DECISION MAKING IN COMPLEX ENVIRONMENTS

Decision Making in Complex Environments

Edited by

MALCOLM COOK
University of Abertay Dundee, UK

JAN NOYES
University of Bristol, UK

YVONNE MASAKOWSKI
Naval Undersea Warfare Center Division, Rhode Island, USA

ASHGATE

Published by
Ashgate Publishing Limited
Gower House
Croft Road
Aldershot
Hampshire GU11 3HR
England

Ashgate Publishing Company
Suite 420
101 Cherry Street
Burlington, VT 05401-4405
USA

Ashgate website: http://www.ashgate.com

British Library Cataloguing in Publication Data
Decision making in complex environments
1. Decision making
I. Cook, Malcolm II. Noyes, Janet M. III. Masakowski, Yvonne
658.4'03

Library of Congress Cataloging-in-Publication Data
Decision making in complex environments / Edited by Malcolm Cook, Jan Noyes and Yvonne Masakowski.
p. cm.
Includes index.
ISBN-13: 978-0-7546-4950-2
1. Decision making. 2. Multiple criteria decision making. 3. Decision making--Data processing. 4. Decision support systems. I. Cook, Malcolm (Malcolm James), 1960- II. Noyes, Janet M. III. Masakowski, Yvonne.

T57.95.D43 2006
153.8'3--dc22

2006021150

ISBN 978-0-7546-4950-2

Printed and bound in Great Britain by TJ International Ltd, Padstow, Cornwall.

Contents

Part 5: Teams and Complex Decision Making

Part 6: Assessment and Measurement

List of Figures

List of Tables

List of Contributors

Corinne Adams graduated from the University of Stirling, Scotland in 2000 with a BSc in Psychology and went on to do an MSc (by research) in Sport Studies. In October 2001, she began working as a Research Assistant at the University of Abertay Dundee, Scotland. In this post she worked on a number of Human Factors research projects covering issues such as command and control, and critical incident management. In 2004, she took up the post of Research Officer within the Scottish Executive. In 2006, she returned to the University of Stirling where she is currently undertaking an MSc in Applied Social Science, with the intention of starting a PhD in October 2007.

Carol Angus is currently the Health Information Analyst within the National Health Service, UK. Formerly working in a University setting, she worked for six years as a researcher within the subject areas of Psychology and Human Factors, working on both commercial and academic funded projects that addressed a wide range of topics including military training, training for terrorism, team work, command and control, organisational psychology and human performance.

Bjørn Tallak Bakken is currently (since 2002) employed as a scientist at the Norwegian Defence Leadership Institute, after working as scientist and project manager at the Norwegian Defence Research Establishment (since 1996). He is currently enrolled in the PhD programme at the Norwegian School of Management (NSM), with a research focus on decision making effectiveness in (military) crisis management operations. Other fields of interest include: operations analysis, game theory, system dynamics, and organisational learning.

Alan Bleakley is an Educational Psychologist at the Royal Cornwall Hospital, Truro, UK. He has an interest in the enhancement of surgical teamwork and decision making, and was instrumental in establishing the hospital's Theatre Resource Management and Team Self-Review programmes.

Olivier Blin, a Researcher-Professor, is board certified in Neurology and Psychiatry and holds a PhD in Pharmacology. He is the head of the Clinical Pharmacology department of the Public Hospitals Marseille, France. He is the Director of the Pathophysiology and Pharmacology of Emotions, Reward and Risk Unit of the Institut des Neurosciences Cognitives de la Méditerranée, a joint organisation between the Centre National de la Recherche Scientifique and Marseille University. He is an elected member of the American Society for Clinical Pharmacology and Therapeutics and of the American College of Clinical Pharmacology. He is also a member of the Scientific Committee of the French Space Medecine Center (MEDES) and the Executive President of the National Workshop for Clinical Pharmacology, the President of the French Association for Biological Psychiatry and the General Secretary for Clinical Pharmacology section for the French Society for Pharmacology and Therapeutics.

Robert Bolia is a Research Scientist in the US Air Force Research Laboratory's Human Effectiveness Directorate. He holds a Bachelor's degree in Mathematics from Wright State University, USA and a Master's degree in Military Studies from the American Military University, USA. His primary research focus is on the effects of collaboration technology on team performance, workload, and situation awareness in tactical command and control environments. He is also a military historian.

James Boyden is a Consultant Anaesthetist at the Royal Cornwall Hospital, Truro, UK. He has an interest in the enhancement of surgical teamwork and decision making, and was instrumental in establishing the hospital's Theatre Resource Management and Team-Self Review programmes.

Berndt Brehmer is Professor of Command and Control Decision Making at the Swedish National Defence College and he was the Vice President of the college, 1998-2004. He came to the college from the Swedish National Defence Research Establishment where he was head of the Division of Human Sciences. Before that he was Professor of Psychology at Uppsala University, Sweden for 17 years. He is a member of the Royal Swedish Academy of Letters, Antiquities and History, and of the Royal Swedish Academy of War Sciences.

Richard Breton is a Defence Scientist at Defence R&D Canada-Valcartier, and an Associate Professor at the School of Psychology, Université Laval, Québec, Canada. He received his PhD in Psychology from Université Laval in 1997. His research interests focus on decision making and situation awareness in military environments and cognitive system engineering.

C. Shawn Burke is a Research Scientist at the Institute for Simulation and Training of the University of Central Florida, USA. Her expertise includes teams and their leadership, team adaptability, team training, measurement, evaluation, and team effectiveness. She has published over 40 journal articles and book chapters related to these topics and has presented at over 70 peer-reviewed conferences. She is currently investigating team adaptability and its corresponding measurement, issues related to multi-cultural team performance, leadership, and training of such teams, and the impact of stress on team process and performance. She earned her doctorate in Industrial/Organizational Psychology from George Mason University, USA, and serves as an *ad-hoc* reviewer for *Human Factors, Leadership Quarterly, Human Resource Management,* and *Quality and Safety in Healthcare*. She has co-edited a book on adaptability and is currently co-editing a book on advances in team effectiveness research.

Gloria L. Calhoun is a Senior Engineering Research Psychologist with the US Air Force Research Laboratory. Currently, she is studying control station design for single operator supervision of multiple unmanned aerial vehicles. Her previous research focused on colour/pictorial formats, as well as multifunction controls, groundwork for today's "glass cockpits". She has also evaluated tactile and spatial auditory displays as well as voice, eye gaze and brain EEG-based controllers.

Richard Clark is Associate Professor of Psychology and Director of the Cognitive Neuroscience Laboratory at Flinders University in Adelaide, Australia. He is recognised for his contributions to Neuropsychology and the related field of Cognitive Neuroscience and the applications of these fields to the understanding of brain and cognitive function in health and disease. His laboratory provided some of the first evidence of brain dysfunction in post-traumatic stress disorder and it has contributed significantly to better understanding of the biological basis in many other psychopathologies, including panic disorder, attention deficit disorder, schizophrenia and head injury. More recently, he has played a leading role in the development of the first international and integrative database of brain and cognitive function. This work has led to enhancement in the evidence base for medicolegal assessment of brain functional disorders in Australia.

Elizabeth Clarke studied Psychology at the University of Adelaide, Australia and subsequently spent a year conducting research in the military environments as a member of the Human Sciences Discipline in the Land Operations Division of Australia's Defence Science and Technology Organisation. She then travelled to Victoria to finish a second degree in Law at the University of Melbourne. She is currently a commercial solicitor working in private practice in Melbourne where she continues to study problem solving techniques commonly employed in conflict situations.

Malcolm James Cook is a Senior Lecturer in Military Human Factors. He has an eclectic career addressing a wide range of Applied Psychology issues in areas such as Command & Control, military targeting, sensor fusion, advanced interface development, intelligence assessment, error management systems, tele-robotics, and mission planning. He is currently interested in differences in performance in complex cognitive tasks with and without familiarity with meta-cognitive methods and prior experience with the information context.

Gemma Cox currently works as a Human Factors Consultant for Human Engineering Ltd, UK. Previously she was a Research Associate, working for the University of Nottingham, UK. Here she worked on two major research projects in the domains of Rail and Aviation. Her current research focuses on the assessment of suitable methodologies for investigating human performance in safety critical contexts and she is currently writing her PhD thesis.

Margaret Crichton is the founder of People Factor Consultants, and is a Chartered Psychologist. She held a post as Research Fellow at the University of Aberdeen, Scotland from 1998-2005. Her primary interests are developing and presenting training in human factor skills such as decision making (especially under stress), situation awareness, and communication in high hazard organisations, for example, offshore oil and gas drilling teams, UK nuclear power production emergency response organisations, and the emergency services. She has also been involved in developing computer-based training for decision making skills improvement, as well as observing and monitoring emergency exercises (full-scale and desk top).

Sidney Dekker is Professor of Human Factors, and Director of Research at Lund University, School of Aviation, Sweden. He gained his PhD in Cognitive Systems Engineering from The Ohio State University, USA, and has previously worked for the Public Transport Cooperation in Melbourne, Australia, the Massey University School of Aviation, New Zealand, British Aerospace, UK, and has been a Senior Fellow at Nanyang Technological University, Singapore. His research interests are system safety, human error, reactions to failure and criminalisation, and organisational resilience. He flies as instructor pilot on the Cirrus SR-20 at Lund University School of Aviation. His most recent books are *Ten questions about human error: A new view of human factors and system safety* (2005) and *The field guide to understanding human error* (2006).

Georgios Detsis is currently Managing Director and R&D Coordinator at ENTELIA Informatics, Athens, Greece. He has a long track record in managing R&D programmes in various fields. He holds a BEng (Honours) in Electronic Systems Engineering and an MSc (Honours) in Intelligent Knowledge Based Systems both from the University of Essex, UK, and an Economics & Management (executive MBA) from Athens Laboratory of Business Administration, Athens, Greece.

Mark Draper is a Senior Engineering Research Psychologist at the US Air Force Research Laboratory's Human Effectiveness Directorate. He received his PhD in Industrial Engineering from the University of Washington, USA where he investigated adaptation to virtual reality technology. Since 1998, he has been exploring the potential for advanced operator interface concepts to improve unmanned air vehicle performance. His current research focuses on decision support interfaces for supervisory control of multiple unmanned systems.

Graham K. Edgar is a Senior Lecturer in Psychology at the University of Gloucestershire, UK, and has previously conducted Human Factors research as a Principal Scientist for BAE Systems. His general interests are in Cognitive Psychology and Human Factors, with a particular interest in situation awareness.

He is a Chartered Psychologist and Associate Fellow of the British Psychological Society.

Helen E. Edgar is a Chartered Psychologist and Associate Fellow of the British Psychological Society. She was previously a Research Project Manager and Principal Scientist at BAE Systems, and is currently doing freelance work in the areas of cognitive psychology and situation awareness.

Eric W. Farmer is Technical Director of Human Sciences in QinetiQ, UK. He is a cognitive psychologist who has worked primarily in the field of aviation human factors. His current research interests include workload measurement and modelling, and the effects of fear on human performance and behaviour.

Philip S. E. Farrell is a scientist employed by the Defence Research and Development Canada and works at the Canadian Forces Experimentation Centre. With a PhD in Mechanical Engineering from the University of Toronto, Canada, he continues to gain expertise in the area of Human Factors: specifically multiple agent interaction. His current research interests include knowledge management, common intent, team performance, and the application of control theory to effects-based thinking and joint fires concepts.

Rhona Flin is Professor of Applied Psychology at the University of Aberdeen, Scotland. She directs a team of industrial psychologists working with high risk industries on the management of safety and emergency response. The group has worked on research projects in civil aviation, the offshore oil industry, the nuclear power industry, as well as in acute medicine. They have recently been designing systems to assess surgeons' and anaesthetists' non-technical skills, and developing techniques to measure safety climate in hospitals.

Han Tin French heads the Human Sciences Discipline in the Land Operations Division of Australia's Defence Science and Technology Organisation. Her research areas, spanning 17 years, include the effects of blast overpressures on humans, collective training and situation awareness in the land environment. She gained her PhD in Physical Chemistry from the University of New England, New South Wales, Australia.

Joseph W. Guthrie, Jr. received his MS in Industrial/Organizational Psychology from the University of Tennessee at Chattanooga, USA in 2003, where he received the award for Outstanding Graduate Student in Industrial/Organizational Psychology. He is now a third year PhD student enrolled in the Applied Experimental and Human Factors Psychology programme at the University of Central Florida and works as a graduate research assistant at the Institute for Simulation and Training. He is currently the lead graduate student on a project sponsored by the Air Force Research Lab investigating the effects of collaboration technology on team decision making and performance. He is also currently involved in projects investigating team performance under stress and how human factors can improve patient safety. His research interests include teams and technology, usability evaluation, multi-team systems, distributed teams, and patient safety.

Peggy L. Heffner works for the Naval Aviation Systems Command, F/A-18 Program Office as the Electronic Warfare Integrated Product Team Lead in the USA. She supports the EA-18G electronic attack variant programme where she combines her interest in technology solutions with her prior experience in human factors and Crew Systems. She received a Bachelor's degree from Embry Riddle Aeronautical University and a Master's degree from the Pennsylvania State University.

Anne Helsdingen is a researcher for TNO, the Netherlands organisation for applied research. Her major research interest is training complex decision making and currently she is working on a thesis on this subject. Other national and international research projects that she is involved in include subjects such as gaming and human behaviour modelling.

Simon Henderson is a Principal Consultant with QinetiQ Ltd in the UK. His work is concerned with understanding and enhancing teamwork and decision making across a wide variety of domains; and he has a particular interest in enabling teams to learn more effectively from their own experience. He has developed organisational learning systems for the UK Armed Services, the Police force, and most recently, the National Health Service.

Adrian Hobbs is a Consultant Anaesthetist at the Royal Cornwall Hospital, Truro, UK. He has an interest in the enhancement of surgical teamwork and decision making, and was instrumental in establishing the hospital's Theatre Resource Management and Team Self-Review programmes.

Alina Holgate is a Lecturer in the School of Psychology, Deakin University, Melbourne, Australia. She is involved in a research programme studying decision making under stress, and instructs in Country Fire Authority Victoria training programmes for volunteer firefighters who aspire to leadership roles.

Erik Hollnagel is Industrial Safety Chair at École des Mines de Paris, France, as well as Professor of Human-Machine Interaction at Linköping University, Sweden. Since 1971, he has worked at universities, research centres, and various industries with problems in the domains of nuclear power production, aerospace and aviation, software engineering, healthcare, and automobiles. He is an internationally recognised expert in the fields of industrial safety, accident analysis, cognitive systems engineering, cognitive ergonomics and intelligent human-machine systems. He has published widely including 12 books: the most recent titles being *Resilience engineering* (2006), *Joint cognitive systems: Foundations of cognitive systems* *engineering* (2005) and *Barriers and accident prevention* (2004). He is, together with Pietro C. Cacciabue, Editor-in-Chief of the international journal of *Cognition, Technology & Work*.

Stig Johannessen is currently working as Senior Adviser in Organisational Questions at Hedmark University College, Norway. He holds a Master's degree in Public Administration from Karlstad University. He has his highest military education from General Staff course, Sweden Defence College. His current research is within the area of leadership and teambuilding.

Elisabeth Jouve is Biostatician in the Centre de Pharmacologie Clinique et d'Evaluation Thérapeutique of the Assistance Publique – Hôpitaux Marseille, France. Her work focuses around the methodology and the statistical analysis in clinical research especially of clinical trials. The development and the assessment of psychometric scales are among her research interests.

Gary Klein is Chief Scientist of the Klein Associates Division of Applied Research Associates (ARA) in the USA. He is one of the founders of the field of Naturalistic Decision Making. In recent years he has extended his work on recognitional decision making to describe problem detection, option generation, sensemaking, planning and replanning. He is the author of *Sources of power: How people make decisions* and also *The power of intuition*.

Thomas Koester, Psychologist and Human Factors specialist, currently works for FORCE Technology and Danish Human Factors Centre. His experience covers R&D projects, course development and teaching of Human Factors. His work with applied human factors and field studies has covered many different safety critical domains including maritime transport, off-shore industry, power plants and hospitals. He has also in close cooperation with industrial designers been engaged in projects on design and optimisation of human-machine interaction in various products such as communication equipment for ships, hospital equipment and agricultural machines.

John N. Kostaras is employed as a Systems Analyst with TIM Hellas, a mobile telecommunications company located in Athens, Greece. He holds a BSc in Informatics from Athens University, an MSc in Telecommunications and Information Systems and an MA in Management Studies with distinction both from Essex University, UK. While employed with INTRACOM SA, he has been involved in a number of EUCLID research projects for defence for which he has written a number of relevant scientific articles. His interests involve human computer interaction (HCI), geographic information systems (GIS), 3D visualisation and Command & Control Systems (C4I).

Sandra Marshall is Professor of Psychology at San Diego State University, CEO of EyeTracking, Inc., and Director of the Cognitive Ergonomics Research Facility at San

Diego State University, USA. For the past 20 years, she has conducted basic research sponsored by the Office of Naval Research, Air Force Office of Scientific Research, and Defense Advanced Research Projects Agency (DARPA). Currently, she directs several major projects, all of which use eye-tracking measures to investigate aspects of cognitive processing and decision making.

Yvonne Masakowski is a Human Factors Engineering Psychologist at the Naval Undersea Warfare Center serving as the Human Systems Integration Lead for the US Navy Littoral Combat Ship ASW MP programme. Previously, she served as CNO Science Advisor to the Strategic Studies Group (SSG) at the Naval War College for which she received the Civilian Meritorious Service Award. Dr. Masakowski has also served as the first Associate Director, Human Factors for the Office of Naval Research Global, London. She has played a pivotal role in the integration of human factors in the development of numerous tactical control products and system designs. Her current research focuses on distributed decision making in netcentric warfare, global situation awareness, and the impact of globalisation on collaboration. She is the recipient of numerous national and international awards including The Cross of Merit presented by the Ambassador of the Czech Republic to the US, Ambassador M. Palous; two medals for Scientific Achievement for the advancement of Human Factors by the Ministère de la Défense Sciences Médicales et Facteurs Humains and the French Navy of the Department Generale Adminstration (DGA) of France. She also received a Medal for Scientific Achievement from The Polish Air Force Institute of Aviation Medicine in 2001.

Graham Mathieson worked at the UK MOD's Defence Science and Technology Laboratory and its predecessor organisations from 1980 until his death at the age of 47 in March 2006. He was married with two children. He graduated in Applied Physics from Strathclyde University and worked for MOD research establishments in a number of roles, developing his skills and qualifications in Operations Research (OR). He most recently worked in the Human Systems team at Dstl where he regarded it as his mission to understand the cognitive sciences sufficiently to facilitate their proper representation in the analysis approaches employed in OR. His considerable intellectual abilities and passion for his subject were recognised by his appointment as a Dstl Fellow in 2002. This chapter represents a fitting testimony to his work in the field, and provides a legacy from which his colleagues can learn and continue his work.

Peter McGeorge is currently Head of the School of Psychology at the University of Aberdeen, UK. After training as a cognitive psychologist at the University of Nottingham where he obtained his PhD in Experimental Psychology, he then moved to Aberdeen to research medical decision making. Since then he has worked on a range of projects concerned with decision making in complex environments (e.g. nuclear, fire ground, offshore oil and gas production) along with various projects on basic aspects of visual attention.

Barry McGuinness is a Senior Principal Scientist at the BAE Systems Advanced Technology Centre (ATC) in Bristol, UK, where he has worked since 1990. For over 10 years he has led the ATC Human Factors Department's capability in the modelling and measurement of situation awareness. In 2005, he initiated work under contract to the US Department of Defense to investigate issues of human trust in information in network-centric military operations. His work has earned him two company awards for innovation. He holds a Master's degree in Applied Psychology from Cranfield University in the UK.

Jim McLennan is a Senior Research Fellow in the School of Psychological Science, La Trobe University, Melbourne Australia. He is a Project Manager in the Bushfire Cooperative Research Centre.

Anne Melia is a Programmer and Human Factors Consultant in QinetiQ, UK. Amongst other projects, Anne has collaborated with the US Military to develop a real time cognitive support tool to augment decision making for fast-jet pilots. She currently undertakes human factors research in the domain of aviation security, specifically the screening of hold-baggage for terrorist threats.

Matthew Mills is a Senior Consultant with the UK National Air Traffic Services. Prior to this he worked for QinetiQ Ltd. He has an interest in enhancing teamwork and decision making through the use of web-based technologies, and has been involved in developing a system of Team Self-Review for the National Health Service.

Tracey Milne is a Human Factors specialist with a physiology background, having gained her MSc in Ergonomics in 1997. She currently works for Tube Lines Ltd, UK as Senior Human Factors Delivery Manager. Prior to this, she worked for the Centre for Human Sciences, QinetiQ for seven years and has an extensive knowledge base in both civil and military arenas providing Human Factors/Ergonomics technical advice on a variety of military and civil projects including European, corporate and applied research. Particular focus has been in the rail, aviation, and automotive domains. Her main areas of expertise are in the user-centred design and evaluation of human-machine interfaces (particularly large screen display technology and simulation), the development and assessment of combined speech and text interfaces, cognitive task analysis and situation awareness measurement.

Thierry Morineau is a University Teacher at the University of South Brittany, France. Previously, he was part of the CNRS/LAMIH laboratory where he worked on Air Traffic Control activities. He is interested in ecological psychology applied to user interface design and decision support systems. His current research focuses on decision making in environmental domain and in neurosurgery.

Jennifer McGovern Narkevicius is President of Human Systems Integration for SkillsNET Corporation and Technical Lead of the US Navy's Enterprise program for human systems integration, Systems Engineering, Acquisition, and PeRsonnel INTegration (SEAPRINT). She is developing the interaction of the human based domains that effect workforce performance applicable to both military and commercial workplaces. She holds a PhD in Cognitive Psychology from the University of Florida, with Master's degrees in Special Education: Gifted and Talented (University of South Florida) and Developmental Psychology (University of Florida). She is completing an MS in Systems Engineering at Johns Hopkins University, USA.

W. Todd Nelson is a Senior Engineering Research Psychologist in the US Air Force Research Laboratory's Human Effectiveness Directorate. He earned an MA and PhD in Experimental Psychology and Human Factors from the University of Cincinnati, USA, and a BA in Psychology from Allegheny College, USA. His primary research interests involve the effects of collaboration technology on team performance, situation awareness, and workload. He is adjunct faculty in the Department of Psychology at the University of Cincinnati, where he serves on doctoral and master theses committees.

Alastair P. Nicholls is a Chartered Psychologist in QinetiQ, UK. He is currently supporting, in the capacity of Experiment Designer, the core capability of NITEworks (a Ministry of Defence and defence industry partnership that empirically assesses the benefits of Network Enabled Capability). Prior to arriving at QinetiQ, Alastair was a post-doctoral Research Associate at the School of Psychology, Cardiff University, Wales.

Edward Nicholson was a 3rd Year BSc Student in the School of Mechanical, Materials and Manufacturing Engineering at the University of Nottingham, UK. He piloted the experiment that was developed further for inclusion in this book which was supervised by Dr. Alex Stedmon and Dr. Sarah Sharples.

Jan Noyes is a Professor of Human Factors Psychology at the University of Bristol, UK. She is a Fellow of the Ergonomics Society and a Member of the British Psychological Society and the Institution of Engineering and Technology (formerly IEE – Institution of Electrical Engineers). In 1999, she was awarded the Otto Edholm medal for her contribution to ergonomics. She has authored around 200 publications including six books, and was awarded the IEE Informatics Premium Award in 1998 for her paper on 'engineering psychology and system safety'. She was also Chair of the 1999 and 2001 IEE People In Control conferences, is on the Editorial Boards of *Ergonomics*, and *Interacting With Computers*, and is a member of the Defence Technology Centre (DTC) on Data and Information Fusion.

Mary Omodei is a Senior Lecturer in the School of Psychological Science, La Trobe University, Australia. She is the Bushfire Cooperative Research Centre Leader at La Trobe University, and the developer of the Network Fire Chief microworld research tool.

Göran Pettersson is a Senior Scientist at the Command and Control Systems Department within the Swedish Defence Research Agency. After finishing studies in Mathematics (BSc), he joined the SAAB Company (1983-6) and worked with real time simulation for the JAS39 fighter system. He was then employed as a scientist at the Swedish Defence Research Agency (1986-95) and was active in the decision support systems area. After some years, he was coordinator for the national defence related decision support programme. He was then a system engineer at SAAB (95-97) and worked with decision support and data fusion within a development programme. He then joined again the defence agency and is now working with research applications that relate to cognitive systems engineering and temporal representations of battlefield data.

Diane Pomeroy is a Human Factors Scientist in the Land Operations Division, Defence Science and Technology Organisation, Australia. She holds a PhD in Psychology from Flinders University, Australia. Her current primary research focus is on use of simulation for individual and collective training within the army.

Alastair Ross is a Chartered Psychologist and Senior Research Fellow in the Centre for Applied Social Psychology (CASP), University of Strathclyde, UK where he teaches on an Honours course in Psychology and Technology, and a Master's course in Psychological Research Methodology. His PhD investigated the implications that causal explanations have for motivation and behaviour. He has worked as a consultant for companies such as BP (British Petroleum), British Energy, Rolls-Royce, British Aerospace, and Unilever.

Robert Rousseau is an Emeritus Professor at Université Laval, Québec, Canada. His current work deals with cognitive issues in evaluation of DSS, decision making in C2, situation awareness and teamwork. He has been a faculty member at the School of Psychology at Université Laval from 1968 to 2003, where he has carried out experimental research on timing and attention in humans and performed applied cognition projects on work automation, information systems implementation and knowledge management.

Heath A. Ruff currently works for Advanced Information Engineering Services, a General Dynamics Company, in support of the US Air Force Research Laboratory at Wright-Patterson Air Force Base. He holds a Master's degree in Engineering from Wright State University. His primary research interests are supervisory control of unmanned systems, humans and automation, and decision support interfaces.

Morten Ruud is currently employed as scientist at the Norwegian Defence Leadership Institute. He is enrolled in the PhD programme at the University of Bergen, with research focus on policy development in dynamic systems with multiple stakeholders. Other fields of interest include: system dynamics, group based modelling, multi-user micro worlds, mental models, sense making and organisational learning.

Eduardo Salas is Trustee Chair and Professor of Psychology at the University of Central Florida (UCF), USA and holds an appointment as Program Director for Human Systems Integration Research Department at UCF's Institute for Simulation and Training (IST). Previously, he was Head of the Training Technology Development Branch of NAWC-TSD for 15 years. During this period, he served as a principal investigator for numerous R&D programmes, including TADMUS, that focused on teamwork, team training, decision making under stress and performance assessment. He has co-authored over 300 journal articles and book chapters and has co-edited 19 books. His expertise includes assisting organisations in how to foster teamwork, design and implement team training strategies, facilitate training effectiveness, manage decision making under stress, and develop performance measurement tools. He is a Fellow of the American Psychological Association, the Human Factors and Ergonomics Society, and a recipient of the Meritorious Civil Service Award from the US Department of the Navy.

Melanie Seymour is a Research Assistant in the School of Psychology at Flinders University, Adelaide, Australia. She holds a BSc with Majors in Cognitive Science, and Honours in Psychology. Previously, she has been employed as Manager of the Cognitive Neuroscience Laboratory, Flinders University and a User Interface Designer. Her research interests are in the field of cognitive neuroscience.

Sarah Sharples (née Nichols) is a Lecturer in Human Factors in the School of Mechanical, Materials and Manufacturing Engineering at the University of Nottingham, UK. She completed her PhD in 1999 on methodological and theoretical issues in the assessment of participants' experiences of virtual environments. She has been a researcher, research manager or grant holder on several industrial, government and EU funded projects, including VIEW of the Future, and a long-term programme of research for Network Rail examining implications, design and implementation of novel interfaces for railway control and use of rail simulation for human factors research. She was co-investigator and project manager on the EPSRC funded project GR/R86898/01 Flightdeck and Air Traffic Control Collaboration Evaluation (FACE). She is a Registered Ergonomist, and her main areas of interest and expertise are Human-Computer Interaction, cognitive ergonomics and development of quantitative and qualitative research methodologies for examination of interaction with innovative technologies.

Bruno Sicard currently serves as Medical Liaison Officer for the US Army Medical Research and Materiel Command, Fort Detrick, Maryland. Previously, he was Head of the Human Factors Department in the French Navy where he applied his dual experience as naval flight surgeon and physiologist to research and applications in naval, combat and many other extreme environments. He is a graduate from medical school in Lyon, France and holds a PhD in Physiology from Marseille, France, and is an Associate Scientist at the Institut de Neurosciences Cognitives de la Méditerranée.

Carys Siemieniuch is a Senior Lecturer in Systems Engineering in the Department of Electronic and Electrical Engineering, Loughborough University, UK. A systems ergonomist for 20 years, with both UK professional and European CREE registration, she has expertise across the full range of systems-related human factors topics, but retains a particular interest in four main areas: capability acquisition systems, knowledge life cycle management (KLM) systems; organisational and cultural aspects of enterprise modelling techniques; impact of situation awareness and individual/team decision making structures and processes on human performance.

Murray Sinclair is a Senior Lecturer in the Department of Human Sciences, Loughborough University, UK. Current research areas concern advanced manufacturing technology: computer-supported co-operative working between the organisation and its suppliers; the design of computer-aided engineering (CAE) systems to optimise activities in manufacture, assembly and field service; human supervision and control in 'flexible' manufacturing environments; knowledge structures and the roles of knowledge within organisations; and issues of corporate governance and the organisational structures needed to support this.

Christopher Smyth is an Engineering Psychologist with the US Army Research Laboratory, Human Research and Engineering Directorate, at Aberdeen Proving Ground, Maryland. He is responsible for research in the area of crew station design for military ground vehicles and holds several patents in the field. He has degrees in engineering and experimental psychology from sundry universities, all earned following a career in the military.

Dag Søberg is currently the commanding officer of the Border Guard Battalion located at Kirkenes in Eastern Finnmark. He has his highest military education from General Staff course, Sweden Defence College.

Lynn Springall is a Senior Human Factors Specialist for National Air Traffic Services in the UK. Her background includes many years in Air Traffic Control (ATC) operations, during which she studied to move into the human factors domain of ATC. She holds a Master's degree in Human Factors and Safety Assessment in Aeronautics from Cranfield University, UK. Her main interest is in the development of future tools and systems for air traffic controllers ensuring that they are efficient, error tolerant and usable.

Alex W. Stedmon is a Lecturer in Human Factors in the School of Mechanical, Materials and Manufacturing Engineering at the University of Nottingham, UK. He is also a Registered Ergonomist and Chartered Psychologist who completed a PhD in Human Factors of speech input for real world and virtual reality applications. He has also been a Consultant and Senior Research Fellow involved in several industrial, government and EU funded projects, including VIEW of the Future and research with Network Rail examining CCTV. He managed work on the EPSRC funded project GR/

R86898/01 Flightdeck and Air Traffic Control Collaboration Evaluation (FACE). He is also a keen motorcyclist and is developing research into motorcycle ergonomics through involvement with the Ergonomics Society Motorcycle Ergonomics Special Interest Group.

Nalini Suparamaniam is a Senior Consultant and Subject Expert at Det Norske Veritas, Norway. A large part of her current work is associated with management consulting within safety, health and environment in the oil and gas industry. She gained her PhD in Human Systems Engineering at Linköping University, Department of Mechanical Engineering, Sweden. She has previously been attached to the University of Central Florida (UCF) where she first began her research activities. Her experiences include work in the military, aviation, rescue and relief services, medical, maritime, and currently, oil and gas industry. She has taught at University of Central Florida, at Linköping University, and is currently engaged at the University of Bergen, Norway. Her approach to address safety challenges in a high reliability industry is through assessing and providing for organisational robustness, technology functionality and usability, and clarifying human interaction with these systems. Her research interests follow her in her career in industry.

Karel Van Den Bosch is a Senior Researcher for TNO, the Netherlands organisation for applied scientific research. He manages and performs national and international projects, involving fundamental and applied research on the training of complex cognitive tasks (e.g. command & control, decision making, crisis management). The objective of his current work is to make simulation-based training more effective by using cognitive software agents (e.g. agents playing the role of team mate, adversary, or instructor). He is investigating how such agents can successfully support training, thus making training more systematic (uniform behaviour of agents), more effective (agents consistently eliciting intended behaviour of trainee), and more efficient (team members need no longer be present during training). Outcomes and products of this research are used to develop autonomous and independent forms of training.

Michael Vidulich is the Technical Advisor for the Warfighter Interface Division of the US Air Force Research Laboratory's Human Effectiveness Directorate. He holds a Master's degree in Experimental Psychology from Ohio State University and a PhD in Experimental/engineering Psychology from the University of Illinois at Urbana-Champaign, USA. His main research interests are mental workload and situation awareness assessment. He is also a member of the adjunct faculty of the Wright State University Department of Psychology, USA, where he has taught since 1989.

Brendan Wallace is a Research Fellow at the Glasgow Centre for the Child and Society, University of Glasgow, UK. Previously he worked on issues relating to safety, risk, accidents and 'human error' at the Centre for Applied Social Psychology, University of Strathclyde, UK. His current research interests include the development of innovative qualitative research methods, and risk and safety issues amongst young people. His latest book is *Beyond human error* (2006) with Alastair Ross.

Linda Walsh is a Theatre Manager at the Royal Cornwall Hospital, Truro, UK. She has an interest in the enhancement of surgical teamwork and decision making, and was instrumental in establishing the hospital's Theatre Resource Management and Team Self-Review programmes.

Alexander Wearing is Professor of Psychology at the University of Melbourne, Australia. He has a distinguished record of research in judgement and decision making, and in economic psychology.

Damien J. Williams is a final year postgraduate student in the Department of Experimental Psychology at the University of Bristol, UK. His research interests focus on the area of risk perception and decision making, with particular application to the provision of risk information.

John R. Wilson is Professor of Human Factors in the School of Mechanical, Materials, Manufacturing Engineering and Management at the University of Nottingham, UK. He previously held posts at the Universities of Loughborough and Birmingham in the UK, and was Visiting Professor at the University of California, Berkeley and University of New South Wales, Australia. At Nottingham, he is Director of the Institute for Occupational Ergonomics and Director of the Virtual Reality Applications Research Team. In addition, he has been part-time Strategic Advisor on Human Factors on secondment to Network Rail since 2001. He has produced over 450 publications, and more than 250 are in refereed books, journals or collections. He was awarded the Sir Frederic Bartlett Medal of the Ergonomics Society in 1995, for services to international ergonomics teaching and research. He has been principal investigator or grant holder on over 50 major grants from Research Councils, government, the European Union and public bodies as well as having carried out research or consultancy for over 100 companies. He is a Chartered Psychologist and a Chartered Engineer, a member of the Peer Review College for both the Engineering and Physical Sciences Research Council and the Economic and Social Research Council, member of Editorial Boards for a number of journals, and is Editor-in-Chief of the journal, *Applied Ergonomics*.

Foreword

As we move into the 21st Century, decision making can no longer be viewed as an isolated event or experience but rather mandates that "decision making" must be assessed within the context of globalisation and the distribution of information.

Globalisation emerged from innovations in technology, economic shifts in trade and political exchange in conjunction with the advent of the digital revolution. Furthermore, the digital revolution serves as an enabler that secures global connectivity, accelerates the pace of information exchange and helps to shape individuals' lives, business strategies and military operations throughout the world.

Therefore, if we are to consider decision making and its impact on a global scale we must do so within the context of the complex digitally connected environment in which decisions are made. Today, we are faced with a critical need to address the ways in which information is generated and distributed to inform, enlighten and shape human decision making in a complex workplace environment.

Advances in science and technology, along with the emergence of globalisation, continue to have an impact on the ways in which information is distributed and decisions are made in a 21st Century environment. A thorough discussion on the topic of decision making must therefore make note of the transition of decision making. We must begin our exploration of this topic by examining traditional theories and move forward to address challenges that must be faced in this new complex environment. This shift in information exchange and its accelerated pace serves as a means of empowering both individuals and nations.

The digital revolution has enhanced access to information and has been linked with the dramatic shifts in economic growth and the empowerment of nations such as China, India and Singapore (Friedman, 2004). The global exchange of information also presents new opportunities and challenges to nations in terms of their business and military strategies. Indeed, each nation's vision for economic, social and technological progress has the potential to transform nations on a global scale. Namely, decision making will extend its reach beyond the local neighborhood and have an impact on a global scale.

Decision making as a topic merits consideration and scrutiny with regard to theories, cultural differences, and tools that may augment it, as these lend support to the process of decision making that may have significant consequences globally. Specifically, globalisation and decision making are inter-connected issues that need to be discussed to achieve a level of understanding with regard to the theories defining decision making, the tools that support effective decision making and the impact of decisions on a global scale. Furthermore, decision making needs to be energised by innovative thinking that fosters cooperation and competition. Today, each nation is faced with a challenge to maintain their respective place in science and technology.

Namely, globalisation levels the playing field and avails opportunities for nations that were previously isolated.

Traditionally, decision making research has focused on theories that support and/or derive pathways to the selection of choices and the potential impacts of these choices. Networks, models and methodologies have illustrated a variety of approaches to decision making. Indeed, intelligent agent networks, embedded with computational models, algorithms and tools, have been shown to provide critical information required by the user to enhance the effectiveness and accuracy of the decision being made. In addition, systems have been developed for querying, probing and planning to facilitate predictive situation awareness.

Today, we are inundated with a plethora of information, emails and ever-changing software. There is a dynamic relationship among humans, computers, expert systems and intelligent agent software that shapes the way we live, conduct business and participate in war and its related activities. It is imperative that we master the critical components of knowledge management and decision making that will enhance and empower the individual and/or nation.

In the 21st Century, knowledge management tools, intelligent agent architectures, robotics and automated systems will facilitate the expert performance necessary to secure effective and accurate decision making. One of the principal metrics of performance on any scale rests on an individual's ability to reduce uncertainty and optimise their decision making. The complexity of the global information landscape presents challenges to the attention management mechanisms of the decision maker. Issues such as information overload and situation awareness will be supported by technologies which will be developed to augment human cognitive processes and support better and more accurate decision making.

There has been significant progress made in the development of technologies that serve to modify data, reduce the clutter and present information/knowledge in a manner in keeping with human information processing. However, there is still a need to be aware of the trade-offs involved between the human decision maker and those automated technologies that support their decision maker. Currently, we are faced with an abundance of information that challenges our attention and cognitive capacities, as well as placing increased demands on time management. The question is, which tools and technologies can we provide in future designs that will effectively support cognitive processes and facilitate effective decision making?

One approach is to examine the ways in which we think about the human-automation relationship. There is a need to strike a balance between cognitive workload and automation. Workload could be considered from the perspective of the joint human-automation time-line and opportunities could be sought to exploit technologies that will augment the human decision maker. We need to explore time-related trade-offs between automation and humans wherein technologies can accelerate the human's ability to assimilate, disseminate and communicate information from a variety of sources.

The combination of an individual's knowledge, experience and expertise and the information that has been analysed and processed by an autonomous system provides for a more robust knowledge which facilitates their judgement and enables them to make the most effective and accurate decision within a complex environment. The emergence of adaptive adversaries on a global scale will mandate the need for immediate situation awareness and will require increased speed and accuracy of decision makers operating within this dynamic environment.

Automated technologies endeavour to extend human cognitive capacities and enhance human performance through a better understanding of limitations in human processing that impede global situation awareness. Current research is focusing on the development of direct brain interfaces that will enable the control of multiple independent channels that will facilitate our ability to optimise human pattern recognition, classification and memory. In addition, nanotechnologies are currently under development that will enhance human strength and performance, provide medical monitoring and intervention and serve as a node in the command and control centre of the future.

Given these advances in automated decision support technologies, we must not make the mistake of ignoring or setting the human decision maker aside. Rather, we must remember that decision making is a human endeavour within the context of events that requires integration of information, consideration of alternatives and consequences for the decisions made. Decision making needs to be considered with a view to the path of information exchange, analysis and the decision maker's strategy for evaluating, assessing and selecting the choices which support consequences intended by their decisions. It is our conjecture that deliberation of decisions and alternatives reflects the decision maker's choices in consequences at the start.

Most of the books written on decision making have examined this topic from one specific theorist's perspective or from a comparative viewpoint of decision theories. The chapters presented in this book reflect a mosaic of theories and approaches to decision making that will prove to be a valuable tool for the reader who is interested in learning and applying decision making theories in their respective research and practice.

This book will take the reader on a journey from the early theories of Plato and Aristotle (Hollnagel) to current day issues such as sensemaking (Klein), network-centric warfare and military command and control (Bolia et al.) and the complexities of team decision making (Salas et al.). The authors examine the challenges presented to decision makers in the complexities of environments from the flightdeck of the future (Cox et al.) to the virtual environments used in planning and decision making (Kostaras and Detsis), as well as providing an analysis of risk taking (Sicard et al.).

This book is a compilation of valuable insights based on years of experience, expertise and knowledge from the major leaders in the field of decision making research. This book also represents the cutting edge in information management decision making and is a must have for readers to achieve an understanding of the impact of decision making in an international community in which nations have joined together in military operations, business, political and collaborative research. It will serve as an essential tool for researchers as it addresses a topic that is both critical and timely with regard to the challenges presented on a global level. Disparate perspectives, cultural influences and approaches to decision making as discussed in these chapters serve to educate, enlighten and afford an opportunity for each reader to gain a new perspective on a field that is continuously evolving. The progress achieved to date in understanding the human mind, its processes and ability to solve problems and make decisions serves as the foundation for research and technological advances that will support decision making in the future.

The discussions and descriptions provided in the following chapters are aimed towards enlightening us all and affording us an opportunity to reflect on our strategies, choices and consequences for our decision making paradigms. Around the globe, civilian and military personnel will face unique challenges in which they will be called upon to make accurate and effective decisions.

The importance of this book is that it provides critical insights and scientific evidence of decision making theories, strategies, approaches and methodologies that will help to shape future decisions and decision makers' paradigms. The war fighter, peacekeeper, and/or business person, et al. will discover new trails to blaze that will minimise uncertainty and empower decision makers around the world.

This book will serve as a valuable tool that can be used as a compass to navigate your way through a maze of information and achieve a comprehensive understanding of a topic that is often complex and confusing. Further, it will light the path to your understanding and the importance of information management and decision making. To quote Clausewitz, who stated that "Imperfect knowledge of the situation… can bring military action to a standstill…", this book helps to provide the informed individual with the way ahead to effective decision making, regardless of the domain.

Yvonne Masakowski

We hope you enjoy it!

Malcolm Cook, Jan Noyes and Yvonne Masakowski
May 2006

References

Clausewitz, Carl von (1832/1993). *On War*. (Originally published in German by Dümmlers Verlag, Berlin; J.J. Graham translation published in London in 1873.) English version: Everyman's Library.

Clausewitz, Carl von (2003). *Principles of War*. Dover Publications.

Friedman, Thomas L. (2004/2006). *The World is Flat: A Brief History of the Twenty-first Century*. Farrar, Straus and Giroux.

PART 1
Characteristics of Complex Decision Making

Chapter 1

Decisions about "What" and Decisions about "How"

Erik Hollnagel

We assign a moment to decision, to dignify the process as a timely result of rational and conscious thought. But decisions are made of kneaded feelings; they are more often a lump than a sum.
Thomas Harris, *Hannibal*, p. 162

Introduction

Research into decision making has traditionally focused on how people – as individuals – choose among alternatives and specifically about how they go about finding the best alternative or making the "right" decisions. This is most clearly expressed in the three assumptions that characterise the rational decision maker, the *homo economicus* (Lee, 1971). The first assumption is that a rational decision maker is completely informed, which means that s/he knows what all the possible alternatives are and what the outcome of any action will be. This presumably includes both short-term and long-term outcomes. The second assumption is that a rational decision maker is infinitely sensitive; hence s/he is able to notice even the slightest difference between alternatives and use this to discriminate among them. One consequence of this assumption is that two alternatives never can be identical, as they will always differ in some way. The third assumption is that the decision maker is rational, which implies that alternatives can be put into a weak ordering and that choices are made so as to maximise something. The weak ordering means that if for three alternatives A, B, and C, the decision maker prefers A over B, and B over C, then the decision maker must also prefer A over C. This in turn requires that there is a common dimension, which can be either simple or composite, by which all alternatives can be rated. This common dimension also enables the decision maker to identify the alternative that has the highest value, hence to maximise his or her decision outcome.

Decision making as an identifiable process

The origins of looking at decision making as an identifiable process can be found in the early history of thinking, possibly beginning with the development of logic. Broadly speaking, a decision is supposed to be the result of rational reasoning, which is exactly what logic is about. The doctrine of the rationalist school of thinking, which goes back to Plato, is that human knowledge can be derived on the basis of reason alone, using self-evident propositions and logical deduction. Aristotle formalised the idea of rules of logic and the notion of a logical proof, by means of which one could determine whether a conclusion was true or false. This established the tradition of rational thinking – hence of making a rational decision – and the requirement that the conclusion or decision must be consistent and logical, that is, that it must be understandable according to some rules or criteria; otherwise it is called irrational. The same ethos is found in the question from Chevalier de Méré to Blaise Pascal about whether to accept a bet for a specific outcome in a game of dice. This was reformulated into a question of whether one outcome was more likely than another, with the logic of rationality dictating that the decision should be to choose just that. (In Chevalier de Méré's problem, one alternative was that at least one six would appear within four throws of one die; the other that there would be at least one double six within 24 throws of two dice. Since the probabilities of the two events are 0.5177 and 0.4913 respectively, the rational decision is to choose the first alternative and forget the other.) This problem was formulated in 1654 and is generally seen as the start of probability theory, which has been of paramount importance to decision making.

The assumptions of the *homo economicus* refer to the nature of decision making as a process, as something that takes place in the mind of the individual decision maker. Even for organisations, collective decision making "boils down" to what individual decision makers do and the collective is expected to behave just as rationally as the individual. Although the assumptions about the *homo economicus* have rightly been criticised as psychologically unrealistic (Edwards, 1954), they are nevertheless entrenched in the architecture of many decision support systems. These generally aim to replace by machines what humans are bad at doing, echoing the compensatory principle associated with the Fitts' list of function allocation (Dekker and Woods, 2002; Fitts, 1951). They thereby preserve the illusion that decision making is a rational process, and that it is only because of some noticeable human shortcomings that rationality does not manifest itself in practice. Yet this endeavour is unlikely to succeed because it falls prey to what Bainbridge (1983) termed the "ironies of automation". In this case it means that we attempt to use technological artefacts to compensate for a function or process that we cannot describe precisely.

Decision making as an activity

It is, however, possible to see decision making as an activity or a phenomenon rather than as a process, thereby replacing the idealistic assumptions about a rational decision maker with a more realistic set of assumptions about decision making as a facet of work. These assumptions are:

- Decision making is not a discrete and identifiable event, but rather represents an attribution after the fact. In hindsight, looking back at a specific event or activity, we can identify points in time where a decision "must" have been made in the sense that the events could have gone one way rather than the other (Hollnagel, 1984). Yet this does not necessarily mean that the people who were involved made an explicit decision at the time, even though in hindsight they may come to accept that they did. This first assumption also points to an interesting similarity between the conceptual status of "decisions" and of "human error", cf. the discussion of the latter in Woods, Johannesen, Cook and Sarter (1994).
- Decision making is not primarily a choice among alternatives. It is very difficult in practice to separate decisions from what is otherwise needed to achieve a decision maker's objectives, that is, what is required to implement the chosen alternative (for example, Klein, Oranasu, Calderwood and Zsambok, 1993). A decision cannot be made without some information about the situation, the demands, and the possibilities of action. Yet the extent (quality and quantity) of that information may indirectly favour one outcome rather than another. Even worse, a lack of information about something may severely curtail the choices that can be made. Similarly, a decision in most cases also requires actions to ensure that the expected outcomes obtain. It is therefore proper to ask whether the term decision making should be restricted to the "moment of choice" or whether it should also cover what goes on before and after.
- Decision making is not usually a distinct event that takes place at a specific point in time, or within a certain time window and which therefore can be dissociated or isolated – even if ever so briefly – from what goes on in the environment. Decision making cannot be decoupled from the continuous coping with complexity that characterises human endeavours (Hollnagel and Woods, 2005). This assumption is superfluous only if decision making is noticeably faster than the changes in the environment, but this condition is rarely made explicit.

The problems arising from the last assumption have been addressed by the theories of dynamic decision making (for example, Brehmer, 1992), although these still see decision making as a distinct process rather than as a facet of human work and activity. There are some really fundamental problems arising from the fact that decision making, whatever it is, takes time and therefore logically requires that the information it uses remains valid while the decision is made. Despite the obvious importance of time – for decision making as well as for human actions in general – few models of human behaviour take this into account (Hollnagel, 2002). Yet rather than getting lost in this fascinating issue, the rest of the chapter will consider the consequences of seeing decision making as an activity rather than a process. This reflects the fact that the problems people have when managing complex and dynamic processes are not so much about *what* to do, as about *how* and *when* to do it. Indeed, decision making is less a question of choosing the best alternative than a question of knowing what to do in a given situation as described, for example, by the school of naturalistic decision making.

This can easily be illustrated by considering the use of procedures in a job. A procedure is an explicit and detailed description of the actions required to achieve a specific purpose, whether it is a recipe for making spaghetti carbonara or the emergency operating procedures to recover from a tube rupture in a nuclear power plant. The procedure is an externalisation of decision making so that the user can concentrate on when to do something rather than struggle with finding out what to do. Deciding on when to do something is, of course, also a decision of sorts, but it is qualitatively different from those that decision theory generally has focused on.

Decisions and actions

It is a consequence of changing the view of decision making from being a separate process to being a facet of work and of acknowledging the paramount importance of time that the three assumptions implied by rational decision making are no longer tenable. The first assumption, complete information, cannot be upheld because the environment is dynamic rather than static. Complete information can therefore only be achieved if all the necessary information can be sampled so fast that nothing changes while the sampling takes place. The second assumption, infinite sensitivity, is untenable for the same reason, namely that it would require time to differentiate among alternatives. The third assumption, weak ordering, must be abandoned because people normally do not have time to consider all the alternatives they have found, even if it is not the complete set. For instance, a study of how senior reactor operators in a nuclear power plant diagnosed disturbances showed that they used a non-compensatory approach. "When people adopt this approach, they reduce the number of alternatives by selecting the most important attributes, instead of performing trade-offs among them. This reduction will be continued until one alternative remains …" (Park and Jung, 2003, p. 210). In practice there is rarely time to match all alternatives against each other, even if a common evaluation criterion could be found. Quite apart form that, numerous studies have shown that people in practice often have problems in making transitive orderings of alternatives (Fishburn, 1991).

When decision making is described as an activity, the relation between time and decision making can be seen as a special case of the relation between time and actions. This relation can be represented by the contextual control model (COCOM; cf. Hollnagel and Woods, 2005), which describes how the ability to maintain control depends on the controlling system's interpretation of events and selection of action alternatives (Figure 1.1). At the heart of the model is a cyclical relation linking *events*, *intentions* and *actions* where in particular the two arcs called "evaluating / assessing the situation" and "choosing what to do" are relevant for the present discussion. Associated to the former is the time needed to evaluate events and assess the current situation, while associated to the latter is the time needed to select the action alternatives that will bring about the desired outcomes. The time needed to accomplish both of these must be seen in relation to the time that is available for carrying out the action, represented in Figure 1.1 as the window of opportunity.

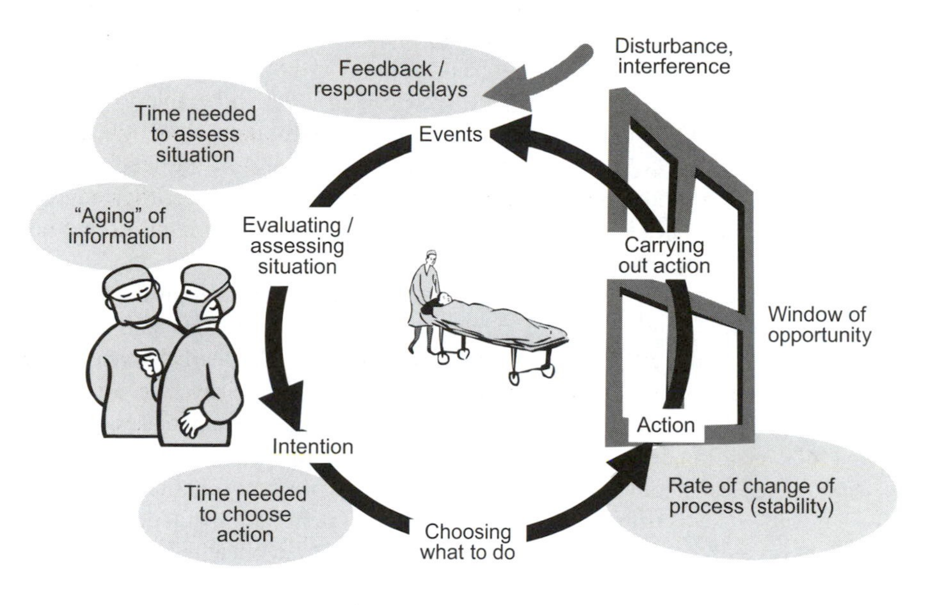

Figure 1.1 The contextual control model applied to decision making

Decision making in the traditional sense can be seen as the interface between the evaluation of the situation and the choice of action, or perhaps as a combination of the two. Based on the COCOM, it is clear that time plays a role in several different ways. The evaluation of the situation is susceptible to delays in feedback or responses from the process or application being controlled, as well as to the aging of information. This in itself depends, among other things, on how long the evaluation takes, which leads to an intricate coupling between the two. The choice of action depends on the stability of the process and on the window of opportunity. In many cases it is more important to do something quickly than to spend time on finding the optimal action. In other cases there may be limited time in which to carry out an action, and an alternative that is quick to implement may therefore be preferred over one that takes longer although that in other respects might have been better.

Balance between feedback and feedforward

In practice people are well aware of these dynamic dependencies and usually try to reduce the time needed to evaluate and choose in several ways. Efficient performance depends on an equilibrium between being proactive (relying on feedforward) and being reactive (relying on feedback), which people usually are able to achieve provided the work environment is reasonably stable. If performance is dominated by feedback control, that is, by mainly reacting to what happens, people may soon find themselves in a situation where they lag behind, which invariably will aggravate any shortage of time. In extreme cases, a dependence on reactions means that it is impossible to make any plans, hence to be prepared for what may come.

In order to avoid this, people try to look ahead in order to be able to respond more quickly. The use of anticipation or feedforward control gains time by reducing the need to make a detailed evaluation of what happens and of the feedback; by being prepared to respond, actions may be taken faster and the necessary resources may be made ready ahead of time. Looking ahead, however, also requires that not all the time is used to evaluate the feedback and that the risks from ignoring part of the feedback are sufficiently small. The disadvantage of relying on feedforward control is that the expectations or predictions may be inaccurate and that the prepared response therefore may be inappropriate. This risk clearly increases the further ahead the decision maker tries to look. If the predictions are inaccurate then the chosen actions may lead to unexpected results, which in turn increases the time needed for evaluation, hence increases the role of feedback. In terms of this balance, decisions can be seen both as buying time – by enabling the controller to be ahead of developments – and as using time. It is therefore important both that it does not take too long to make the decision and that the actions are carried out at the right moment and with the right speed.

Efficiency-thoroughness trade-off

One way of reducing the time needed to make a decision is to be sufficiently rather than completely thorough in the evaluation of the situation and in choosing what to do. This is a well-known phenomenon in decision theory and has over the years been referred to by names such as muddling through (Lindblom, 1959), satisficing (March and Simon, 1993) or recognition-primed decisions (Klein, Oranasu, Calderwood and Zsambok, 1993). Trading thoroughness for efficiency, an efficiency-thoroughness trade-off (ETTO) principle (Hollnagel, 2004), makes good sense since there is never infinite time or infinite resources available for the decision. Actions must be taken so that they fit into the window of opportunity, and for that to happen shortcuts and trade-offs must be made.

This efficiency-thoroughness trade-off is a common feature of human performance on the level of individuals and of organisations alike. On the individual level the ETTO principle can be found both in the characteristics of cognitive functioning as well as in how people go about their work. Tversky and Kahneman (1974), for instance, demonstrated that people rely on a small number of heuristics or mental shortcuts that reduce the complex tasks of assessing probabilities to simpler judgemental operations. Functionally equivalent principles can be found on the level of activities, in work and at home. Rules of thumb such as "looks OK", "not really important", "normally OK, no need to check this now", or "it will be checked by someone else later" are probably recognised by everyone. The reason why people behave in this way is that they try to be efficient by only being as thorough as they believe is necessary. The criterion for making the trade-off is, however, not fixed but depends on the context. For instance, if the external or internal pressure to complete a task or meet a deadline is very high, people will lower their demands to thoroughness, that is, they are willing to take greater risks. Although this strategy on the whole is successful and therefore must be considered normal, the outcome may every now and then differ from what was expected, hence be classified as a failure. Yet the failure will be one of acting rather than one of reasoning.

Decision failure modes

By shifting the perspective of decision making from being a question about *what* to do, to become a question about *how* to do it, the importance of the rationality of choice is diminished. Instead of being isolated as a distinct process, decisions become part of the activities by which people – and systems – stay in control of what they do and thereby become a natural part of how people cope with complexity. An added benefit is that instead of seeing decisions as being either right or wrong, we can describe them in the same ways as other actions, specifically in how they can fail. Just as it is pointless to describe human actions in the binary categories of errors and correct actions, it is similarly unproductive to describe decisions only as right or wrong.

In the field of human reliability and socio-technical accident analysis, failure modes are characterised in terms of the possible manifestations or phenotypes (Hollnagel, 1993). This produces a definition of eight distinct failure modes as shown in Table 1.1.

Table 1.1 Human failure modes

Failure Mode	Categories
Timing (when)	Action performed too early or too late
Duration (how long)	Action performed too briefly or for too long
Distance (how far)	Object/control moved too short or too far
Speed (how fast)	Action performed too slowly or too fast
Direction (where to)	Action performed in the wrong direction
Force/power/pressure (how strong)	Action performed with too little or too much force
Object (which object)	Action performed on wrong object
Sequence (which order)	Two or more actions perfomed in the wrong order

Since a decision is just like any other type of activity, it can also go wrong in the same ways, hence be described using the same principles. The simple-minded classification of a decision as being correct or incorrect is thereby replaced by a more detailed account of how decisions can fail, which must be the necessary starting point for understanding why it failed. Even if we focus only on the activities that constitute the choice as such, which is perfectly reasonable in cases where a deliberate consideration of alternatives precede an action, it is clear that a decision can fail in a variety of ways.

- A decision can be wrong because it is made at the wrong time, either too early before the necessary information was available or too late when the opportunity for action has disappeared. The latter can happen either if the decision maker started too late, or if too much time was spent in considering the options and alternatives.
- A decision can be wrong because it is made too quickly, that is, because not enough time was spent in finding and considering alternatives. Note that this failure mode (duration) is conceptually distinct from the one above (timing). It is a common finding that people try to save time by making an efficiency-thoroughness trade-off (for example, Hollnagel, 2004).
- A decision can fail because it considers the wrong object (that is, the wrong alternative), specifically because it excludes the right alternative(s). This is less straightforward than the above categories, since the determination of which alternatives were right and wrong often cannot be made *a priori*. In some cases choices are based on "irrational" criteria such as an imperious immediacy of interest or basic values (Merton, 1936), which may lead to the wrong alternative being preferred.
- A decision can finally fail if it is taken out of order or sequence. This recognises that a decision is not a single and isolated activity, but part of a set of activities that for practical reasons have to be organised or ordered. Indeed, few real-life decisions only involve one choice; most involve multiple choices that are related in one way or another, for instance as pre-conditions or post-conditions. The choices may also refer to processes that have different temporal characteristics (rate, speed) and scope (extension, time horizon), and thereby be mutually incompatible.

Each of these failure modes may, of course, occur for a number of reasons referring both to situational factors and to more inherent characteristics of human activity. A decision can, for instance, be made too quickly because of a communication failure, an inadequate procedure or plan, missing information or limited resources, time pressure, and so on. While it is quite possible to propose a set of generic causes or antecedents (for example, Hollnagel, 1998), a specific analysis should not be done without referring to the characteristics of the domain and situation.

If we widen the scope and look at decision making as an activity or as part of meaningful actions, all the failure modes of Table 1.1 become relevant. We can therefore replace the normative criterion of making the right decisions by the more detailed analysis of whether the decision is made at the right time, with the right duration,and so on, and implemented in a way that achieves the desired objectives (using the right force, in the right direction, with the right speed, and so on). This means that decision making, instead of being just the choice among alternatives, takes its natural place in the activities needed to achieve the overall objective of staying in control. Thus rather than looking at decisions in isolation, one by one so to speak, we have to look at decisions as part of the ongoing activity as emphasised, for example, by the contextual control model.

Consequences

Changing the view of decision making from focusing on *what* to focusing on *how* also changes the issue of decision support completely. One consequence is that decision automation is no longer an issue, since decision making cannot be automated without ceasing to be decision making. The reason for that is that whereas automation is feasible for situations that are highly regular, hence can be analysed completely in advance, decision support is needed for situations that are irregular and unpredictable, which makes prior analyses difficult or impossible. The support must first and foremost be closely integrated with the task and therefore be continuous rather than discrete. In that sense many, if not all, aspects of interface and interaction design become issues of decision support.

A further consequence is that discussions about "intelligence" in decision support must also change. If decision making is an activity rather than a process, then the intelligence clearly cannot be in the support as the embodied process but must reside in the decision maker (as an individual or a group). We should therefore strive to support intelligent decisions and intelligent implementations rather than to build intelligent support systems. In that sense the implementation issues – *how*, *when*, and so on, rather than *what* – become issues of maintaining control and of regaining control if it is lost, rather than of supporting decision making as a mental process.

References

Bainbridge, L. (1983). "Ironies of automation". *Automatica*, 19, 775-779.

Brehmer, B. (1992). "Dynamic decision making: Human control of complex systems". *Acta Psychologica*, 81, 211-241.

Dekker, S. W. A. and Woods, D. D. (2002). "MABA-MABA or Abracadabra? Progress on Human-Automation Co-ordination". *Cognition, Technology & Work*, 4(4), 240-244.

Edwards, W. (1954). "The theory of decision making". *Psychological Bulletin*, 51, 380-417.

Fishburn, P. C. (1991). "Nontransitive preferences in decision theory". *Journal of Risk and Uncertainty*, 4(2), 113-34.

Fitts, P. M. (1951). *Human engineering for an effective air navigation and traffic-control system.* Ohio State University Research Foundation, Columbus, Ohio.

Hollnagel, E. (1984). "Inductive and deductive approaches to modelling of human decision making". *Psyke & Logos*, 5(2), 288-301.

Hollnagel, E. (1993). "The phenotype of erroneous actions". *International Journal of Man-Machine Studies, 39,* 1-32.

Hollnagel, E. (1998). *Cognitive reliability and error analysis method – CREAM.* Oxford: Elsevier Science.

Hollnagel, E. (2002). "Time and time again". *Theoretical Issues in Ergonomics Science*, 3, 143-158.

Hollnagel, E. (2004). *Barriers and accident prevention*. Aldershot, UK: Ashgate.

Hollnagel, E. and Woods, D. D. (2005). *Joint cognitive systems: Foundations of cognitive systems engineering*. Boca Raton, FL: CRC Press / Francis & Taylor.

Klein, G. A., Oranasu, J., Calderwood, R. and Zsambok, C. E. (1993). *Decision making in action: Models and methods*. Norwood, NJ: Ablex.

Lee, W. (1971). *Decision theory and human behaviour*. New York: John Wiley & Sons.

Lindblom, C. E. (1959). "The science of 'muddling through'". *Public Administration Review*, 19, 79-88.

March, J. G. and Simon, H. A. (1993). *Organizations* (2nd Ed.). Cambridge, MA: Blackwell Publishers.

Merton, R. K. (1936). "The unanticipated consequences of social action". *American Sociological Review*, 1 (December), 894-904.

Park, J. and Jung, W. (2003). "The requisite characteristics for diagnosis procedures based on the empirical findings of the operators' behavior under emergency situations". *Reliability Engineering and System Safety*, 81, 197-213.

Tversky, A. and Kahneman, D. (1974). "Judgment under uncertainty: Heuristics and biases". *Science*, 185, 1124-1131.

Woods, D. D., Johannesen, L. J., Cook, R. I. and Sarter, N. B. (1994). *Behind human error: Cognitive systems, computers and hindsight*. Columbus, Ohio: CSERIAC.

Chapter 2

Corruption and Recovery of Sensemaking During Navigation

Gary Klein

Introduction

The purpose of this chapter is to examine the activity of navigation as a form of sensemaking. In particular, the chapter examines the process of getting lost and then recovering. Navigation can be described as the process of getting from one location to another. But the same tactics of sensemaking that let people understand how things fit together can also result in navigation errors, leading people to become lost. By studying how the process of sensemaking can corrupt our understanding of events, we may be able to learn more about the way people seek to understand events in the decision making process. The cycle of getting lost and then recovering – getting found – is a window into the process of sensemaking.

The nature of sensemaking

The concept of sensemaking came into prominence with the publication in 1995 of the book *Sensemaking in Organizations* by Karl Weick. Weick noted that sensemaking is often triggered by a surprise, such as a data element that does not fit with the accepted interpretation of events. Thus, sensemaking is most visible when predictions break down. From this surprise, people may look back to realise that there were other discrepant cues that now fit into place; Weick refers to this as a retrospective examination of elapsed experience.

Leedom (2002) has drawn on the writings of Weick to compile a set of conditions that influence sensemaking:

- The nature of the problem is itself in question: what requires attention and adjustment is unclear, shifting, or intertwined with other concerns.
- Information is problematic: there is doubt regarding what information is needed, and how it is to be collected, filtered or categorised.
- There are multiple, conflicting interpretations: reported facts and their significance can be read in different ways.
- Differences of value orientation exist: lacking adequate objective criteria, decision makers rely on personal and cultural values to read significance into a situation.
- Goals are unclear, multiple or conflicting.

- Time, combat resources and attention are limited.
- Contradictions or paradoxes appear.
- Roles and responsibilities are unclear: decision makers are unsure as to what mission success means and how to measure it.
- Poor understanding of cause-effect relationships.
- Symbols and metaphors are used in confusing ways.
- Key decision participants are fluid: different staff members are entering or leaving the situation as a function of problem redefinition or staff shift rotation.

Weick also makes the point that sensemaking is focused *on* the cues extracted from a continuous flow of events; sensemaking is also focused *by* these cues as the person selects and defines the cues and highlights implications that might be missed.

Klein, Phillips, Rall and Peluso (forthcoming) defined sensemaking as "a process: the deliberate effort to understand events". It is typically triggered by unexpected changes or other surprises that make us doubt our prior understanding. Sensemaking can also be initiated in the absence of surprise, as when goals change and a person realises the value of thinking more deeply about some issues.

Information processing accounts of sensemaking have tended to describe a mechanical sequence of inputting data and applying transformation rules, and obtaining inferences. These bottom-up accounts miss the importance of a parallel top-down process that defines the cues, guides the information seeking, and organises the inferences. Klein et al. (forthcoming) have presented a data/frame theory of sensemaking that portrays the iteration of top-down and bottom-up processes.

Data/frame theory

In the data/frame theory of sensemaking (Klein et al., forthcoming), understanding is achieved when the data and events can be fitted into a frame such as a story, a script or a map. However, people also use the frame to define the data. Therefore, we posit an interaction between the data and the frame, with neither having precedence. We need the data to retrieve and construct appropriate explanatory frames. We need the frames in order to define the cues and separate relevant cues from noise.

The process of sensemaking is to understand a situation against a noisy background where the cues themselves are not given but must be constructed. The signal stream contains irrelevant signals, incorrect signals, obsolete signals, circular reporting, accurate signals that come from sources that cannot be trusted, inconsistent signals, sets of signals that are too complex to link together, signals that appear to be irrelevant but are actually important, subtle signals that are difficult to notice, signals that may be deliberately misleading, as well as relevant and accurate signals. Sensemaking requires the person to determine which signals to discard, and which to emphasise.

Noise is not simply the background for a "real" situation. Noise has its own reality. Part of sensemaking is in identifying what is signal and what is noise. Erroneous reports have to be interpreted as such, and explained away, or they will corrupt the sensemaking. Irrelevant data have to be perceived as such and given little attention except the attention needed to dismiss them.

Furthermore, we see sensemaking as deliberate, a mental manipulation of cues in order to derive inferences, as opposed to a recognitional pattern matching process. Unconscious activities such as pattern matching play a role in sensemaking as do many other cognitive processes, and unconscious recognition of situations and events result in situation awareness – we are treating sensemaking as the deliberate reasoning that takes place when a person has to expand or revise the way events are understood.

Klein et al. (forthcoming) have also identified different forms of sensemaking, as shown in Figure 2.1. Sometimes we elaborate the frame as we learn more about a situation. Other times we try to puzzle out the significance of data that are inconsistent with the frame we are using. We may explain away these data, or we may take them seriously and revise or replace the frame. Sometimes we have to compare alternative frames, and sometimes we may truly be in the dark, seeing some sort of frame for data that we cannot automatically connect.

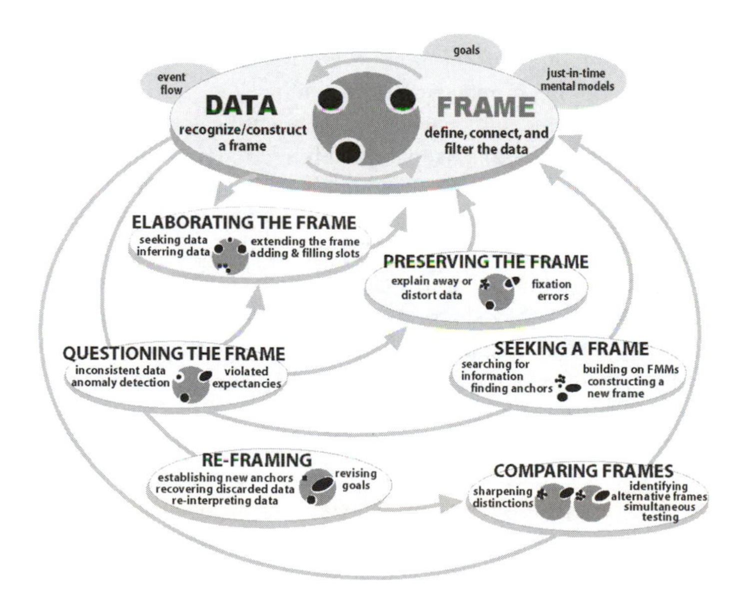

Figure 2.1 Data/frame model of sensemaking

The starting point in Figure 2.1 is a recognitional match between data and frame, as described in Klein's recognition-primed decision model (Klein, 1998). This is usually not a conscious process, and therefore is not a form of deliberate sensemaking.

Based on this match, relevant data are attended to, and the person has retrieved or constructed some sort of frame (for example, a story, script or map).

The specific nature of the frame can have a great influence on problem solving activities. Alan Brunacini, the chief of the Phoenix Fire Department, has stated that the way an on-scene commander sizes up the situation in the first five minutes determines how the fire will be fought for the next five hours. Chi, Feltovich and Glaser (1981) showed that more skilled physics students adopted a deeper representation of paper and pencil problems than novices. They framed the same problems in terms of laws of physics that corresponded to formulae that could be applied. These formulae were serving as scripts about how to carry out a sequence of actions. The frames/scripts provided the higher-level students with a functional understanding of the problems – an understanding in terms of operations that could be carried out to solve the problems. In contrast, the novices adopted representations based on physical features of the problems that did not correspond to formulae they could use.

Klein, Calderwood and MacGregor (1989) distinguished between two types of activities: elaborating one's situational understanding (facet two in this model of sensemaking), and shifting one's understanding (facet four), and demonstrated a "situation assessment record" that represented the evolution of understanding in terms of whether new messages led to an elaboration or a shift from the existing account of the situation. Figure 2.1 has expanded on this distinction by further distinguishing the activities of questioning the frame and either disregarding the contrary data or modifying or replacing the frame.

The process of sensemaking depends on the identification and interpretation of new data. In many cases, a frame is identified based only on three or four key data elements, which can be thought of as anchors. Sensemakers identify a small set of anchors that are promising cues and inferences, as the basis for constructing the frame. Anchors can be thought of as "holds" in rock climbing – bases for constructing a story or script (or for traversing a route, in climbing). They are firm cues with high information value – they shape the nature of the story/script/map. Klein and Crandall (1995) hypothesised that because of memory limitations, only three to four anchors can be sustained at once. The sensemaker needs to identify a promising set of anchors and work from there to construct a story. The story or explanation is woven together from the anchors and from additional cues that are pertinent.

However, some new data will not fit well into the frame, leading sensemakers to *question the frame*. These data can be explained away, as shown in Figure 2.1 by the aspect of *preserving the frame*. One way to do this is to elaborate the frame so that its general structure is preserved. For example, Ptolemy framed the orbits of the sun and moon and planets as circles. Observations showed that this simple frame was inaccurate. In order to maintain a frame of circular orbits, Ptolemy had to devise a complicated system of epicycles. The Ptolemaic system survived until Galileo and Copernicus and Kepler suggested a new frame – an orbiting earth, and elliptical orbits instead of circular ones. At this point, astronomers *compared* frames, and settled on the Copernican view rather than the Ptolemaic view.

Therefore, we have a balancing act as discrepant data are explained away by increasing the complexity of the frame, until a point where the complexity becomes too unwieldy and the frame is rejected in favour of another that is simpler. We posit a strategy whereby the initial data elements are critical for eliciting a frame that will be used to guide further information search and interpretation. The succeeding data elements will be less influential than the initial one, except in the case where the person detects an anchor that is so discrepant and so credible – a framebreaker. The identification of a framebreaker results in the rejection of one frame for another. Frame replacement can also be triggered by the weight of evidence, or by the unwieldy nature of the revised frame.

Spreading corruption

Although some distortions are inevitable, they usually do not pose problems. On occasion, however, one or more of the anchors that a person is using to make sense of events can be inaccurate. When this happens, the sensemaking processes that worked so well in routine circumstances can become corrupted, as the frame is twisted to accommodate flawed data, and data are twisted to fit the frame. The erroneous frame leads to distortions in the data, which supports fixation on the frame and selective attention to irrelevant data and further distortions of new data, a process we can describe as "spreading corruption". The initial confusion corrupts the data seeking and interpretation and therefore builds on itself. Inferences derived from the inaccurate belief are tainted, and taint other inferences. Each of the aspects of sensemaking shown in Figure 2.1 can contribute to the breakdown of understanding when the process of spreading corruption takes over.

Corruptions are inevitable, given our buggy mental models of the world, our inaccurate stories about how things happen, our flawed scripts, our mistaken maps. Given the number of beliefs we hold in a given domain, it is essentially impossible to eliminate inconsistencies. Corruptions are also created because of erroneous data. And sometimes, the situation itself may change so that, through no fault of our own, we come to believe in the validity of cues and representations that are inaccurate. When the degraded cues are the ones we are depending on to anchor our representation, the impact is particularly strong.

Syrotuck's classic work *Analysis of Lost Person Behavior* (1977) provides some excellent examples of how people got profoundly lost in the wild – lost without any idea of how to recover.

The experience of getting profoundly lost is highly stressful:

> People who are lost may experience differing types of reactions. They may panic, become depressed, or suffer from 'woods shock'. Panic usually implies tearing around or thrashing through the brush, but in its earlier stages it is less frantic. Most lost people go through some of the stages. It all starts when they look about and find that a supposedly familiar location now appears totally strange, or when they start to realise that it seems to be taking longer to reach a particular place than they had expected. There is a tendency

> to hurry to 'find the right place'. 'Maybe it's just over that little ridge.' If things get progressively more unfamiliar and mixed up, they may then develop a feeling of vertigo, the trees and slopes seem to be closing in and a feeling of claustrophobia compels them to try to 'break out'. This is the point at which running or frantic scrambling may occur and indeed outright panic has occurred.
>
> Regardless of how well and healthy a person seems to be when rescued, there is almost always some degree of shock. Even people who, while lost, appeared to use good judgement with no suggestion of overt panic, exhibit what we like to call 'woods shock'. Many persons found mobile and well will seem to converse in a completely normal manner. Only upon close questioning does it become evident that they are unable to remember where they spent the first night, whether they had any water to drink, or whether they crossed the river yesterday – or maybe the day before (p. 11).

Syrotuck identifies the confusion factors that lead people to get lost in the woods: confusing intersection of roads or trails, often unmarked; trails obliterated by snow fields, rock slides, overgrowth, lack of use, or multiple routes; subtle changes in terrain; similarity of terrain features such as meadows, lakes or rocky outcrops.

We can use the example of getting lost in the woods as a metaphor of what happens during sensemaking when beliefs and frames are corrupted.

Sarter and Woods (1995) performed research on mode errors in aviation. With advanced information technology, modern cockpits offer pilots a variety of modes, rather than requiring pilots to operate separate, stovepiped systems. One risk is that it may not always be clear to the pilot which mode the system is in. If the pilot makes an error about mode, it can be possible to operate the system for some time, making inputs and interpreting system responses, all the while being in the wrong mode. This type of mode error is an example of the way corruption can creep into sensemaking and be masked and sustained.

As the inaccuracies increase, evidence is explained away by "bending the map" (Gonzales, 2001) or explaining away the discrepancies. The phenomenon of fixation is sometimes seen as a wanton sustainment of faulty beliefs, but we can also see fixation as a natural consequence of spreading corruption. Similarly, confirmation bias (Mitroff, 1981; Tolcott, Marvin and Bresnick, 1989) may really be the application of corrupted cues and stories. In fixation and confirmation bias people are seeking data that confirm what they already believe rather than looking for counter-indicators that could test their beliefs.

The data/frame theory would view these phenomena as consequences of using frames to identify and represent data. For example, scientists typically try to confirm their theories by demonstrating how their theories can account for various phenomena – scientists use their theories to guide the way they set up experiments. The same process holds for ordinary people trying to make sense of events. The process becomes more visible when it breaks down. The apparatus we use to make sense of the world, the story/script/map, has become flawed and as we use it, we bend the map and further distort the story. In navigating a hiking trail we may expect to see a stream and do not find one – perhaps it dried up, or was diverted since our map was produced. We spot a hill that is not supposed to be there – maybe it is another hill further away – we know how difficult it is to judge distances accurately. In one extreme case, a hiker was seen smashing his compass against a rock because it was giving him information that was so discrepant with his beliefs that it was driving him crazy.

Sensemaking during navigation

To explore the phenomenon of sensemaking, I conducted a very informal investigation of navigation. I asked a small, unselected sample of people, all with college degrees and many with advanced degrees, to generate incidents in which they had become lost. I was primarily interested in how they got lost while driving their cars but some of the incidents involved land navigation. The informants described how they came to be lost, how they realised they were lost, and how they "got found" again. This call for examples generated 14 incidents that were examined individually for ideas and hypotheses. No attempt was made to code these data because their collection was not systematic. Instead, the incidents were examined individually for ideas and hypotheses, to see how these incidents either reinforced or extended the data/frame model of sensemaking.

Levels of being lost

I was able to distinguish four levels of being lost: not lost, mildly lost, seriously lost, and profoundly lost. If we know exactly where we are with high precision, we are not lost. Most of us are mildly lost most of the time, in terms of not being able to specify with high accuracy where we are. For example, few of us can, at this instant, point to the front door of our house or office with great accuracy. This is not a problem because we usually possess very good routines for reorienting. We do not feel lost because our inadequate knowledge of our orientation does not prevent us from carrying out functions. So we can say that we are mildly lost – not exactly locked into a map, but not suffering any consequences either.

The next level is being seriously lost – disoriented and temporarily blocked from carrying out important functions until we can reorient. When we lose our rental car in a large parking lot, we can be seriously lost if we are unable to locate it after wandering around for 20-30 minutes. The level of "lost-ness" is related to the proportion of routines we can still perform. As the routines and affordances diminish, the person is moving from being mildly lost to being seriously lost. We do not have enough reliable anchors to be able to proceed, and we will have to take steps to repair the flaws and increase the reliable anchors.

The final phase is being profoundly lost. Here, we are unable to repair the flaws in our story, script, or map, and are unable to obtain additional reliable anchors. We may only have a single anchor available to us, perhaps none at all. People in the wilderness who become profoundly lost often die within a few days, sometimes in less than a day (Syrotuck, 1977).

Most of us do not venture very deeply into wilderness terrain to reach the level of being profoundly lost. However, even the simpler task of automobile navigation offers a range of opportunities to become lost. Consider the task of travelling to an unfamiliar city and attempting to get from one place to another. We are trying to fit new data into a map, to make these consistent with the anchors. Because of the paper artefact of a map, we can have many more anchors than in other types of sensemaking.

When we think of how we navigate to drive to a destination, it usually seems like a simple matter of following the directions, making the correct turns, and being careful. And most of the time, that is what happens. We know where we are, we know our destination, and we know our route.

Navigation gets much more exciting when the directions do not work and we end up lost. That is a true test of sensemaking. We lose some or all of the anchors – we no longer know where we are, and therefore we cannot know our route. We may not even be sure about our destination – we may not trust the person who gave us the directions. The navigation incidents I collected described anchors that help us maintain our direction and notice deviations. The next section describes the cues and anchors available to us for navigating. After reviewing these, we will cover the ways that these cues and anchors can become corrupted.

Range of cues and anchors

In navigation we usually have a wide range of cues and anchors to help guide us. Generally, we only use a subset of these cues. The more we prepare, the less likely we are to get seriously lost. The things we know will often include:

- Our current location when we started out
- Our destination
- The route we are planning to take
- The places where we are going to shift directions
- For each shift point, we can know:
    - The preparations (for example, the previous street)
    - The error markers (indications that you have missed the turn)
    - The possible areas of confusion (anticipations about potential errors in shifting directions, or in following a street that may not continue in a linear fashion)
- The recovery routes (other roads we can get to without trouble)
- The danger zones (places from which it will be difficult to recover should we make an error, such as getting on the wrong expressway during rush hour)
- The ways to increase flexibility (such as using streets and staying off expressways)
- Speed accuracy trade-offs (using expressways rather than local streets)

- Compass headings and landmarks
- A sense of scale – time/distance relationships – to tell if we should be reaching a given street by now and have not

With all of these safeguards, navigation is usually successful. But the process of navigation also has its troublesome elements. For example, giving directions seems straightforward, and web-based mapping programs not only provide directions once we enter our originating point and destination, but also offer the choice of the fastest route, the shortest route, and even a route that avoids freeways. Yet what we really need in trying to find our way to an unknown location is the direction with the least amount of confusability, and the software mapping programs do not generate that. The incidents of navigational failures contained a large assortment of breakdowns as described in the next subsection.

Corruption of cues and anchors

There are many ways we can entertain erroneous cues during navigation:

- Faulty directions:
  - Right/left reversals
  - Inaccurate scale
  - Wrong street names
  - Flat-out errors
- Street names change:
  - Street signs can be turned around
  - Streets can be renamed after the map was published
- Lines on a map are misleading. Web mapping services may show streets going through areas when these streets may be temporarily or permanently blocked
- Duplicate street names, as when neighbouring municipalities use the same street name for different streets
- Numbering discontinuities, as when neighbouring municipalities fail to maintain sequential numbers for a street and we do not realise we have travelled from one municipality to another because they have grown to be contiguous, as in Boston
- Subtle differences:
  - We see the correct name but it refers to an avenue or road, not a street (for example, the various Peachtrees in Atlanta)
  - The streets are not at right angles, so we struggle to maintain a straight course
- Mismatched designations:
  - We know the route number but the street signs tell the names of the roads
  - The signs tell us the final destination of a highway and we do not know where on the map to look to find this city
- Road blockage:

  - Construction forces us to abandon the route we planned
  - An accident to some other cars prevents us from making the turn we need
  - We are blocked by a sign that prohibits left turns, or some other unexpected constraint
- Obscuration of landmarks:
  - Night-time driving
  - Fog
- Street signs that are difficult to locate or read:
  - Too small
  - Placed obscurely
  - Hidden behind a large truck
  - Lost in the glare as we drive towards the sun
  - Unlit at night
  - Covered by snow
- Map unreadability, as when we turn the map upside down to orient ourselves, and now the words are unreadable
- Unexpected highway type, as when we fail to realise that we are entering a limited access roadway
- Map scale: one of the most interesting aspects of corruption is due to the difficulty of gauging the scale of the map

To expand on only one of these topics, map scale, we may be able to match up the street names we are seeing with the ones shown on our map – if we knew where we were. But if we are lost, we may be looking in the wrong place, and not finding the observed streets. Maps do not work well when we need them the most. In some cases, we look at the wrong map (if we have printed out several) because, being lost, we did not know which map to consult. We need a scale sense to know where to look on a map, or to judge if we have gone too far, or other types of activities that depend on time/distance models.

Example 1. Dayton to Chicago by way of Detroit.

This incident involved a trip from Dayton, Ohio to Chicago, Illinois. The navigator checked a map and determined that the route, as planned, was simple: West on I-70 to Indianapolis, take the 465 beltway north around Indianapolis, go north when reaching the I-65 highway and take that towards Chicago (see Figure 2.2).

However, traffic was heavy and moving fast on the 465 beltway north of Indianapolis. The navigator had non-prescription sunglasses, and when the driver said "I-69 – isn't this our exit", he agreed, confusing I-69 and I-65. Trying to be careful, he decided to check, and looked at the map. But he looked at the I-65 highway, which is where he thought they were.

He was unable to find any of the roads they intersected on the map (because he was looking in the wrong place, on the I-65 route), but he explained away this anomaly as the poor level of detail available on the map. The navigator could have easily spotted those same roads along the I-69 route, but then he would not have been lost.

He could have found the intersecting roads if he conducted an extensive information search of the entire map. This type of exhaustive search was simply impractical.

After continuing north for half an hour, the navigator noticed a sign giving the number of miles to Detroit. The sign did not mention Chicago. This struck the navigator as odd. He had also been growing uneasy about not seeing a single familiar cross road. The sign pointing the way to Detroit was the framebreaker.

The navigator studied the map more intently because he now had two anchors – the 465 beltway, and Detroit as a destination. He noticed that there was another highway I-69 leading off 465, and connecting to interstate I-94 that heads east to Detroit. He also recognised several of the cross roads they had passed earlier, now that he realised his error, and diagnosed how he had gotten on the wrong interstate. It was all a matter of knowing where to look. The use of an erroneous frame (believing that he was on I-65 instead of I-69) misdirected attention and information seeking.

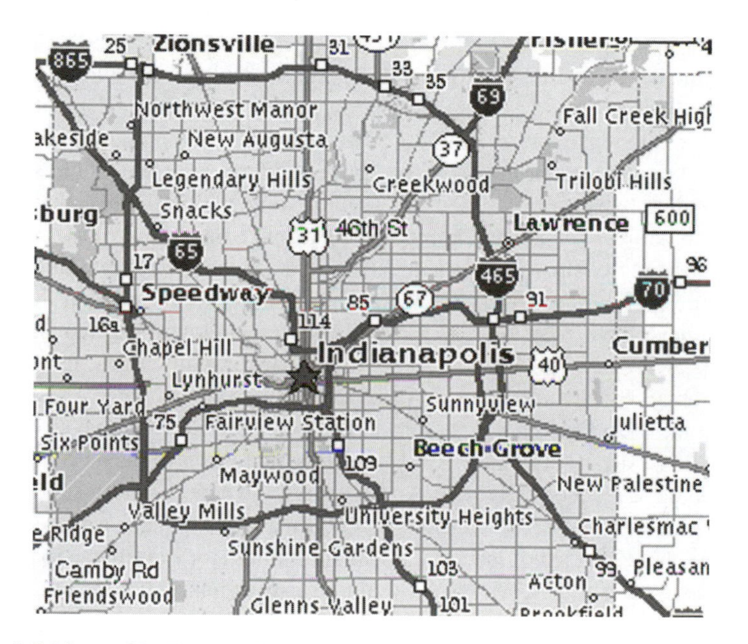

Figure 2.2 Map of Indianapolis

Realising that we are lost

Getting lost is only a part of the sensemaking challenge. In following unfamiliar routes, we commonly make small errors or become momentarily confused. In most cases we immediately realise the error and correct it. On some occasions we fail to notice that we are getting lost, and persevere until our chance for recovery has diminished. This section examines the functions of detecting problems, and the process of using knowledge shields to become trapped in fixation.

Problem detection

The incidents I collected showed how critical it is for people to decide that they are lost, they are confused, and their sensemaking has failed. We all know the experience in driving someplace that if we just keep going, perhaps the landmark or the street we are looking for will appear over the next hill or at the next stoplight. The further we go before admitting confusion, the harder to recover. If we catch the confusion early enough, we can backtrack, or at least start monitoring our progress more carefully to enable us to backtrack later. A more typical pattern is to press on, to continue bending the map. In the worst cases, people become profoundly lost and cannot even backtrack. Weick (personal communication, 25 June 2002) has distinguished between effective sensemaking ("I can keep the action going") and ineffective sensemaking ("I'm getting stuck"). The navigation incidents illustrated this distinction. People can keep the action going for a long time even as their understanding has become seriously corrupted.

One of the hardest parts of getting found is realising that we are lost. (See Klein, Pliske, Crandall and Woods, 2005, for a description of problem detection.) If we are having trouble reading or even finding street signs, how long do we keep driving before we acknowledge a problem? If we turn around too soon, we will miss the next landmark and perhaps get lost. Often, we wait too long, making it much harder to get found. We may press on, accumulating more evidence that we are lost. Sometimes we use a subtle indicator – perhaps a negative cue, as when we pass a landmark and say "the fellow giving me directions would have surely told me about this – since he didn't, I must have gone too far". Or a designated landmark such as an overpass has not appeared and does not seem likely to appear for many miles. In most of the incidents I obtained, problem detection was triggered by a framebreaker, such as the sign giving miles to Detroit in Example 1. But often we do not have any dramatic cues to tell us we are lost. We just have the feeling that we have been driving forever – we should have reached it by now. And we have the accumulating set of nagging discrepancies, such as a perceived ambiguity about a turn we made, a sense of having gone too far on a given leg of the route, local anomalies (for example, a road passing through a deserted stretch when it should not be), street names that do not match the map.

Knowledge shields

Problem detection does not always succeed. Perrow (1984) used the term "de minimus explanations" to describe how people explain away inconvenient data and fail to attend to early cues that something is wrong with the way they are understanding events. Feltovich, Coulson and Spiro (2001) have described a wide array of "knowledge shields" that enable us to explain away anomalies.

Feltovich, Coulson and Spiro (2001) observed cardiac surgeons in simulated scenarios, where the surgeons adopted an erroneous diagnosis. Frequently, the surgeons discounted new data that contradicted the misdiagnosis. They explained these data away. Feltovich et al. distinguished more than 20 different types of arguments, or "knowledge shields", that the surgeons used to discard the discrepant new data.

Not all of the knowledge shields identified by Feltovich et al. (2001) are examples of flawed reasoning. Some are normally adaptive reasoning that can be misapplied to important signals. For example, one knowledge shield is to treat the discrepant information as a special case that does not apply, or to decide that the discrepant phenomenon only applies under extreme conditions that are different from the current conditions. Other knowledge shields, such as argument from authority, can sometimes be appropriate. Some of the knowledge shields, such as relying on a flawed or buggy mental model, are not really about the form of the reasoning at all. Regardless of the nature of the knowledge shields, their effect is to delay the realisation that sensemaking has become corrupted by explaining away inconvenient data.

Applying the concept of knowledge shields to navigation errors, if a highway sign seems to give us inappropriate information, perhaps there is another route, involving a highway not relevant to us. If an overpass does not appear, perhaps we just have not gone far enough. The appearance of an unexpected landmark may just indicate that the person giving us directions was not very compulsive. Ambiguity about a turn is pervasive, and a sign that we are vigilant after all. The sense that we have already gone too far can be countered by the need to be sure that we are not almost there. Local anomalies can be ignored if we are in unfamiliar terrain. The mismatch of street names to our map is not a dead giveaway of being lost. The map may not be sufficiently complete to capture all the streets. Clearly, there are many available ways to dismiss the initial indications that we might be getting lost.

Fixation

Knowledge shields help to sustain fixation on an incorrect hypothesis. Because we rely on the story/script/map to direct our attention, interpret signals, and filter signals, it is very disorienting to give up this frame. The phenomenon of fixation is not simply due to laziness and inertia. De Keyser and Woods (1990) described fixation as the failure to revise a mistaken explanation despite the opportunity to do so. The data/frame theory would hypothesise that people ordinarily rely on frames, and use these frames to direct their search for new data. When the frames turn out to be flawed, we treat persistence in their use as fixation. When the frames are reasonably accurate we do not notice how discrepant data are dismissed.

Smith, Giffin, Rockwell and Thomas (1986) have conducted a study of how experienced pilots perform fault diagnosis, given a complex malfunction during a simulated flight. The analysis provided by Smith et al. includes a frame in the form of a diagnosis script, with two slots: possible causes, and expectations about instrument readings and other indicators. Smith et al. found 18 variants of this script. Pilots who missed the diagnosis generally had or knew the necessary knowledge, but got fixated on a wrong hypothesis, and as a result, ignored data that suggested something else. Another group of pilots who missed the diagnosis failed to judge if their expectations from a script were accurate. Smith et al. concluded that for most of the errors, the initial hypothesis was incorrect and the pilot fixated on that working hypothesis, and failed to look at competing hypotheses.

The effect of fixation was demonstrated by Bruner and Potter (1964), who used the paradigm of presenting a visual stimulus in a blurred form, and gradually improving the focus until the subject was able to recognise it. Bruner and Potter found that if they had college student participants who did not see the stimulus in its most blurred form, they had significantly faster recognition. In other words, the incorrect hypotheses they formed upon first seeing the stimulus in its most blurred form was preventing them from making sense of it as it came more into focus.

The fixation in Example 1 does not seem like a deliberate attempt to maintain faith in a flawed frame. Rather, the fixation was a simple result of using a frame to direct attention, which is one of the functions of frames. We call this fixation when we are working with a flawed frame. We call it efficient attention management when our frames are accurate.

Defences against knowledge shields

There are several possible actions we can take if we suspect that our belief system is becoming corrupted. One is to be alert to conditions where a routine is altered, as a stimulus to become more alert, to increase the level of monitoring for disconnects. We can also try to anticipate problems that might introduce corrupted beliefs, or to be more sensitive to reduced margin of error as a danger sign.

Thus, in Example 1, the navigator should have looked at the map in a different way. Instead of passively tracing the routes to be taken, he should have looked for potential confusions along the way. He could have tried to anticipate where he could get lost and how he would know so that he could recover.

Another tactic is to try actively to construct alternative stories to explain events, in order to increase flexibility and reduce reliance on anchors that might be untrustworthy. Cohen, Freeman and Wolf (1996) have described an exercise of deliberately rejecting the current story as an account of events, in order to become aware of alternative stories.

A third tactic is to increase the data flow – this is different from increased vigilance of the existing data. It means seeking more information, and perhaps being more careful to catalogue data that might otherwise be discarded. These data might turn out to be relevant if the frame is replaced. They might also contribute to the confusion, adding more events to be tracked, adding more noise, and adding more coincidental connections that would have the appearance of patterns. Klein (2004) discusses additional practices for managing uncertainty.

The defences against knowledge shields can help us prevent spreading corruption from becoming too damaging. The next section discusses the activity of recovering once we have already become confused.

Recovering from corrupted understanding

In many settings, guidance is given in point-to-point navigation but not to recovery. Yet recovery is the greater challenge. Thus, in observing simulated Army helicopter missions (Thordsen, Klein and Wolf, 1990), my colleagues and I were struck by a common event. A helicopter would be picked up by enemy radar, and the crew would receive an electronic warning that they were in danger of being engaged by a missile. The crew would then initiate strenuous avoidance manoeuvres until they broke the radar lock. At that point, they were often fairly disoriented about where they were on the map, and spent some time piecing together their location. They did not appear to have received any training in recovery, in "getting found", even though they frequently became lost in trying to carry out the complex missions.

Recovering – shifting rather than elaborating a frame – becomes harder the more we are committed to the existing frame. The more we use knowledge shields to preserve our corrupted beliefs, the more we will bend the map and the more difficult it will be to recover. Recovery means selecting a new frame, rather than trying to elaborate and preserve the previous frame.

The process of getting found may be more difficult for novices than for experienced decision makers. In our research with fireground commanders (Klein, Calderwood and Clinton-Cirocco, 1986), we found that trained professionals were prepared to shift their understanding of events – they spotted early signs of a problem and started tracking a different and more accurate account. This facet of sensemaking, shifting one's understanding, is most in line with Weick's account of sensemaking as a response to an anomaly or to confusion and uncertainty. In Figure 2.1, the recovery process requires a person to question the existing frame, and to reframe the data and cues, replacing the flawed frame with a more accurate one.

Strategies for recovery

Once we accept being lost, how do we reorient? If we are lucky, we just missed a turn, and by reversing our direction we will come to it. That is not really being profoundly lost. Being profoundly lost means the emotional feeling of helplessness as we look at a map and have no clue about where we are.

To reorient we can drive around in circles, hoping to catch a familiar street. That is a pretty desperate activity, although I am sure we have all been reduced to this state on occasion. We can rely on an intuitive sense of which direction we should be heading in, and that sometimes works – it works better when we have been using a map than when we just rely on directions about turning right and left. We can try to diagnose why we got lost – where we made the fateful mistake. Or we can use some blend of these tactics, working with a vague sense of what direction we have to take, a suspicion about where we went wrong, a few assumptions about landmarks such as that elevated expressway over there, perhaps a dim memory that we passed this street earlier. This is the true test of sensemaking – discovering where we really are.

And sometimes we do not even try to discover our current location. We spot a workable street on the map, we make an assumption about how to travel in order to run into it, we keep scanning to make sure we are not going in a wrong direction, and we grit our teeth and keep going.

If we realise the difficulty in time, before reaching the stage of profound confusion, then it is reasonable to diagnose the nature of the corruption and make a more moderate change rather than starting all over. We can:

- Press on and hope to find a familiar street or landmark
- Look for "catching" features (for example, if we get to this highway we know we have gone too far)
- Retrace steps to the last known anchor
- Ask directions, although this requires us to self-locate, if we are calling someone for help
- Verify the anomalies
- Pore over the map to find our location
- Generate and test hypotheses (for example, we should be coming to a certain street very soon)
- And/or diagnose the anomaly

Example 2. Flying blind.

Klein, Phillips, Rall and Peluso (forthcoming) described a navigation example in which an amateur pilot took a simple cross-country flight. The pilot built a flight plan for the 45-minute duration of his flight. He determined the heading, course, planned airspeed, way points for each leg of the trip, diagram of the destination airport, and so forth. He performed his pre-flight routine, which included calibrating the altimeter and the directional gyro (DG). He was familiar with the terrain in the area around the airport from which he took off. Some of the visual markers did not match the map, but he interpreted these types of discrepancies as the inevitable ways that maps become obsolete. For example, one village extended right up to a small airport – obviously the village had added a new subdivision in the past few years. But about 30 minutes into the flight he began to feel that he was not on his flight path. However, he did not have any clear landmarks because all the small towns and villages looked similar.

His hardest judgement was to determine if he was still on his course or if his intuition was right and he was off course. His instruments had been telling him he was on course, so he should have been okay. Nevertheless, he decided to check his DG against the compass, and discovered that his DG was about 20-30 degrees off. That was the framebreaker, shifting him into a different mode of sensemaking. Instead of trying to explain away small discrepancies, he now had to figure out where he was. He could no longer trust the DG. He had to start over and establish his location. He had a rough idea of where he was because he had been heading south, drifting farther east than he had planned. He knew that if he kept flying south he would soon cross the Ohio River, which would be a clear landmark that he could use as an anchor to discover his true position.

Once he reached the Ohio River, he could see that the configuration of factories on the river did not match the map at his planned crossing point. He flew up the river for a short distance and came to a bend (another good anchor) that had power plants and factories with large smokestacks – still more good anchors. He noticed a railroad crossing that was crossed by hi-tension power lines. He noticed some high-tension power lines just where he expected to find them, given his new hypothesis about where he was. On the map, the high-tension lines intersected with a railroad crossing, and sure enough, that is what he saw on the ground. He then followed the power lines directly to his destination airport.

In this example, we see several of the sensemaking activities involved in recovery. The pilot started with a good frame, in the form of a map and knowledge of how to use basic navigational equipment. Unknown to him, the equipment was malfunctioning. Nevertheless, he attempted to elaborate the frame as his journey progressed. He encountered data that made him question his frame – question his position on the map. But he explained these data away and preserved the frame. Eventually, he reached a point where the deviation was too great, and where the topology was too discrepant from his expectations.

He knew that he was heading south, and would eventually cross the Ohio River, and he prepared his maps to check his location at that time. The Ohio River was a major landmark, a dominating anchor, and he hoped he could discard all of his confused notions about location and start fresh, using the Ohio River and seeing what map features corresponded to the visual features he would spot. He also had a rough idea of how far he had drifted from his original course, so he could start his search from a likely point. If he had tried to reorient earlier, he probably would have failed because he simply did not have any useful anchors. The small villages and farm fields did not provide useful anchors – they were quite hard to distinguish from each other.

If decision makers have to re-frame their understanding, Figure 2.1 shows that they may want to re-examine the data they have already interpreted using a different frame, and data they may have discarded. In fitting a new frame to the data – a new map – there is a good chance that some data that were discarded, or possibly put in temporary storage, may be relevant. Weick (1995) has talked about this as a retrospective activity, looking back at events and reconsidering them. Obviously, data that were discarded may no longer be fully or even partially accessible.

Hikers are advised *not* to try to diagnose what went wrong if they have reached the point of profound confusion. At that point, the corruption has spread so far that they cannot depend on any of their beliefs. Attempting to diagnose how they got lost may be a poor idea – the spreading corruption of cues and anchors has gotten so widespread that attempted diagnosis is unlikely to help, and may make things worse. They know at least one of their beliefs is wrong, but which one?

The best strategy is usually to retrace their steps. Failure to do so almost ensures that they will become profoundly lost. However, retracing depends on having a good memory of recent route choices.

If they cannot backtrack, they are encouraged to move into "zero-based sensemaking", starting fresh with data that are immediately at hand in order to avoid the influence of corrupted beliefs. They have to throw out their existing beliefs – these beliefs are too contaminated by spreading corruption. They have to begin again, using only those cues and anchors they are sure they can trust. If they select the wrong anchor, then they bend the map to conform to that anchor (so, that must be the hill over there, and that stream must be this river – must have dried up a lot) and turn being seriously lost into being profoundly lost. Once they are safely recovered, they can look at the map and then try to diagnose the problem. But this strategy may be easier to describe and prescribe than to use. The concept of zero-based sensemaking may be psychologically impossible. Therefore, the advice may reduce to a caution to be wary of interpretations and to try to depend only on anchors that are firmly established, such as north and south (if a compass is available) and east and west (if the movement of the sun can be studied and fixed against an unambiguous distant landmark).

People may vary in their ability to apply different recovery strategies. Goldin and Thorndyke (1982) studied individual differences in navigational styles and abilities, contrasting people with different levels of spatial ability. People with greater visual memory, spatial orientation and visualisation will be more successful in learning a new environment through navigation or from a map. Participants with a visual/perceptual style approached the navigation task differently than those with a verbal/analytical style. It might be useful to conduct research on simulated navigation tasks, entering incorrect data in a garden path type of scenario to see how the navigation was corrupted and how people recovered. Such a design could be used to contrast the reactions of individuals with different navigational styles.

Conclusions

The act of navigation, particularly the way confusions are generated and then resolved, provides us with some additional features of sensemaking. The concept of spreading corruption describes how sensemaking can fall apart as one misconception leads to another and another. In many cases, the only thing that can stop this downward spiral is to encounter a framebreaker – a data element that firmly contradicts the prevailing frame. However, the incident of the hiker who smashed his compass because it was confusing him shows our ability to reject contradictory data that should serve as a framebreaker.

The cycle of corruption and recovery can also have some benefits. Despite the frustrations of having to cope with corrupted beliefs, it is possible that this corruption/

recovery cycle is valuable in helping us build richer mental models. By being confronted with the inadequacies of mental models, their bugginess or shallowness, we are both motivated and directed to strengthen the way we understand events.

This process may account for our determination to find our own way after getting lost, rather than asking for help. Perhaps only by getting found, only by personally recovering, can we correct our mental map and learn the true layout and scale of our surroundings. Only by getting seriously lost, and emotionally experiencing the confusion, can we give up shallow frames and mental models and replace them with richer ones.

In fact, we may resist changing outmoded mental models unless we have clear evidence of their inadequacy, and a strong experience of frustration in trying to use them. Such evidence and experience of failure can be a platform for growth.

Acknowledgements

I would like to thank Jenni Phillips and Debbie Peluso for their helpful review of a draft of this chapter, and for their collaboration in all of our studies of sensemaking. The research reported in this paper was supported by contracts with the Army Research Institute for the Behavioral and Social Sciences (Contracts 1435-01-01-CT-3116 and1435-01-01-CT-3116).

References

Bruner, J. and Potter, M. C. (1964). “Interference in visual recognition”. *Science*, 144, 424-425.

Chi, M. T. H., Feltovich, P. J. and Glaser, R. (1981). “Categorization and representation of physics problems by experts and novices”. *Cognitive Science*, 5, 121-152.

Cohen, M. S., Freeman, J. T. and Wolf, S. (1996). “Meta-recognition in time-stressed decision making: Recognizing, critiquing, and correcting”. *Human Factors*, 38, 206-219.

De Keyser, V. and Woods, D. D. (1990). “Fixation errors: Failures to revise situation assessment in dynamic and risky systems”. In A. G. Colombo and A. Saiz de Bustamente (eds), *Systems reliability assessment* (pp. 231-251). Amsterdam: Kluwer.

Feltovich, P. J., Coulson, R. L. and Spiro, R. J. (2001). “Learners’ (mis)understanding of important and difficult concepts: A challenge to smart machines in education”. In K. D. Forbus and P. J. Feltovich (eds), *Smart machines in education*. Menlo Park, CA: AAAI/MIT Press.

Goldin, S. and Thorndyke, P. (1982). “Simulating navigation for spatial knowledge acquisition”. *Human Factors*, 24, 457-471.

Gonzales, L. (2001). “Land of the lost”. *National Geographic Adventure* (Nov/Dec), 84-96, 155-156.

Klein, G. (1998). *Sources of power: How people make decisions*. Cambridge, MA: The MIT Press.

Klein, G. (2004). *The power of intuition.* New York: Doubleday.

Klein, G. A., Calderwood, R. and Clinton-Cirocco, A. (1986). "Rapid decision making on the fire ground". In *Proceedings of the Human Factors Society 30th Annual Meeting*, 1, 576-580.

Klein, G. A., Calderwood, R. and MacGregor, D. (1989). "Critical decision method for eliciting knowledge". *IEEE Transactions on Systems, Man, and Cybernetics*, 19(3), 462-472.

Klein, G. A. and Crandall, B. W. (1995). "The role of mental simulation in naturalistic decision making". In P. Hancock, J. Flach, J. Caird and K. Vicente (eds), *Local applications of the ecological approach to human-machine systems* (Vol. 2, pp. 324-358). Mahwah, NJ: Lawrence Erlbaum Associates.

Klein, G., Phillips, J. K., Rall, E. and Peluso, D. A. (forthcoming). "A data/frame theory of sensemaking". In R.R. Hoffman (ed.), *Expertise out of context: Proceedings of the 6th international conference on naturalistic decision making.* Mahwah, NJ: Erlbaum.

Klein, G., Pliske, R. M., Crandall, B. and Woods, D. (2005). "Problem detection". *Cognition: International Journal of Cognitive Science*, 7, 14-28.

Leedom, D. K. (2002). *Important issues in sensemaking research* (White Paper). Vienna, VA: Evidenced Based Research, Inc.

Mitroff, I. I. (1981). "Scientists and confirmation bias". In R. D. Tweney, M. D. Doherty and C. R. Mynatt (eds), *On scientific thinking*. New York, NY: Columbia University Press.

Perrow, C. (1984). *Normal accidents: Living with high-risk technologies*. New York, NY: Basic Books.

Sarter, N. B. and Woods, D. D. (1995). "How in the world did we ever get into that mode? Mode error and awareness in supervisory control". *Human Factors*, 37(1), 15-19.

Smith, P. J., Giffin, W. C., Rockwell, T. H. and Thomas, M. (1986). "Modeling fault diagnosis as the activation and use of a frame system". *Human Factors*, 28(6), 703-716.

Syrotuck, W. G., with Syrotuck, J. A. (ed.) (1977). *Analysis of lost person behavior: An aid to search planning*. Mechanicsburg, PA: Barkleigh Productions, Inc.

Thordsen, M. L., Klein, G. A. and Wolf, S. (1990). *Observing team coordination within Army rotary-wing aircraft crews* (Contract MDA903-87-C-0523 for the U.S. Army Research Institute, Aviation Research and Development Activity, Ft. Rucker, AL). Fairborn, OH: Klein Associates Inc.

Tolcott, M. A., Marvin, F. F. and Bresnick, T. A. (1989). "The confirmation bias in evolving decisions". In *Proceedings of the 1989 Symposium on Command-and-Control Research* (pp. 232-238). McLean, VA: Science Applications International Corporation.

Weick, K. E. (1995). *Sensemaking in organizations*. Thousand Oaks, CA: Sage Publications.

Chapter 3

Time and Design in Decision Making Environments

Göran Pettersson

Background

There have been many sources and influences on this work. Planning and temporal representation of data were discussed in the Artificial Intelligence (AI) research during the 1970s and 1980s. Inspiration has been found in temporal logic described in papers such as Allen (1983, 1984). My own experience as test engineer of software functions in a JAS 39 Griffin PM (Presentation and Maneuvering) simulator during the mid 1980s and later, during the mid 1990s, as system engineer at the avionics department at SAAB AB, have had a high impact on the motivation and engagement to proceed and intensify efforts to continue the investigation of the characteristics and effects of time. Interviews and communication with people from the Swedish Air Force staff have, to a large extent, also contributed.

Introduction

The purpose of this chapter is to introduce time and relations between time and action, as means in the system design process. One part of the chapter focuses on characteristics and effects of time and human perception and understanding of time (Hollnagel, 2002a). The effects of time on design has not yet fully been learned and understood. The other part of the chapter deals with how time can be used as a means to organise and distribute cognitive tasks in cockpit environments.

Higher demands on effectiveness and safety have forced system designers to improve aircraft performance and to integrate more automation and sophisticated support systems. Technical advances have been adapted and implemented with the aim of improving system effectiveness. Problems with high workloads have been met with extended automation. However, many advanced concepts have failed to reach the desired level of safety and effectiveness. Design principles have failed to give relevant support and intentions with extended automation have not always been successful. The consequence has been that system interaction complexity, in dynamic contexts, has increased and also exceeded human capability.

A consistently shared situation awareness between the human operator and the technical system is a necessity to enable an effective *joint system*, described by Hollnagel (2002b). An effective joint system, in this context the integrated aircrew-avionics system, is important to reach enough safety and effectiveness. In dynamic system environments time and temporal restrictions always influence tasks and activities, that is, time to move, sense, think and act. The impact of time is rarely explicitly estimated when new functions are integrated in dynamic decision support systems. In air combat contexts conventional design approaches have often been applied for development of the next generation support systems. The spatial moving map has been the base for almost all kinds of combat and missions. It has then been natural to use spatial data, such as location, range, distance, velocity and direction as the base for development of the support system. A spatial foundation is important as a starting point for design concepts. However, the spatial description is continuously changed through impact of movement and actions that occur at different times. As a consequence of the effects from these actions temporal relations between plans and actions are important design complements. In Pettersson et al. (1998, 1999) a temporal display concept is proposed, as a supplement to the conventional spatial tracked target data representation. The display concept is based on temporal action zones where own or hostile action capability is visualised. A first simulator study (Linde et al., 2001), indicated improved effectiveness, but also a decreased safety.

Time or temporal aspects on design of decision support systems have more or less been forgotten during decades of research, development and construction for applications in real world contexts. In dynamic contexts time is not only a continuous change that creates history. It is also a resource and means that, together with management of movement and action, can be used to model and estimate possibilities, impossibilities or other restrictions. The feeling of time depends on mental and contextual factors and is also based on subjective and individual experience. Humans show in general low capability to estimate relations between own and other environmental temporal actions. In both military and civilian applications it is the "occurrence" of actions and events that should be estimated and rated when success, effectiveness and safety are the dominating design goals.

Humans and time

There does not exist a coherent and common view on the nature of time. The experience or perception of time depends on the context. Time has always also impacted on the mental process of cognition and understanding that is distributed within the joint system. Each context generates different requirements on temporal restrictions and conditions. Sometimes a clock or schedule is enough to meet required demands. In other cases a decomposition of a mission must be done, in terms of plans and actions, to meet requirements (Figure 3.1).

This affection can be described as psychological factors such as a) engagement in and motivation for the task or mission in progress, b) risks and uncertainties, c) fear, d) importance of mission, e) mental workload which also affects situation awareness (Endsley, 1999). Besides psychological affection there is also affection from physiological factors such as a) noise, b) fatigue c) heat and cold and d) boredom. However, it is hard to find reports and research that have seriously investigated the impact of these factors on perception of time and the consequence of temporal affection on decision support systems.

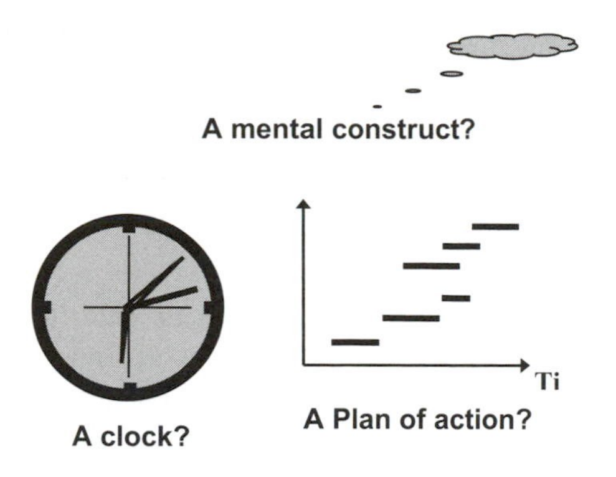

Figure 3.1 What is time?

Tears

It is important to realise also that effectiveness, risk and safety must be integrated in the discussion about time and action. Time, effectiveness, action, risk and safety (TEARS), shown in Figure 3.2, are related to each other.

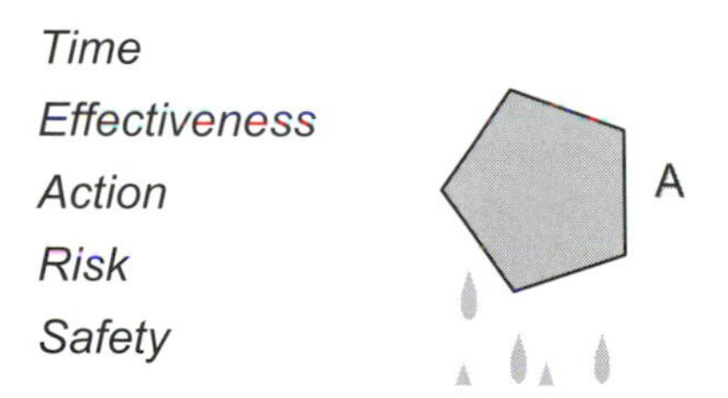

Figure 3.2 The optimal shape of the TEARS pentagon?

But safety and effectiveness are more consequences of how much time and actions are available and present in the contextual situation. The desired goals are to reach and maintain a high effectiveness and a high level of safety. However, it may be hard to maintain high safety at the same time as high effectiveness, and vice versa. The available time during the mission phase may be spent either on efforts to maintain effectiveness or safety. Too much time spent on actions that support high safety may result in low effectiveness. On the other hand, too much attention focused on activities that enable high effectiveness can result in reduced safety. The consequence can be that the risk of mistakes increases. Risk can be defined as the probability of a mission failure or that the mission goal cannot be reached.

Time, safety and effectiveness

Safety can in many cases be improved if more time is spent on control of the task or mission in progress. Humans mostly act as if it was more important to save time for future use than to improve safety by instant use of time. Humans are mostly not aware of, or do not believe or care about, the fact that safety is affected by the amount of time that is used on control of the task or mission that is in progress.

Perception and effects of time and action

The impact of time on system performance must be modelled and used in system design. It is important to understand the qualities of time, and the effects of the interaction between time and action, before the concept can be applied in conceptual models and design. The characteristics of time and influence of time on humans in real world contexts are the effects that must be known, and the only ones that can be used. Measurements of influence from temporal restrictions on human participants in laboratories are of limited value. Affection of temporal restrictions on human perception of time can only be detected and measured in the correct and real context.

Temporal Plan and Action Model

Symbols and other information that are visualised on displays in cockpit systems are projections of objects and activities in the environment and also components that contribute to the characteristics of the situation. According to Hollan et al. (2000), cognition is distributed in the environment. Consequently should the processing of cognitive tasks also be distributed in the environment? A cognitive task is a task with the capability to be able to contribute to the creation and modification of the current construct. The current construct is distributed in the cockpit environment. The majority of temporal tasks, plans and actions, are also cognitive tasks and can be distributed within the joint system. The characteristic of a temporal task, beside the cognitive property, is that it is affected by temporal conditions and restrictions. Actions that have impact on movement are cognitive tasks, since movement induces changes in the perceptual flow from the environment, and consequently also has impact on the cognition. Important temporal actions or tasks (see Figure 3.3), are tasks like: *Assignment of plan/action priorities*, *Definition of plan/action content*, *Monitoring of effectiveness*, *Monitoring of alternative plans*, *Temporal plan computations*, *Plan evaluation*, *Plan evaluation and assessment*, *Selection of plan*. They will be described in the following sections.

Control models

Mission planning has previously mainly been done before the start of missions. There were many reasons for that. Pre-mission planning routines have many aims: a) put the aircrew into context, b) prepare the aircrew for expected situations, and c) update the support systems with the most recent intelligence reports. The old reasons to have pre-mission planning will probably also be valid in the future, but during special conditions it may be necessary to extend the planning or re-planning capability to be a facility applicable in real time during execution of missions in progress. Sets of plans with associated sets of actions must be equipped with relevant content and also have been assigned relative priorities. This must be enclosed to enable effective recipes that result in successful accomplishment of missions. To each action there is associated a temporal description that determines under which conditions and restrictions the action will or can be executed. These conditions must be calculated and updated continuously during execution of a mission.

Contextual control models

Contextual control models (COCOM) are suggested by Hollnagel (2002b), and are closed loop models where selection of actions is based on the current understanding of a situation. The model is general and should be applicable in any context.

The cyclical Temporal Plan and Action Model (TPAM– see Figure 3.3), proposed here, is a combination of COCOM and the ECOM (Extended COntrol Model) model, proposed by Lindh (2003) and then modified to apply a temporal representation of the environment. Such a modification should be relatively easy to implement as software in a computer. Different contexts, of course, require relevant plan and action sets. Plan content, temporal conditions and equations must also be adapted to the current context. The TPAM describes how focus of attention moves between and spends time on different cognitive tasks during execution of a mission. The amount of time spent on each task during an execution cycle relates to changes and differences in the environment. The phases in the model are described here.

Update of current construct

The current construct is the latest updated representation of the environment, and that constitutes a base for the selection of the next plan or action. The construct can be defined as a shared representation between the technical system and the human operator.

Selection of plan

At this phase of the TPAM cycle the *current construct* is the base for the selection of a new plan or if the current plan shall continue. The selection is based on what the system knows about the environment and which next plans will best be able to solve the problem or are believed to meet the desired mission goal.

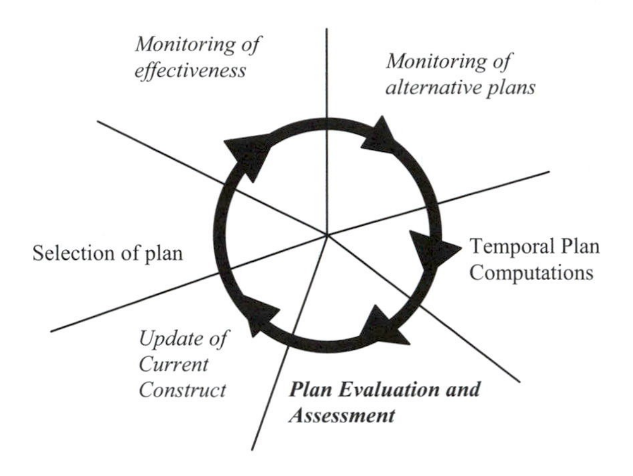

Figure 3.3 Temporal Plan and Action Model (TPAM)

Monitoring of effectiveness

To ensure that the selected plan contributes to the success of the plan goal, it is important to monitor the execution of the current plan. If the current plan does not support the mission in progress, then focus of attention moves to other parts of the TPAM cycle, and a new plan can be selected for execution. In Pettersson (2002) a method for estimation of *plan effectiveness* is implemented in a simple PC environment. The aim of the implementation was to test the validity of the selected plan monitoring approach. The algorithm estimates temporal relations between own and hostile actions during the monitoring phase of the TPAM cycle. If the own planned sequence of actions can be executed with success before threatening hostile assumed actions, then the own plan is probably effective. An *action success matrix* can be created where each element, ASM(i,j) is a temporal numerical estimation of the relationship between the i:th own and the j:th hostile action.

ASM(i,j)= Advantage (OwnA(i),HostileA(j))
i=1,M; j=1,N

The ASM matrix is an estimation of to what degree own actions can be executed before the relevant set of hostile assumed actions. The *Advantage* function uses a temporal logic, described in Allen (1983) as a tool for calculation of relations between own and hostile actions. The matrix is a base to be used for update of the current construct and can be used as support for the selection of the next plan and action.

Monitoring of alternative plans

Alternative plans must be compared with the execution of the selected plan that is in progress. If there are alternative mission plans that can more effectively contribute to ensure that the plan and/or mission goal is reached, then the pilot must be informed.

Temporal plan computations

The current and all alternative plans must continuously be updated with the environmental data from all data sources.

Plan evaluation and assessment

Evaluation of the plan's feasibility and validity in current context and situation must be done continuously during the whole mission or mission phase. Different amounts of time are used on each temporal task in the cyclic model when the situation is changing.

Temporal design guidelines

The design goals in cockpit environments must reach a sufficient level of effectiveness and safety. The impact of time on system design has rarely been applied in real world contexts. Only in very few cases has time been used and then more as additional information than as a major design guide. Since time is not normally used as a guiding factor in design, it is now time to introduce a set of temporal guidelines.

- TGL1 (Temporal restriction): If time is a constraint in the current context in question, then design must be avoided that requires high efforts of human interpretation of environmental issues.
- TGL2 (Temporal perception): Design must improve and support perception and understanding of temporal relations between actions in the environment.
- TGL3 (Instant action capability): Design that supports awareness of action variety and applicability in the current situation.
- TGL4 (Future action improvement): If and how preconditions for an action can be reached or not.
- TGL5 (Time to act and move): Activities and movement in dynamic system environments require time. This must be considered and modelled. The available time to perform actions can, within limits, be moderated through changes in speed or altitude.
- TGL6 (Temporal effectiveness): The joint system functions should be designed and distributed with respect to their temporal and cognitive processing complexity.

Cogni-temporal processing

How shall control of plans and actions be performed in dynamic contexts and by whom? Should it be *the one who is most qualified*? The problem is that the most qualified manager can shift depending on the situation and environmental issues. The optimal solution may be a temporal optimisation that should make it possible to sense and find where the narrow paths are located in every situation. The computer system must be a support to humans' understanding of the environment, in which the computer system is also an integrated part. The goal is to facilitate a consistently shared functional situation awareness; an awareness that is extended to include, not only awareness of states of the outside world, but also of the state and capabilities of the other part of the joint system. This does not imply that both parts must have identical representations, but they must be consistent.

During pre-flight mission planning preparation of the mission, plans and actions are assigned priorities. These priorities are based on pilot experience, intelligence reports and risk acceptance.

The effective content of actions and plans is constructed from a plan and action library during the mission planning process. During real time processing of plans and actions, their temporal applicability are calculated and updated by the system computer. Sensor and other mission data, and online intelligence reports that are available from the network, are main input sources to the temporal calculations. This should probably best be executed by the computer, without intervention from the pilot, because of the human limitations concerning time and temporal task processing.

Monitoring of alternative plans should be based on collected data of the environment, which are then further processed in the *temporal plan computations* phase. The reason is that temporal relations, between plans and actions, give effective information about environmental objects threat capability and own action capability. The task is processed in parallel by both the pilot and the computer.

Plan evaluation should be designed with the aim of engaging both the pilot and the computer together, to avoid the risk of unexpected and incomprehensible proposals from the computer. The computer updates the internal temporal relations and updates the display. The pilot notes the changes on displays and updates the current mental construct of the situation. The shared representation between pilot and display/computer should be more consistent when the impact from time and action are considered.

The *selection* of next plan or action is based on the current construct or understanding of the situation. An understanding of the qualities and possibilities of the available set of plans and actions must be included in the update of the current construct of the situation.

At first it seems obvious that *selection* is a cognitive task that should only be done by the pilot. But certainly situations will occur when the pilot has lost control of the TPAM-loop. If it can be verified that the pilot has lost control or it is obvious that the pilot does not have enough time to perceive, act and estimate the consequences of actions in the environment, then appropriate plans and actions can be selected and executed by the computer. Selection and execution of selected tasks is a question of effectiveness and safety.

Conclusions

It is known that environmental conditions have an impact on how humans perceive time and that cognition is affected by stress factors and other conditions. Such conditions many not always be classified as stress factors but are often parts of the context and located in the environment. Humans' problems in managing time have rarely been considered in system design processes. With aspects from the concept of TEARS (Time, Effectiveness, Action, Risk and Safety) a set of temporal guidelines is proposed, with the aim of improving system performance for a joint pilot-support system in a complex dynamic air combat context. These guidelines should improve effectiveness or safety, depending on where efforts are focused. Focus on high effectiveness may result in less safety and high focus on safety may induce lower effectiveness. The consequence will be that it is important to find a balance between guidelines that enable effectiveness and those that offer higher safety. It is also a question of how much risk or effectiveness is required to make it possible to deal with expected threats or demands. How this balance will be found is not known yet and will also shift depending on contextual demands and factors.

The result of a primary test (Linde et al., 2001) indicated that support that presented hostile action capability gave the consequence that improved effectiveness but decreased safety. However, should these results be further investigated in more extended simulator studies and for real world situations? Temporal design guidelines as proposed in this chapter can be estimated and adapted in each context and application. Humans, from birth to death, are affected by and forced to manage time. But is this a lesson learned today in the current modern and technological world?

References

Allen, J. (1983). "Maintaining knowledge about temporal intervals". *Communications of the ACM*, 26(11), 832-843.

Allen, J. (1984). "Towards a general theory of action and time". *Artificial Intelligence*, 24(2), 123-154.

Endsley R. M. (1999). "Situation awareness in aviation systems". In D. J. Garland and V. D. Hopkin (eds), *Handbook of aviation human factors*. Mahwah, NJ: Lawrence Erlbaum Associates.

Hollan, J., Hutchins, E. and Kirsch D. (2000). "Distributed cognition: Toward a new foundation for human-computer-interaction research". *ACM Transactions on Computer-Human Interaction*, 7(2), 174-196.

Hollnagel, E. (2002a). "Time and time again". *Theoretical Issues in Ergonomics Science*, 3(2), 143-158.

Hollnagel, E. (2002b). "Cognition as control: A pragmatic approach to the modeling of the joint cognitive system". Available online: http://www.ida.liu.se/~eriho/

Linde, L., Strömberg, D., Pettersson, G., Andersson, R. and Alfredsson, J. (2001). "Evaluation of a temporal display for presentation of threats in air combat". In *Proceedings of CSAPC – 01* (pp. 137-146), Neubiberg, Germany.

Lindh, J. O. (2003). "Cognitive aspects of network enabled operations". In *Proccedings of 21st International System Safety Conference* (ISSC21), (pp. 422-431) Ottawa, Canada.

Pettersson G. (2002). "A temporal action based approach to decision support in air combat". In *Proceedings of 34th Annual Congress of the Nordic Ergonomics Society*, Vol. II, (pp. 651-657) Kolmården, Sweden.

Pettersson, G., Axelsson, L., Jensen, T., Karlsson, M. and Malmberg, A. (1999). "Multi-source integration and temporal situation assessment in air combat". In *Proceedings of IDC-99*, Adelaide, Australia.

Pettersson, G., Strömberg, D. and Roldan-Prado, R. (1998). "Temporal decision support and datafusion in BVR-combat". In *Proceedings of Eurofusion98*, Malvern, UK.

Chapter 4

Risk and Decision Making

Damien J. Williams

Introduction

The ability to identify and avoid harmful situations is necessary to the survival of all organisms. The reason why humans have come to dominate this planet has been credited to our exceptional capacity for decision making (Hastie and Dawes, 2001). The development of this ability can be attributed to our ancestors who were engaged in activities that necessitated effective decisions for survival and passing on their genes, for example, choosing a mate, choosing a watering hole, choosing a place for shelter, and so on. Put simply, those who made ineffective/inappropriate decisions often died, while the successful ones reproduced and passed on their decision making capabilities.

What becomes immediately obvious in these situations is that risk or the perception of risk plays an important role in the decision making process. For instance, the selection of a mate required the assessment of the risk associated with choosing one potential mate over another in terms of the likelihood of successful reproduction. Another example could be the selection of a place to shelter whereby an assessment of the environment would indicate the risk of approaching a cave and inhabiting it. Those who neglect to make assessments of risk would be more likely to make ineffective decisions; moreover, those who make erroneous assessments of risk are also likely to make sub-optimal decisions. It is apparent that an individual's assessment of risk has implications for decision making; however, there are numerous barriers to achieving an effective assessment of risk.

Risk in everyday decision making

An important feature of any decision is the degree of uncertainty associated with future outcomes. Uncertainty can arise because of a lack of or incomplete knowledge (for example, intrinsic factors – "I'm not sure if I left my wallet on the bus?"), or incomplete information (for example, external factors – "It's not certain if the sunshine will hold out for the rest of the week") (Scholz, 1983). What is more, if uncertain outcomes are costly it is typical to talk about *risk*. Consequently, risk is an inherent part of everyday life and is a fundamental consideration in "a vast range of decision making situations, from allocating wealth to safeguarding public health, from waging war to planning a family, from paying insurance premiums to wearing a seatbelt, from planting corn to marketing cornflakes" (Bernstein, 1998, p. 2).

A particular everyday instance could be the decision to purchase one of two possible brands of margarine, which relates to the issue of personal health. Choice must be made based on various sources of information: value for money, nutritional content (for example, fat content), personal experience (for example, palatability), and recommendations from health experts. This information removes much of the uncertainty and enables an informed decision. Moreover, knowing that any negative consequences would not, in most cases, be immediate and could be rectified/reversed, means the situation could be considered "low-risk". However, situations in which outcomes are associated with a greater degree of uncertainty are considered higher in risk. For instance, the decision to take a journey in uncertain, bad weather would be considered "high-risk", as it is difficult to predict the impact of the weather despite reports from the Met Office. This is further complicated when considering the possible modes of transport, which cause variations in the degree of perceived risk; some would perceive flying or even taking the train as being riskier than taking the car in these conditions. This illustrates that risk is context specific and implies that it cannot be *fully* understood by simply applying information from one situation to another.

Any consideration of risk usually arouses negative connotations such as the possibility of loss of life, money, time, and so on. However, in many uncertain situations, risky decisions have resulted in positive consequences. For instance, had the founders not taken risky decisions, many of the defining enterprises of our time (for example, Apple, Microsoft, BAE Systems, Coca-cola, and so on) might never have come to fruition. Hence, risky decisions can result in outcomes much better or much worse than those of less risky options (Baron, 2000). In essence, risk is an inherent part of the decision making process, as we constantly balance risks in order to make decisions (Wickens and Hollands, 2000): the over-arching factor is the effectiveness of the risk assessment on which decisions are based.

Risk perception and decision making

Intuition and even personal experience would suggest that risk perception influences decision making, and while many consider this to be the case (for example, Siegrist, Gutscher and Earle, 2005), empirical research has often failed to acknowledge adequately the role of risk perception in the decision making process. A classic example can be found in the framing literature (which will be dealt with later in more detail). While risk or uncertainty is manipulated as part of the methodology, little reference is made to the role of risk perception in describing the counterintuitive findings (see McNeil, Pauker, Sox and Tversky, 1982; Tversky and Kahneman, 1981). Consequently, it has been generally assumed that such exogenous variables directly influence decision making, which gives an invalid account of the decision making process.

This situation has since been addressed by Sitkin and Pablo's (1992) mediated model of the determinants of decision making under risk. In essence, the model posits that the effect of a variety of exogenous variables (for example, framing effect, cognitive biases) are mediated by two causal mechanisms – risk perception and risk propensity – that are believed to regulate cognitive processes such as information gathering and sensemaking (Pablo, Sitkin and Jemison, 1996; see, Keil, Wallace, Turk, Dixon-Randall and Nulden, 2000; Simon, Houghton and Aquino, 1999; Sitkin and Weingart, 1995). Moreover, Keil et al. and Sitkin and Weingart's findings suggest that risk perception is the more influential mediator in this process. Not only does this support the intuition that humans assess risk when making decisions, but also infers that risk perception should be considered when investigating the effects of exogenous variables on decision making (Sitkin and Weingart).

Cognitive heuristics

Human decision making typically takes place under a constant barrage of information. Gathering information is critical for effective decision making (Cooper, Folta and Woo, 1995); however, due to limitations in cognitive capacity it is often difficult to process accurately all the information (March and Simon, 1958). While this may be true of low-risk situations such as the margarine example, high-risk situations are typically defined by a lack of information or knowledge and therefore are not necessarily affected by this limitation. However, the psychological state created by uncertainty is very painful, which can affect the ability to make important judgements and operate in risky situations (Schwenk, 1984). According to cognitive dissonance theory (see Festinger, 1957), such cognitive discomfort or tension (dissonance) motivates the individual to reduce the pain and achieve a state of consonance. This can be achieved through the use of simplification strategies, referred to as heuristics (Schwenk, 1986), which repress awareness of uncertainty and create a simplified view of the situation (Michael, 1973). Hence, in situations of high risk these processes can reduce difficult mental tasks (for example, assessing probabilities and predicting values) to simpler judgemental ones (Tversky and Kahneman, 1974). Moreover, in low risk situations they can make the available information more manageable, thus reducing the cognitive load (Marsh, 2002).

In general, heuristics are quite useful; they are cognitively economical, robust in dynamic situations (Hastie and Dawes, 2001), and enable satisfactory choices to be made quickly and with incomplete information by exploiting the way information is presented (Todd and Gigerenzer, 2000). Gigerenzer, Todd, and the ABC Research Group (1999) described heuristics as *fast*, in the sense that they do not require much computation, and *frugal*, as they only use available information. However, when they are misapplied they can lead to severe and systematic biases (Nisbett and Ross, 1980) that influence the search for information and subsequent interpretations, often resulting in less rational, less comprehensive decision making (Barnes, 1984). This is particularly pertinent when making complex and uncertain or risky decisions (Schwenk, 1984). For instance, invalid estimates of risk can arise because the negative outcomes and uncertainty associated with a decision are disregarded (Hogarth, 1987;

Xie et al. 2003). Hence, given that the consideration of risk requires the estimation of values, the importance of heuristics in risk perception is undeniable as heuristics are associated with the evaluation of numerical values such as frequency, likelihood and probability, which are objective measures of risk.

Biases in decision making

When making judgements under uncertainty, the four most common heuristics/biases are: availability, representativeness, anchor and adjustment, and overconfidence. A further issue of interest that has already been identified as prevalent in the decision making literature and relevant to the understanding of risk perception is the framing effect. While this phenomenon is rarely referred to as a bias, the systematic effect it has on decision making suggests that it could be consider as one. As such, what is intended is a brief introduction to each of these biases along with an illustration of how they influence the perception of risk in the decision making process.

Framing effect

The way in which a situation is presented can affect perceptions of risk (Gordon-Lubitz, 2003). What is more, perceptions of risk are particularly vulnerable to framing effects, which influences the way individuals approach a risk, and *biases* the decisions they make (Edwards and Elwyn, 2001). The term "framing effect" has traditionally been used to refer to the type of framing identified by Tversky and Kahneman (1981) known as risky choice framing; however, a number of alternative types of "framing" have been identified with different underlying mechanisms that also serve to bias the decision making process, namely: *attribute framing* and *goal framing*. Each of these will be briefly described in turn (see Levin, Schneider and Gaeth, 1998 for an in-depth review of the literature).

Risky choice framing involves the manipulation of risk through the variation of probability/likelihood information. For instance, in the positive frame, a certain and risky choice will be presented in terms of gains (for example, lives, money, and so on) and the negative frame in terms of losses. The general finding is that in the positive frame, people demonstrate a bias toward risk-averse behaviour choosing the certain option over the risky option, and in the negative frame, a bias toward risk-seeking behaviour choosing the risky option over the certain option (see Tversky and Kahneman, 1981).

Attribute framing is when a particular characteristic or attribute of an object or event is manipulated and described in a positive or negative light (for example, project funding allocation based on completion rates for previous projects; see Dunegan, 1993, 1995). Under such circumstances, there is a bias toward choosing the positively framed alternative. This form of framing demonstrates that risk perception is not an essential feature of the framing bias, which suggests that while many decision making situations require the consideration of risk, in others this is not necessarily an immediately obvious consideration.

Goal framing refers to instances when an action or behaviour is framed in terms of gains or losses (for example, credit card use when given a message stressing the losses associated with not using it or the gains from using it; see Ganzach and Karsahi, 1995). Here, people typically demonstrate a bias toward the negative frame (losses). While the effect of risk is inherent in this type of framing, attempts to explain it as the redefinition of the situation in terms of the unstated but implicit risk have been unsatisfactory (Levin et al., 1998). An alternative perspective is that goal framing may operate via the confirmatory bias, which suggests that people choose the risky, negative frame, as it is congruent with negative attitudes.

What is generally apparent from the literature is that subtle changes in the way risky situations or alternatives are presented has the potential to influence decision making. What is more, through Sitkin and Pablo's (1992) model and the supporting empirical research, it is evident that the effects of the framing bias (risky choice framing especially) are moderated by perceptions of risk.

Availability bias

The availability bias concerns how readily information, thought to be relevant to the current decision or examples of similar situations, can be recalled – its cognitive availability (Freudenberg, 1993). Perceptions of risk might be based on events that occur frequently, which are easier to imagine and recall than rare events and are therefore considered more likely. For instance, the public's opposition to the building of nuclear power plants can be attributed to instances such as Chernobyl or Three Mile Island, despite occurring some time ago. Such incidents, often highly documented and sensationalised in the press, are easily imagined and recalled, and result in an increased perception of risk associated with the nuclear industry in general. While this fear could be considered irrational, as there is an abundance of material indicating the safety of nuclear power, people still consider this industry to be risky because of the available memories of past failures.

The availability bias does not always carry with it negative consequences for risk perception and decision making. For instance, when it is formed from experience or an adequate evaluation of relevant information, it may be an accurate basis from which judgements can be made (Redmill, 2002). Problems occur when correspondence is assumed between the ease with which specific events can be imagined or recalled and relative frequency (Hogarth, 1987). This can be influenced by subtle factors unrelated to first-hand experience, such as familiarity, recency, salience, and vividness, which can be highly misleading (Slovic, Fischhoff and Lichtenstein, 1977) and pose a barrier to effective assessments of risk.

Representativeness

Representativeness is a subjective judgement typically employed when people are required to judge the likelihood that an object or event (A) belongs to class or process (B). Kahneman and Tversky (1972) proposed that people answer such questions by assessing the degree of similarity between A and B. More specifically, the bias could be understood as operating in a similar fashion to the activation of a stereotype (Tversky and Kahneman, 1974). For instance, when walking down the street at night and you are confronted by a youth (object) who has a shaved head and facial piercing, information that will be compared with the representation/stereotype of a mugger (class) held in long-term memory to judge whether the individual is likely to be a mugger so as to take the necessary course of evasive action.

This bias may introduce two kinds of systematic error into judgements. First, it may give undue influence to variables that affect representativeness. One such variable is "the law of small numbers" whereby people's intuitions about random sampling lead them to believe that small samples are as representative of the population as very large samples (Tversky and Kahneman, 1971), thereby exaggerating the confidence placed in conclusions based on small samples. This bias has been used to describe the use of a limited number of information sources to assess a potentially risky situation, resulting in an erroneous perception of risk (as in the hypothetical mugger scenario above). It is typical for those with a belief in the law of small numbers to have a reduced perception of risk (see Houghton et al., 2000; Simon et al., 1999). Secondly, and likely a consequence of the first, individuals may neglect important information (for example, prior probabilities; see Tversky and Kahneman, 1973). For instance, individuals who utilise a limited number of positive information sources and neglect negative counter-evidence are unable to carry out an appropriate evaluation of a risky situation, leading to an underestimation of risk and the formulation of overly optimistic conclusions (Kahneman and Lovallo, 1993).

As with the availability bias, representativeness depends on retrieval from long-term memory. Consequently, it is plausible that availability may play a role in representativeness. For instance, the representation accessed for comparison may simply be the most "available" one. A further limitation of judgements based on representativeness is that the evaluation of similarity often fails to reflect the underlying statistical and casual structure of the objects/situations being judged (Hastie and Dawes, 2001). Assessments of risk based on representativeness are likely to be erroneous as relevant information is often ignored, which leads to an insufficient evaluation of the situation, ultimately resulting in sub-optimal decision making.

Anchor and adjustment

The process of decision making usually begins with an initial judgement of the situation being faced. This initial value, or anchor, may be evoked by the formulation of the problem, the partial computation of available information (Tversky and

Kahneman, 1974), or in uncertain situations, seemingly trivial factors may have profound effects (Hastie and Dawes, 2001). For instance, the availability bias can influence the "anchor" process as the most readily available information can be quickly and easily recalled. Following this the anchor is evaluated and revised based on supplementary information, referred to as adjustment. However, these revisions are typically insufficient and the process is prone to underadjustment, or "primacy effects" such that the anchor not only serves as the starting point, but also biases the search for additional information (Hastie and Dawes).

This can best be illustrated using an example. Flying is generally considered a high-risk activity. This anchor could be based on the recall of available information regarding the safety of air travel as inferred from news stories such as the recent report of an Iranian military transport plane that crashed into a 10-storey apartment building in Tehran, killing 128 people. Any additional safety information indicating the safety of air travel compared to other forms of more risky modes of transport would result in little change to the initial level of perceived risk. Indeed, Stewart (2004) noted that:

> You can tell someone who is afraid of flying that the chances of being killed in a plane crash is roughly 1 in 4,000,000, but being presented with that statistical reality probably will not make that person feel any better about flying (p.368).

As such, perceptions of risk are likely to be governed by the initial values, and the processing of further information will only result in small revisions, which may be smaller than are justified by supplementary information (Schwenk, 1984).

Overconfidence

An over-arching bias of cognitive processing is that people tend to be overconfident about their judgements, particularly those based on the use of heuristics (Fischhoff, Slovic and Lichtenstein, 1977). This can have a number of effects on risk perception and the assessment of risk, which may not be mutually exclusive:

- Risk assessment is accepted because the reasons for confidence are easily recalled (see availability bias) (Fischhoff et al., 1977)
- Insufficient adjustment of initial assessment of risk after receiving new information (see anchoring and adjusting bias) (Simon et al., 1999)
- Unequal treatment of confirming evidence over disconfirming evidence (see representatives bias) (Henrion and Fischhoff, 1986)
- Prevents the realisation of the limits of one's knowledge and how much additional information is needed to make a correct assessment of risk (Tversky and Kahneman, 1974)

In essence, the overconfidence bias would lead to a reduced perception of risk (Russo and Schoemaker, 1992; cf Simon et al., 1999) and to the erroneous conclusion that an outcome is not risky.

Overconfidence could be understood as the final step in the process of attempting to bring a degree of certainty to an uncertain situation. While this is not in itself a flawed process, it is only legitimate if possible risks are acknowledged. However, this may not always be the case as people often have difficulty thinking about and resolving risk/benefit conflicts and one way of resolving this is through denial (Barnes, 1984). Risk denial can be further understood in terms of unrealistic optimism (see Klein and Weinstein, 1997; Weinstein, 1987). This suggests that individuals believe that they are less prone to risk or the outcomes of a risky choice, therefore denying the presence/effect of risk. What is more, purposefully denying the presence of risk may lead to an unjustified inflation in expectations of success, termed an "illusion of control" (Langer, 1975). This occurs when an individual is overly confident that (their) skill has a greater impact on performance than chance, even in situations where chance plays more of a decisive role. Consequently, individuals exhibiting this type of behaviour could very quickly take a risky situation and make it less recoverable and more risky. Hence, overconfidence appears to be mediated by the processes of risk denial and illusion of control, which could be seen as creating an inappropriate impression that the individual is less susceptible to the negative consequences associated with a risk. This could subsequently give rise to underestimations of the level of risk, and result in inappropriate behaviour.

Conclusions

In this chapter, the influence of risk perception on decision making has been considered. It is evident that a consideration of risk is inherent in most of the decisions we make. What is more, only when an accurate assessment of risk occurs, can we make optimal decisions. However, there are many factors both intrinsic (that is, information processing limitations and subsequent simplification strategies) and extrinsic (that is, the way in which situations are presented) that influence the assessment of risk. While there is no escaping many of these factors, it is evident that in order to understand this most vital of cognitive functions – decision making – it is necessary to consider the role of risk and risk perception, and understand its influence on information processing.

References

Barnes, J. H., Jr. (1984). "Cognitive strategies and their impact on strategic planning". *Strategic Management Journal*, 5(2), 129-137.

Baron, J. (2000). *Thinking and deciding* (3rd Ed.). Cambridge: Cambridge University Press.

Bernstein, P. L. (1998). *Against the gods: The remarkable story of risk*. Chichester, UK: Wiley.

Cooper, A. C., Folta, T. B. and Woo, C. (1995). "Entrepreneurial information search". *Journal of Business Venturing*, 3(3), 107-120.

Dunegan, K. J. (1993) "Framing, cognitive modes, and image theory: Toward an understanding of a glass half full". *Journal of Applied Psychology*, 78, 491-503.

Dunegan, K. J. (1995). "Image theory: Testing the role of image compatibility in progress decisions". *Organizational Behaviour and Human Decision Processes*, 62, 79-86.

Edwards, A. and Elwyn, G. (2001). "Understanding risk and lessons for clinical risk communication about treatment preferences". *Quality in Health Care*, 10, 9-13.

Festinger, L. (1957). *A theory of cognitive dissonance*. Stanford: Stanford University Press.

Fischhoff, B., Slovic, P. and Lichtenstein, S. (1977). "Knowing with certainty: The appropriateness of extreme confidence". *Journal of Experimental Psychology: Human Perception and Performance*, 3, 552-564.

Freudenberg. W. R. (1993). "Risk and recreancy: Weber, the division of labour, and the rationality of risk perceptions". *Social Forces*, 71(4), 909-932.

Ganzach, Y. and Karsahi, N. (1995). "Message framing and buying behavior: A field experiment". *Journal of Business Research*, 32, 11-17.

Gigerenzer, G., Todd, P. M. and the ABC Research Group (1999). "Simple heuristics that make us smart". Oxford: Oxford University Press.

Gordon-Lubitz, R. J. (2003). "Risk communication: Problems of presentation and understanding". *Journal of the American Medical Association*, 289(1), 95.

Hastie, R. and Dawes, R. M. (2001). *Rational choice in an uncertain world: The psychology of judgment and decision making*. London: Sage Publications.

Henrion, M. and Fischhoff, B. (1986). "Assessing uncertainty in physical constants". *American Journal of Physics*, 54(9), 791-798.

Hogarth, R. M. (1987). *Judgement and choice: The psychology of choice* (2nd Ed.). New York: John Wiley and Sons.

Houghton, S. M., Simon, M., Aquino, K. and Goldberg, C. B. (2000). "No safety in numbers: Persistence of biases and their effects on team risk perception and team decision-making". *Group* and *Organisational Management*, 25(4), 325-353.

Kahneman, D. and Lovallo, D. (1993). "Timid choices and bold forecasts: A cognitive perspective on risk-taking". *Management Science*, 39(1), 17-31.

Kahneman, D. and Tversky, A. (1972). "Subjective probability: Judgment of representativeness". *Cognitive Psychology*, 3(3), 430-454.

Keil, M., Wallace, L., Turk, D., Dixon-Randall, G. and Nulden, U. (2000). "An investigation of risk perception and risk propensity on the decision to continue a software development project". *The Journal of Systems and Software*, 53, 145-157.

Klein, W. M. and Weinstein, N. D. (1997). "Social comparison and unrealistic optimism about personal risk". In B. P. Buunk and F. X. Gibbons (eds), *Health, coping, and well-being: Perspectives from social comparison theory* (pp. 25-61). Hillsdale, NJ: Lawrence Erlbaum.

Langer, E. J. (1975). "The illusion of control". *Journal of Personality and Social Psychology*, 32(2), 311-328.

Levin, I. P., Schneider, S. L. and Gaeth, G. J. (1998). "All frames are not created equal: A typology and critical analysis of framing effects". *Organizational Behavior and Human Decision Processes*, 76(2), 149-188.

March, J. and Simon, H. (1958). *Organizations*. New York: Wiley.

Marsh, B. (2002). "Heuristics as social tools". *New Ideas in Psychology*, 20, 49-57.

McNeil, B. J., Pauker, S. G., Sox, H. C. Jr. and Tversky, A. (1982). "On the elicitation of preferences for alternative therapies". *New England Journal of Medicine*, 306, 1259-1262.

Michael, D. (1973). *On learning to plan and planning to learn*. San Francisco, CA: Jossey-Bass.

Nisbett, R. and Ross, L. (1980). *Human inference: Strategies and shortcomings of social judgment*. Englewood Cliffs, NJ: Prentice-Hall.

Pablo, A. L., Sitkin, S. B. and Jemison, D. B. (1996). "Acquisition decision making processes: The central role of risk". *Journal of Management*, 22(5), 723-746.

Redmill, F. (2002). "Some dimensions of risk not often considered by engineers". *Computing & Control Engineering Journal*, 268-272.

Russo, J. E. and Schoemaker, P. J. (1992). "Managing overconfidence". *Sloan Management Review*, 23, 7-17.

Scholz, R. W. (1983). "Introduction to decision making under uncertainty: Biases, fallacies, and the development of decision making" (pp. 3-18). In R. W. Scholz (ed.) *Decision making under uncertainty: Cognitive decision research, social interaction, development and epistemology*. Amsterdam: Elsevier Science Publishers.

Schwenk, C. R. (1984). "Cognitive simplification processes in strategic decision making". *Strategic Management Journal*, 5, 111-128.

Schwenk, C. R. (1986). "Information, cognitive biases and commitment to a course of action". *Academy of Management Review*, 11(2), 298-310.

Siegrist, M., Gutscher, H. and Earle, T. C. (2005). "Perception of risk: the influence of general trust, and general confidence". *Journal of Risk Research*, 8(2), 145-155.

Simon, M., Houghton, S. M. and Aquino, K. (1999). "Cognitive biases, risk perception, and venture formation: How individuals decide to start companies". *Journal of Business Venturing*, 15, 113-134.

Sitkin, S. B. and Pablo, A. L. (1992). "Reconceptualizing the determinants of risk behaviour". *Academy of Management Review*, 17(1), 9-38.

Sitkin, S. B. and Weingart, L. R. (1995). "Determinants of risky decision making behavior: A test of the mediating role of risk perceptions and propensity". *Academy of Management Journal*, 38(6), 1573-1592.

Slovic, P., Fischhoff, B. and Lichtenstein, S. (1977). "Behavioral decision theory". *Annual Review of Psychology*, 28, 1-39.

Stewart, A. (2004). "On risk: perception and direction". *Computers & Security*, 23(5), 362-370.

Todd, P. M. and Gigerenzer, G. (2000). "Précis of 'Simple heuristics that make us smart'". *Behavioural and Brain Sciences*, 23(5), 727-780.

Tversky, A. and Kahneman, D. (1971). "The belief in the 'law of small numbers'". *Psychological Bulletin*, 76, 105-110.

Tversky, A. and Kahneman, D. (1973). "Availability: A heuristic for judging frequency and probability". *Cognitive Psychology*, 5, 207-232.

Tversky, A. and Kahneman, D. (1974). "Judgment under uncertainty: Heuristics and biases". *Science*, 185, 1124-1131.

Tversky, A. and Kahneman, D. (1981). "The framing of decisions and the psychology of choice". *Science*, 211, 1453-1458.

Weinstein, N. D. (1987). "Unrealistic optimism about illness susceptibility: Conclusions from a community wide sample". *Journal of Behavioural Medicine*, 10, 481-500.

Wickens, C. D. and Hollands, J. G. (2000). *Engineering psychology and human performance* (3rd Ed.). Upper Saddle River, NJ: Prentice-Hall.

Xie, X., Wang, M. and Xu, L. (2003). "What risks are Chinese people concerned about?" *Risk Analysis*, 23(4), 685-695.

Chapter 5

Extreme Risk-taking and Decision Making

Bruno Sicard, Elisabeth Jouve and Olivier Blin

Introduction

Any action, whenever the outcome is uncertain, requires a certain amount of risk-taking. Decision making requires one to compare, evaluate and manage risks. Therefore, decision making and risk-taking are closely related: optimal decision making depends on a balanced risk-taking behaviour. Only extreme risk-taking or total avoidance of risks negatively affects the decision making process and the outcome of an action. The risk homeostasis theory describes how the optimal level of risk is determined by expected benefits and costs of risky and cautious behaviour options. The risk homeostatic mechanism is often compared to a thermostat that regulates the fluctuations in "temperature" (risk level) that may occur (Wilde, 2002).

Our hypothesis was that every individual has his/her thermostat set to a certain level of risk propensity, *risk proneness trait*, adapted to his/her routine environment, and that this level can be temporarily modified, *risk proneness state*, by external or internal factors: environment, drugs or other. Therefore, the safety of actions that relies on the individually adapted thermostat regulation (risk propensity level) should be affected by any sudden significant change in the setting of the thermostat. A trivial illustration of this issue would be: why in certain conditions (sleep deprivation, drugs) will an individual drive through a red light whereas s/he usually stops? Does his/her risk propensity level differ and thereby influence his/her hazardous behaviour?

To study risk propensity we developed and validated a visual analogue scale, EVAR (EVAluation of Risks) (Sicard, Jouve, Blin and Mathieu, 1999). EVAR is composed of 24 items distributed among five factors: *self-control* (self cont), *danger-seeking* (dang), *energy* (energ), *impulsiveness* (imp), and *invincibility* (invinc). Risk proneness evaluated by EVAR is partially correlated with the Zuckerman Sensation Seeking Scale (SSS) (Zuckerman, Eysenck and Eysenck, 1978). Like SSS, some EVAR factors are negatively correlated with age (Sicard, Jouve, Couderc and Blin, 2001).

Like the Zuckerman Sensation Seeking and Cloninger Novelty Seeking traits (Gerra et al., 1999; Zuckerman, 1993) risk proneness could be related to dopamine activity in the mesolimbic system, which is also involved in the human decision making process (Egelman, Person and Montague, 1998). This hypothetical biological model of risk proneness could explain why high risk-takers are predominantly found in adolescent and young adult populations, the majority being males (Michel, 2001). Gonadal hormones, like testosterone, potentiate the dopaminergic system and, with the immaturity of the monoaminergic system, are suspected to underlie adolescent impulsivity (Chambers and Potenza, 2003). The difficulty in decision making observed in the Partial Androgen Deficiency of the Aging Male (PADAM) syndrome and the decrease of risky behaviour with age (Baudier, Guilbert and Gautier, 2002) may also be related to a decrease of free testosterone (Tordjemann, 2001) which influences dopaminergic activity. The relationship between drug use and road accidents could also be partially explained by a change in risk proneness induced by a variation of dopamine activity occurring with ethanol, THC (tetrahydrocannabinol), opioids, and so on (Koob and Le Moal, 1997). Violent behaviour or suicide, found to be proportional to the volume of tobacco consumed and its association with illicit drugs in a survey involving 3,800 adolescents (Binder, 2003), could also be related to this dopamine based biological model.

Whereas risk proneness level was found to be stable for an individual in routine conditions, as expected (*risk proneness trait*), we observed significant short-term intra-individual variations with various stressors (*risk proneness state*). A study of ten naval aviators involved in military special operations with tense flight stress, sleep deprivation, shift lag and fatigue showed an increase in EVAR *impulsiveness* scores whereas risk proneness was steady in a control group (Sicard, Jouve and Blin, 2001). During a combat survival training situation, 30 pilots who had to avoid capture were submitted to sleep deprivation and fatigue, while half of them benefited from caffeine intake (Coste et al., 2002). Caffeine influences dopaminergic activity through its adenosine receptor antagonism (Svenningsson et al., 1999). Pilots given the placebo presented significant variations in risk propensity whereas caffeine had a significant counter-measure effect on EVAR scores.

In similar occupational environments (commercial and military aviation), we also found significant inter-individual differences in risk propensity levels that were unrelated to age; military pilots presented higher scores in all EVAR factors except *impulsiveness* (Sicard, Taillemite, Jouve and Blin, 2003).

These results are thus congruent with our hypothesis that risk proneness is set at a certain level for an individual, with a risk profile most likely adapted to the routine environment of that individual (*risk proneness trait*), and probably related to dopamine mesolimbic activity. Around this pre-set "temperature" the thermostat is susceptible to short-term variations due to various psychophysiological and drug stressors. In line with the "risk thermostat" theory, are decision makers engaged in risky activities using hazardous activities to regulate their risk propensity levels?

To address this issue, we conducted a study on volunteer BASE jumpers to evaluate their risk propensity levels variations with jumps. BASE jumpers are considered very high risk-takers, jumping with a parachute from buildings, antennas, bridges, cliffs, and so on. They are socially tolerated even if their activity is not regulated and at times unlawful when practised from restricted areas such as wildlife-protected land, urban or industrial zones or private property.

The hypothesis was: if these high risk-takers have a risk proneness "thermostat" set at a high level, does BASE jumping quench their thirst for risk and set the thermostat back to a lower level? Otherwise stated, as suggested by risk thermostat theory, are BASE jumpers' behaviours (jumps) and risk proneness levels correlated?

Method

We conducted an anonymous survey on three participants and recorded 20 jumps (17 from cliffs and three from a factory chimney, average height was 215 meters, range 90-300) (Sicard, Jouve, Couderc and Blin, 2005). The volunteers were male BASE jumpers, aged 25 to 31 years (mean 28). They completed EVAR (EVAluation of Risk) and Bond and Lader (Bond and Lader, 1974) visual analogue scales to assess their risk propensity and mood and alertness before and after the jumps. We also used the SSS to evaluate their Sensation Seeking trait, in five dimensions: disinhibition (DIS), thrill and adventure seeking (TAS), experience seeking (ES), boredom susceptibility (BS), general (GEN) (Zuckerman, Eysenck and Eysenck, 1978). Jump difficulties were rated from routine, to uncommon and exploratory (the most stressful).

With such a small number of participants, due to the confidentiality and sometime forbidden nature of this activity, our study is mainly descriptive and does not claim to be representative of all BASE jumpers, but it is the first scientific study conducted with this extreme risk-taking population. Comparison by t-test was conducted for EVAR and Bond and Lader scores. Correlations between risk proneness dimensions and Bond and Lader items were evaluated by the Pearson coefficient. We used, as a reference, non-high-risk-takers, namely a group of 85 participants, aged 18-34 years (mean 26.2).

Results

We observed a pattern with higher SSS Thrill and Danger Seeking (TAS) scores in BASE practitioners, whereas others SSS factors were lower, compared to our reference population (Figure 5.1).

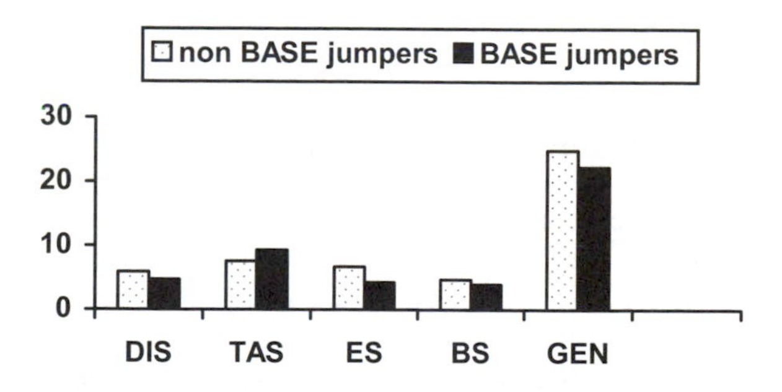

Figure 5.1 BASE jump Sensation Seeking Scale (SSS) scores compared to general population

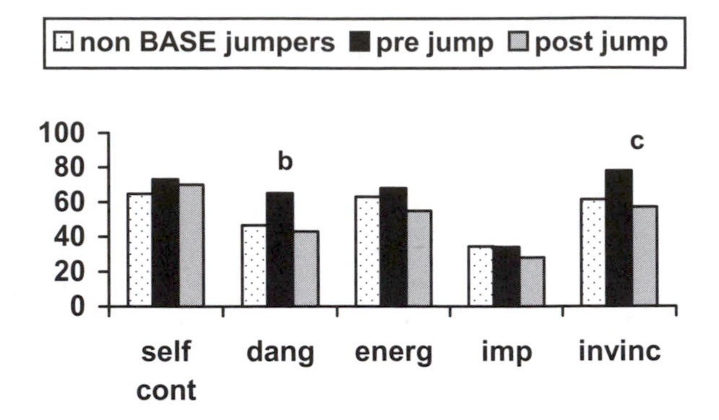

Figure 5.2 BASE pre and post jump and non BASE population EVAR scores comparison between BASE jumpers pre and post jump scores b p<0.01, c p<0.001

Our participants' risk profile displayed significant short term variations (*risk propensity state*) with jumps (Figure 5.2), with a trend that brings back the "thermostat" (risk level) to what could be expected from non BASE jump young adults.

In Figure 5.3 EVAR scores are reported in relation to the difficulty of jump (routine, uncommon or exploratory). Except for *self-control* scores during uncommon jumps, all EVAR scores displayed a trend toward reduction after the jumps, compared to pre-jump levels, and the lowest scores were obtained with the more uncertain, exploratory, "risky" jumps.

Among the five EVAR risk factors, *impulsiveness* was less affected by variations. Statistically significant correlations were found between EVAR and Bond and Lader scales. *Self-control* was correlated with quick witted, *danger seeking* with tense and attentive. *Energy* was correlated with alert, excited, strong, energetic, quick witted, tense and attentive. *Impulsiveness* and *invincibility* dimensions displayed the same correlation as *energy* except for the item tense/relax for *impulsiveness* and strong/feeble for *invincibility*.

Discussion

The objective of this study was to evaluate risk propensity variation before and after BASE jumps. Our hypothesis was that in extreme risk conditions (BASE jump), the risk propensity level, according to risk homeostasis theory, would be high before the jumps and decrease after the jumps: jumps would "quench the thirst" for risk in these high risk-takers who, once satiated, would present a lower proneness for risk.

Compared to a slightly younger population, our participants displayed a lower Sensation Seeking profile, which was expected (SSS scores are negatively affected by age), with the exception of a trend toward higher *thrill and danger seeking*. Therefore the SSS pattern of our BASE jumpers fits the expected profile of young adventurous individuals who engage in high-risk sports and is congruent with previous work showing that students interested in mountain climbing scored higher in TAS than other students (Straub, 1982).

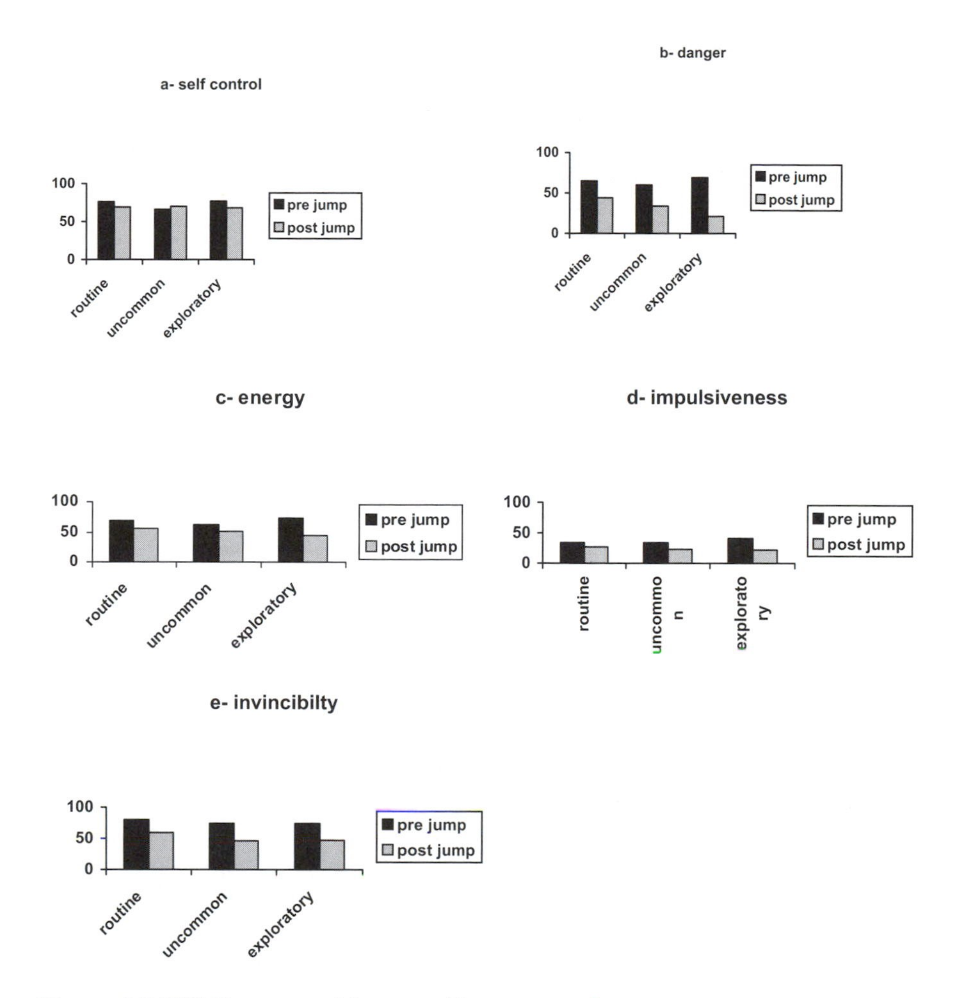

Figure 5.3 EVAR scores with type of jumps: routine, uncommon, exploratory

The risk propensity profile of the BASE jumpers displayed an expected high risk-level, compared to the general population profile, except for the *impulsiveness* dimension. It is interesting to note that military and commercial pilots also present a different risk profile, with the common exception for *impulsiveness* scores, which is a trait considered to be undesirable in the decision making process. Moreover, in both activities (BASE jumping and flying an aircraft) the decision making process quality is critical for survival.

Our BASE jumpers' risk profile displayed significant short term variation (*risk propensity state*) following jumps, with a trend to bring the thermostat (risk level) down to what would be expected from young adult non BASE jumpers. Hence, BASE jump activity seems to satisfy their need for risk-taking and the more stressful the jump (exploratory jumps) the more satiated they seem to be. The results of this descriptive study are congruent with our hypothesis and previous results on risk proneness variations (Sicard, Jouve and Blin, 2001).

Decision making can be affected by risk propensity that seems to be also self-regulated at the individual level. But many questions need to be addressed including the biological model of risk-taking (related to mesolimbic dopamine activity) or the regulation of the risk "thermostat" at the collective level.

References

Baudier, F., Guilbert, P. and Gautier, A. (2002). "Prévenir la violence". *Concours Medical*, 124(12), 826-831.

Binder, P. (2003). "Tabac, cannabis et investissement relationnels". *Revue du Praticien Médecine Générale*, 17(607), 428-344.

Bond, A. and Lader, M. (1974). "The use of analogue scale in rating subjective feelings". *British Journal of Medical Psychology*, 47, 211-218.

Chambers, R. A. and Potenza, M. N. (2003). "Neurodevelopment, impulsivity, and adolescent gambling". *Journal of Gambling Studies*, 19(1), 53-84.

Coste, O., Batéjat, D., Van Beers, P., Piérard, C., Lagarde, D. and Beaumont, M. (2002). "Caféine à libération prolongée: une nouvelle forme galénique de caféine d'intérêt militaire". *Médecine et Armées*, 30(2), 143-149.

Egelman, D. M., Person, C. and Montague, P. R. (1998). "A computational role for dopamine delivery in human-decision making". *Journal of Cognitive Neuroscience*, 10(5), 623-630.

Gerra, G., Avanzini, P., Zaimovic, A., Sartori, R., Bocchi, C., Timpano, M., Zambelli, U., Delsignore, R., Gardini, F., Talarico, E. and Brambilla, F. (1999). "Neurotransmitters, neuroendocrine correlates of Sensation Seeking Temperament in normal humans". *Neuropsychobiology*, 39, 207-213.

Koob, G. F. and Le Moal, M. (1997). "Drug abuse: Hedonic homeostatic dysregulation". *Science*, 278, 52-58.

Michel, G. (2001). "Prise de risque chez l'adolescent: les facteurs de vulnérabilité". *Revue du Praticien Médecine Générale*, 15(525), 238-242.

Sicard, B., Jouve, E. and Blin, O. (2001). "Risk propensity assessment in military special operations". *Military Medicine*, 166, 871-874.

Sicard, B., Jouve, E., Blin, O. and Mathieu C. (1999). "Construction et validation d'une echelle analogique visuelle de risque (EVAR)". *L'Encéphale*, 25, 622-629.

Sicard, B., Jouve, E., Couderc, H. and Blin, O. (2001). "Age and risk-taking in French naval crew". *Aviation, Space & Environmental Medicine*, 72, 59-61.

Sicard, B., Jouve, E., Couderc, H. and Blin, O. (2005). "Propension au risque chez les base jumpers". *Médecine Aéronautique et Spatiale*, 46, 28-32.

Sicard, B., Taillemite, J. P., Jouve, E. and Blin, O. (2003). "Risk propensity in commercial and military pilots". *Aviation, Space & Environmental Medicine*, 74, 879-881.

Straub, W. F. (1982). "Sensation seeking among high and low-risk male athletes". *Journal of Sport Psychology*, 4, 246-253.

Svenningsson, P., Le Moine, C., Fisone, G. and Fredholm, B. B. (1999). "Distribution, biochemistry and function of striatal adenosine A_{2A} receptors". *Progress in Neurobiology*, 59, 355-396.

Tordjemann, G. (2001). "Sexologie, l'andropause". *Concours Medical*, 123(19), 1300.

Wilde, G. J. S. (2002). "Does risk homeostasis theory have implications for road safety?" *British Medical Journal*, 324, 1149-1151.

Zuckerman, M. (1993). "P-Impulsive sensation seeking and its behavioral, psychophysiological and biochemical correlates". *Neuropsychobiology*, 28, 30-36.

Zuckerman, M., Eysenck, S. and Eysenck, H. J. (1978). "Sensation seeking in England and America: cross-cultural, age, and sex comparisons". *Journal of Consulting and Clinical Psychology*, 46, 139-149.

PART 2
Areas of Application

Chapter 6

Human Requirements in Automated Weapons Systems

Jennifer McGovern Narkevicius and Peggy Heffner

Introduction

Humans are evolving slowly. Not so for the systems around the human. These systems can be designed to support decision making, situation awareness, reduced design induced error, and increased operational effectiveness. The costs associated with these human system integration (HSI) improvements are low, especially if introduced early in the programme and carried throughout the acquisition process.

Automated systems are designed more and more frequently into weapons systems. Inclusion of automation provides the potential for improving human performance by reducing errors and enhancing both decision making and situation awareness. Aviation systems clearly are an important technical area for automated systems (Kanki, 2001). These systems are, of necessity, complex and are developed in adherence with the systems engineering principles discussed below.

Systems engineering follows a fairly rigorous sequence of events forming a detailed and documented or documentable process. These events help ensure that the concerns of all the appropriate and applicable disciplines are considered in the design trade-offs made throughout the development of complex systems and systems of systems (where systems are developed by amalgamating multiple complex systems). The phases, illustrated in Figure 6.1, provide checks and balances for decision making throughout the design. The use of clearly defined exit criteria helps decision makers assess the programmatic risks (cost, schedule and performance) associated with proceeding in the selected development path. The process also allows opportunities to inject improvements or design changes based in intelligent flexibility in the design trade space.

The systems engineering process discussed above is predicated on diligent and complete requirement specification. Automated systems require this same careful identification and definition of requirements. However, requirements definition for integration of humans into complex automated systems continues to be an issue. The potential benefits of automation are countered by the very real costs of the increased design complexity that is required to accommodate the automated system and the increased potential for human error brought about by operation of an improperly designed automated system. In addition, automated systems are embedded in the increasingly complex structures of distributed decision networks.

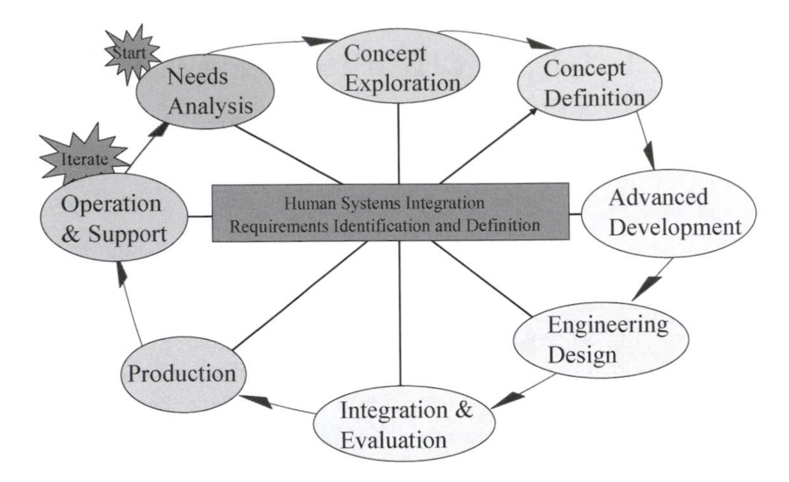

Figure 6.1 Iterative phases of the systems engineering process

Distributed Command and Control (C^2) systems also require the detailed requirements definition that is warranted by complex systems (Sheridan, 2002). There are automated systems embedded in the C^2 systems that take in information from distributed, remotely located automated systems. This nesting of composite systems inside complex systems provides opportunity for requirements to be overlooked, incorrectly captured, or captured at cross purposes. This is especially true for the human performance requirements that will be similar in general appearance but differ in critical functions across systems.

To make requirements definition relevant to systems under development for use by human users, the requirements or operators, maintainers, and other users (such as supervisors) and their place in the system must be carefully documented and utilised throughout the overall system development. Chapanis (1996) reminds us that all systems are developed for human use. There are no unpopulated systems. This clearly applies to automated systems (as well as "unpopulated" ones).

Transformation through technological advances allows many weapons systems to be networked. These networked systems have the potential to generate new warfighting capabilities and possibilities for still to be determined new capabilities. The requirements of these networked systems are not the summation of the requirements of the original individual, stand alone systems. Rather, there will be that summation as well as an amalgamation of new networked requirements (and their derivatives) to be defined, designed to, explored through concepts of operations and analyses of alternatives, and met with design decisions.

Requirements definition

Requirements definition begins with recognising an operational capability is needed (Chapanis, 1996) and must include the needs of the users. It is essential to consider not only the immediate needs driving the system under development or modification, but also to consider the application and use of the system with respect to other systems with which it must interact. This is even more important in networked systems that must work together, preferably seamlessly, to achieve a greater capability than the sum of the individual systems' capabilities. As a discipline, systems engineering provides a framework within which to approach this requirements definition of the system under development. It also provides the framework within which more complex (or expansive) systems can be considered.

Definitions of needs and of requirements are essential in any systems engineering acquisition programme (Martin, 1997; Rechtin and Maier, 1997). The need illustrates the desired capabilities, accomplishments or achievements. The required performance of the system comes from achieving these desires. These requirements must be identified to determine what possible solutions to bring forward in an effort to meet those requirements. Requirements for weapons systems are easily documented for hardware and software but the determination and application of requirements for users is more challenging. Tools, processes and procedures are necessary to apply to users' requirements in engineering acquisitions (Booher, 2003).

Because performance of a system depends on the operator as well as the hardware and software (Chapanis, 1996), it is necessary to translate from the requirements of the overall system to useful, successful human performance in support of that system completing that mission. The primary tools for successful integration of human requirements into systems acquisition and engineering include models, use cases, and requirements management. These tools are necessary to integrate human user requirements and their concomitant features.

Highly networked systems will have nested sets of user requirements based on the capabilities of the entire networked system. Use, early in the systems engineering acquisition process, the tools in the systems engineering acquisition process and follow through with requirements management tools will allow the nested requirements to be incorporated into systems designed to improve situation awareness, decision making, collaborative work, and enhanced operator performance throughout the system.

Use Cases

Use Cases describe what the system under development must do to achieve the mission from the users' perspective. This focus is at the high level of the system. Use Cases focus on the user as the definition of the scope of the project. They can be used to scope how models would (or should) be developed to ensure that how the user will use the system is included in programme decision making. Because of the focus on the users' perspective of the functions of the system, the Use Case maintains focus throughout development.

Use Cases are at a low enough level of granularity that they can be used to describe a weapons system and to describe the networked C^2 system in which that weapons system must operate. The Use Case will facilitate the development of information flow across the C^2 platform and will highlight nodes of information glut (or dearth) that will reduce the performance of the C^2 system and the performance of the weapons system associated with the network.

Use Cases should be developed to help select portions of the operational space to be more fully explored in modelling. They provide a consistent set of scenarios to explore throughout development and operation.

Models and modelling

While the requirements detail *what* a system must be able to do to be considered successful, good requirements do not dictate *how* a system must work or operate. It is quite difficult to get from the what of the requirements to the how of design. One useful tool is modelling of potential solutions to the requirements. Modelling can provide a means for asking and answering questions about functional allocation and tasks assignment across the three major elements of the system: hardware, software, and human users. Modelling requires an understanding of the mission requirements and the means to allocate those requirements within possible solutions. Models must be valid, verifiable, and accurate (Kanki, 2001).

Modelling tools provide an economical, low risk means of exploring potential solutions in the trade space without negative effects on programme cost, schedule, or performance. These tools also provide the means to generate a large pool of potential solution options. Then candidate solutions can be further evaluated and final solutions chosen more freely from the available options rather than selecting, in effect, technical "variations on a theme".

It is feasible (and necessary) to model the automated system and to allocate functions to the automation software, the hardware, and to the human user. Modelling also provides a platform to quickly reallocate functions and observe the effect of different allocations on overall system performance. Models can also be developed from networked distributed systems (such as C^2 entities). Again, it is possible to alter the allocation of functions across the distributed network and determine the optimised way to work within the network.

It is equally necessary to model the elements and entities of distributed C^2 systems. The interactions of the component systems within the C^2 system can be modelled and functions can be allocated to those entities to observe the effects of different allocations on the behaviour and success of the network. Distributed systems also require modelling. These models must incorporate the element systems and the distributed network with its associated connectivity to explore fully the trade space. But more importantly, modelling distributed systems more fully illustrates unintended consequences (both beneficial and unbeneficial). Modelling may also reveal potential, unanticipated enhancements that are an outgrowth of the distribution of systems and their integration.

Requirements management

Requirements management tools allow designers and others associated with the development of the systems under design to ensure that all identified requirements (hardware, software, and user) are documented and are traceable throughout development. These tools ensure that requirements that are difficult to allocate are not dropped. These tools keep all the requirements on equal footing, ensuring that user requirements are not deleted in the face of technical challenges. This is especially essential in automated systems where user requirements make demands that may be difficult to sort out in software architecture development.

Systematic enterprise approaches

Finally, successful implementation of the requirements definition process including human user requirements is predicated on a systemic or enterprise level approach to capturing and incorporating these requirements. This enterprise approach must include a process to be followed by users at all levels of the enterprise. The implementation of the process must be represented in policy that illustrates the roles and responsibilities of various entities throughout the enterprise for completing the process at each level. Finally, these processes and policy must be consistent with Systems Engineering as practised in the enterprise.

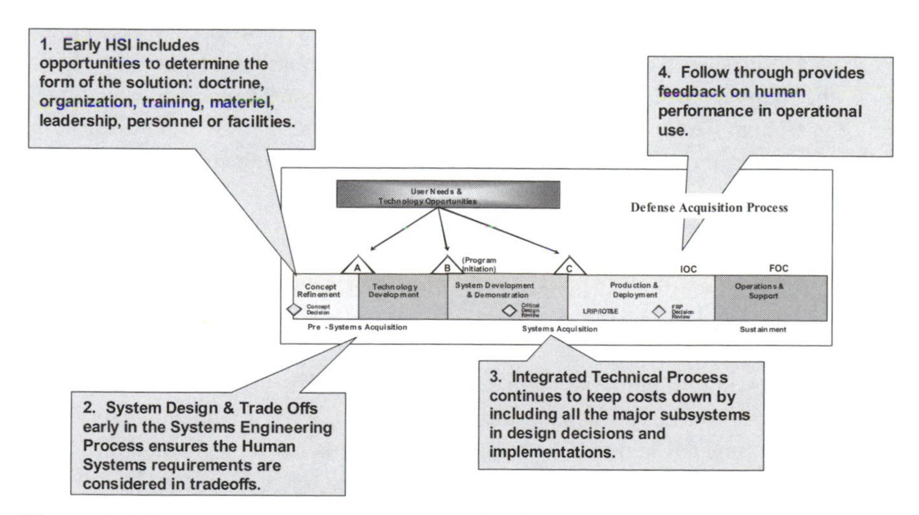

Figure 6.2 Notional injection points for HSI in the systems engineering process

The US Navy has established the Systems Engineering, Acquisition, and Personnel Integration or SEAPRINT Programme. SEAPRINT mirrors the Army's MANPRINT Programme as the Enterprise HSI approach. SEAPRINT establishes the policy, architecture, and processes for Navy HSI. Further, the SEAPRINT Programme continues to explore toolsets that support successful integration of humans and their requirements throughout the enterprise systems engineering acquisition processes. SEAPRINT provides a process, linked to all levels of the enterprise through an integrated architecture. Further, a policy illustrates the required elements which the programme must perform and the Systems Commands guidelines further delineate how the policy will be met by the acquisition portions of the enterprise. These tools (policy, process, architecture, and guidelines) will allow successful inclusion of HSI (and its elements' technical requirements considerations) in systems engineering acquisition, enhancing the use and utility of HSI tools throughout the systems engineering process and across the enterprise organisation. Figure 6.2 illustrates notional injection points for HSI. SEAPRINT specifically identifies "non-traditional" areas for HSI throughout the process.

Example E/A-18G and human systems integration

The US Navy has developed a renewed interest in making the sailor the centre of the Navy. This will strengthen war-fighting capabilities by including the user of weapons rather than focusing solely on the physics of the weapons themselves. The HSI thrust has brought the user requirements to the forefront. This shift of focus requires a modification to the processes used to acquire war-fighting equipment. These changes in focus will include moving to the *integration* of humans as integral parts of the war-fighting system rather than the insertion of humans to a developed system, as has historically been the approach.

This HSI focus will require an integration of tools from across disciplines. These disciplines will be diverse and will include a number of sub-disciplines. From the human focused areas tools will come from Manpower, Personnel, Training, Human Factors, Safety, and Health. Tools will also come from those more traditional disciplines of hardware engineering, software engineering and systems engineering.

The US Navy continues initiatives to compile and integrate processes, tools and techniques from these various human centred disciplines. These activities work to identify, validate, verify, and integrate the tools and their outputs. This effort will ensure that the information and data applicable to design and exploration of the trade space are useful.

In the E/A-18G programme, the outcome of the HSI approach has affected the development of this highly complex automated system. While in development the E/A-18G programme has included a strong reliance on modelling and simulation, Use Cases, and requirements' management. This highly automated, networked system will allow support of distant conflicts with precision, speed, and accuracy (the need for this is highlighted in Pettre, 2003).

Conclusions

As the place of the human portion of advanced technology and more networked systems continues to mature, the necessity for delineating the human requirements as part of the overall systems engineering requirements definition activity will become clearer. Further, the importance of getting those requirements defined correctly and accurately will be evident. The processes and policies that support the development of these human oriented technology solutions will continue to evolve as well. It is the complete and successful integration of HSI processes, policies and architectures into the systems engineering practice that will result in successful automated and networked systems that reach the capabilities required of operational performance.

HSI has a significant impact on systems engineering by including the human requirements into the engineering process, strengthening the ability of the system to meet capability requirements. By ensuring that the human fully participates in the work of the system, highly complex systems, especially automated and networked systems, will succeed. HSI will result in better systems. To achieve this fully, HSI must become integral to the systems' engineering practice from the beginning of programme development. The means to achieve this integration of HSI is through implementation of policy, process, and architecture and tools. These policies and processes will include integrated toolsets to attain, complete, and accomplish HSI throughout the engineering life cycle.

The Navy's acquisition community is finalising these policies and processes and their concomitant architecture. The continued development of highly complex, automated, networked systems will continue placing increasing demands for modelling, Use Case analysis, and requirements management tools in the systems engineering acquisition of weapons systems. The E/A-18G programme is an example of the HSI process effectively supporting development of next generation, automated war-fighting capabilities. Other successes are close behind. As more complex, networked systems are developed, this approach will become more necessary and will shake out the tools needed to support the HSI efforts.

Acknowledgements

The authors would like to thank CRD N. Sutton and Mr T. Witte, both of Naval Air Systems Command and Ms N. Dolan of Office of Chief of Naval Operations for their support of this effort.

References

Booher, H. R. (2003). *Handbook of human systems integration.* New York: Wiley.

Chapanis, A. (1996). *Human factors in systems engineering.* New York: Wiley.

Kanki, B. G. (2001). "Automation in the workplace: Lessons learned in aviation operations". *Human Factors and Ergonomics Society Computer Science Technical Group Bulletin*, 8(1).

Martin, J. N. (1997). *Systems engineering guidebook.* Boca Raton: CRC Press.

Pettre, M. (2003). "Close air support from afar". *Journal of Electronic Defense*, 26(6), 40-45.

Rechtin, E. and Maier, M. W. (1997). *The art of systems architecting.* Boca Raton: CRC Press.

Sheridan, T. B. (2002). *Humans and automation: System design and research issues.* New York: Wiley.

Chapter 7

Automation and Decision Making

Jan Noyes

Introduction

Over the last couple of decades, highly computerised and automated systems have become firmly established in many spheres of industry. Civil aviation has been a particularly significant progenitor and recipient of these technical advances, as typified by the standard "glass cockpit" in today's commercial aircraft. Many of the tasks which flightdeck crew would have previously carried out have now been automated for some or all of the flight, for example, flight path guidance, detection and diagnosis of malfunction. Accordingly, the introduction of automation has led to superior productivity, efficiency and quality control. However, introducing automation also brings with it a number of difficulties. These include loss of situation awareness, and "trust" in the system exemplified by a general belief in the accuracy and authority of computers. Further, there is a deskilling of the crew when carrying out routine operations, resulting in them being less qualified to cope with emergency-type situations when they arise. Indeed, Moray (2001) reported one crew member's belief that only the most experienced individuals should be allowed to fly the Airbus A340 because of this.

In summary, the degree of automation in complex systems such as those found on the civil flightdeck continues to pose a problem. Too much automation, and the human operator is not "in the loop" when failures and malfunctions occur. Making decisions thus becomes problematic as crew are not fully aware of the situation. Too little automation, and the benefits of complex systems remain unrealised. The challenge for system design concerns the development of systems, which provide an appropriate level of automation for a particular situation at a given time (see Kaber and Endsley, 2004, for a recent review).

Flightdeck automation

The primary intention behind flightdeck automation has been the increase of flight safety and efficiency. In many ways, this has been successful. For example, there has been an overall decrease in accident rates concurrent with the introduction of more advanced technology. However, there is also evidence that while the overall numbers of crew errors have been reduced, the advent of automation has generated a new class of errors, of a more fundamental nature, for example, omission and commission errors, arising as a direct result of the automation.

Omission errors arise when operators fail to take appropriate action in response to a problem, because the automated decision aids have failed to inform them of an imminent system failure. An example of this type of error occurred during the China Airlines incident (National Transportation Safety Board, 1986). The Boeing 747 aircraft suffered a slow loss of power from one of the outer engines; however, the autopilot corrected for the resultant yaw, masking the anomaly, until it reached the limit of its compensatory abilities. At that point, the plane rolled and entered a vertical dive of 31,500 feet before it could be recovered. The crew were unaware of the problem before this happened, and did not have time to make a diagnosis or plan a suitable course of action.

In the China Airlines situation, control of the aircraft had been delegated to the autopilot; and because of this, the crew failed to monitor the condition of the engines. This is indicative of another possible basis for error that has been termed "automation bias" (Skitka, Mosier and Burdick, 1999). It has been suggested that a convergence of attitudinal traits, that is, cognitive laziness, diffusion of responsibility, and belief in the authority/accuracy of computers, leads to the occurrence of omission errors as described above, and also commission errors.

Commission errors result from crew following the directives of the automated monitor, when the more reliable non-automated instruments suggest that the automated aid is not taking the correct course of action. An example of this can be seen in the New Zealand DC10 accident. This occurred whilst the aircraft was en route to the South Pole and involved the plane crashing into Mount Erebus in Antarctica, killing all 257 people aboard. The aircraft's navigational systems had been programmed with an incorrect position, yet the crew trusted them and therefore failed to notice how external visual information was inconsistent with their model of where they were (Green, 1990). Both omission and commission errors emanate from the presence of automation, which is biasing decision making processes by providing a replacement for vigilant information seeking and processing.

Technology is now reaching a point of maturity where decision support systems can be designed that utilise the information available on the modern flightdeck to provide an "intelligent cockpit-assistant". Flight crew judgement and decision making have been cited as a key area of weakness (Jensen, 1995). This is true for commercial transport flight crew as well as general aviation pilots. A decision support system could help the crew in determining the best course of action in a given situation. It could present the crew with an assessment of the situation, the options open to them, the probable outcome of each option and the actions required to achieve an outcome. Further, it could provide the crew with an advisory facility by providing immediate feedback on the outcome, or predicted outcome, of their actions – something which is often missing on present flightdecks. All of this would be provided in the context of the current aircraft's situation and crew task(s). At Bristol, we have been addressing some of these issues. Our approach has been to start by looking at the process of decision making as a whole in order to identify parts of the process that are fallible, with the eventual aim of finding ways of supporting crew decision making in those areas. As part of this work, we have been developing an Integrated Decision making

Model (IDM; Donnelly, Noyes and Johnson, 1998); a description of this and some of the experimental work contributing to its development will be reported here.

The flightdeck environment

The civil avionics environment comprises a unique setting for decision making: it is complex, dynamic, subject to distractions, time pressure and at times, information overload. Further, there is a safety critical element; making the right decisions is vital to ensuring the safety of the crew and passengers, and the "health" of the aircraft. Many models of human decision making have been developed; the most successful of which tend to be the more recent ones based on naturalistic decision making (NDM; Klein, 1993). The NDM theories apply directly and particularly well to the avionics environment. They help our understanding of how crew act under certain conditions, how they respond to situations, and how they make decisions. A review of the decision theories developed over the last 50 years or so indicates the following key issues:

1. **Assessment of the situation**. Information is either sought or automatically presented. In both cases, the crew will match the situation to previous experience and look for the closest fit. If the situation is not familiar, more information is sought to improve situation awareness (Endsley, 1995).
2. **Awareness of the situation (mental representation)**. The crew must have an understanding of the situation in order to make a decision. If information or events violate this representation, they must reassess the situation to gain a clearer picture.
3. **Knowledge of the appropriate course of action**. The crew will take action or monitor the situation according to their mental representations. Training, experience and procedures will dictate appropriate action(s).
4. **Awareness of potential consequences of action(s)/inaction**. The crew will perform some type of mental simulation or evaluation of their actions to assess outcomes and check for expectancies. Feedback is received on expectancies, goals, information input, and so on.

There are other characteristics of decision making which are essential to understanding decision making in this domain. First, and perhaps most important, is that crew begin with a high situation awareness rather than acquiring it, unlike other experienced decision makers such as fire-fighters. This is an important reversal since the potential for error occurs when situation awareness degrades (that is, when the pilot's situation awareness differs from the real situation), as opposed to when a situation is not correctly assessed. There is also a time factor involved when assessing a situation, which may not be significant if the situation is already known. Secondly, the use of mental representation (Klein, 1993) is necessary for effective decision making. Similar to situation awareness, this will comprise: 1. Perception of cues/knowledge of what is happening; 2. Comprehension of cues/knowledge of rules governing situation; 3. Projection of future developments/ knowledge of consequences, or expectancies.

These different levels also relate to Rasmussen's Skill-Rule-Knowledge theory (Rasmussen, 1993). It appears that many decision making researchers have come to the same conclusions about information processing, but have named it differently. The crew member has a hierarchy of goals and sub-goals, which correspond to Rasmussen's different levels of information processing. Higher goals, for example, maintaining safe flight, are equivalent to the knowledge level while lower goals, for example, immediate flying tasks, are related to the skill level.

Procedures set out by the authorities and by the operators govern most situations on the flightdeck. Therefore, experience is essential in matching the information and cues to a familiar situation, in order to know which procedure is appropriate. This implies that there is a potential for error when the pilot believes the situation to be something it is not, corresponding to Rasmussen's view of changing between levels of processing, where errors can occur in knowing when to switch levels.

Finally, the decision process is continuous, and may be considered as a series of goals, information processing and actions to achieve these goals. The actions themselves may consist of smaller problems or decisions, with sub-goals. Feedback is essential in this process and may be the most important feature of aviation decision making.

The Integrated Decision making Model (IDM)

The review of decision making theories led to the development of the IDM, shown in Figure 7.1. There are three paths that the crew may take in making a decision, which are as follows:

- Path 1. If there is not enough information, or the situation is complex, the individual may seek more information to clarify their representation of the situation.
- Path 2. If the crew are satisfied with the representation, they may form intentions to act.
- Path 3. There will be effects and consequences of the crew's actions, or failure to act.

The model identifies points in the decision process where errors may be made. It is suggested these may be used as intervention points for decision support in order to prevent errors or to help recover from them. These intervention points are shown in Figure 7.1 and explained in more detail here:

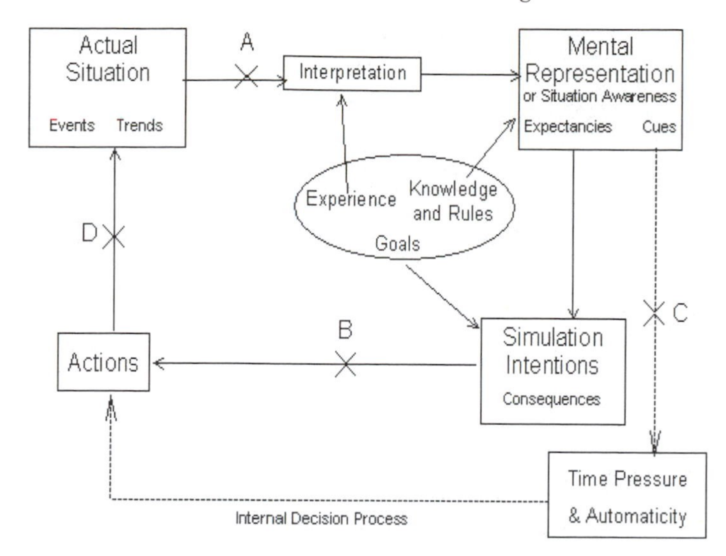

Figure 7.1 Integrated Decision making Model (IDM)
Note: The internal decision making process is indicated by the dashed line.

A – The crew's mental representation/SA and the difference between this and the actual situation play a key role in the decision process.
Situation awareness degrades due to poor information or misinterpretation. This might therefore lead to an error in the situation assessment stage of the model.
Intervention involves presenting information to the pilot in a better format, or inferring what has been missed and re-presenting it in a different way. This is an error-prevention strategy, which cannot guarantee to maintain or recover situation awareness.

B – The crew may not realise the consequences of a course of action, due to inexperience or misrepresentation.
Intervention involves informing the pilot of the consequences of their actions, either when the actions are unsafe, or by inferring intentions. This is an error-tolerant strategy. The system must wait for actions and then assess whether they are unsafe or unintentional. This does include an important feedback element, which may allow the crew to restore situation awareness.

C – The crew may not consider the consequences of a course of action, due to time pressure and/or automaticity. They might erroneously assume that a particular situation has been encountered before, and will therefore not seek to explicitly consider the consequences of his/her actions. This route might also be taken when there is not enough time mentally to simulate events.
Intervention is the same as B. Although the two are essentially different errors, the result is the same.

D – Erroneous actions may go unnoticed due to distraction or lack of feedback. If the crew fail to gain sufficient feedback following an action then their mental representations may not be reinforced, and further necessary action might not be taken.

Intervention involves providing feedback on actions as with B and C. Feedback may be given when actions are unsafe, or by inferring pilot intentions, when actions are unintentional. If the crew are distracted, they may not notice feedback information and so an effective "attention-getter" would be needed. This strategy is also error-tolerant and situation awareness may be restored if information is presented in the right way.

Validation of the model

An important aspect of decision making, and one which often leads to error, is the use of heuristics, that is, short-cuts. The two main conditions under which this route is taken are time pressure and automaticity (when actions or situations become routine) as outlined in C above. The latter seems to be the key to the versatility of human decision making, but can also be its downfall. Hence, it was decided to "test" the IDM with regard to time pressure and automaticity. An experimental study was undertaken to examine the actions of decision makers under these conditions. It was hypothesised that people making decisions do not consider the full consequences of their actions when under time pressure and acting automatically, and therefore will make more errors under these conditions.

In order to test the model, a decision making task was devised. This comprised a process control simulation of an alcohol distillation plant, and is shown in Figure 7.2. The primary task was to produce as much distillate as possible with a purity as high as could be achieved. The participants had three controls with which to regulate the distillation of alcohol according to quantity and quality. They were presented with a pictorial representation of the process control simulation shown in Figure 7.2 along with eight parameters, which must be monitored. These parameters change colour as they reach threshold levels, to provide warning advice: amber indicates that a parameter is at an undesirable level and red warns of a dangerous level. The participants needed to control the fuel flow into the burner, the cooling fan speed and the dump valve. They had to maintain a specified purity level and flow-rate; these were possible to attain temporarily, but not to maintain permanently, thus requiring continuous input to the controls. However, these controls do not directly affect the purity and flow rate, but are second-order controls, thus providing the desired complexity and dynamicism. There was also a secondary, event-handling task where participants were asked to respond to a message. Since there were two possible responses, this required a simple decision to be made by the user.

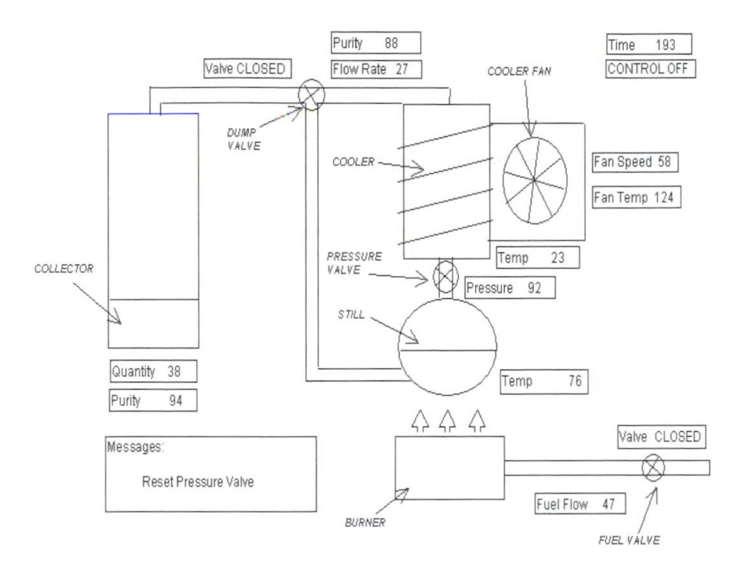

Figure 7.2 Screen view of process control simulation

The time allowed to respond to the event was controlled so that the participants were placed under time pressure. Hence, time pressure was simulated by giving the decision maker a fixed time in which to respond to an event. In the control condition, the participant had five seconds to respond, while in the experimental condition, they had only two seconds. Similarly, the frequency and type of event were controlled such that the response became increasingly automatic. Automaticity can be simulated using two similar events of varying frequency of occurrence: the more frequent event and its response become automatic, while the less frequent event becomes non-routine. Thus the ratio of the occurrence of the two events is a measure of automaticity. In the control condition, the ratio was 1:1, while in the experimental condition, the ratio was 7:1.

Participants were 18 males and 11 females: 14 in the control condition and 15 in the experimental condition (high time pressure and automaticity). Their mean age was 24.9 years, with a standard deviation of 2.09 years. At the end of the experiment, a questionnaire was completed by each participant to provide information relating to the time pressure and automaticity conditions, and participants' performance and motivation.

Results indicated that participants under time pressure made faster decisions, although there was no difference in times taken to respond when the number of events was much higher. It appeared that the automaticity criterion was perhaps too simple and that the decision chosen to represent the automaticity path in the IDM was in itself automatic, thus needing no consideration of consequences. The objective results did show that under time pressure or distractions, the decision maker's actions could be automatic and this might imply that there are two kinds of automaticity, simple and complex. This could relate to Rasmussen's different levels of skill/knowledge. The event-handling used in the experiments would have dealt with "simple automaticity", that is, skill/rule level decisions (yes/no decisions) which are simple enough not to warrant any real consideration of consequences under any conditions, but which, under conditions of time pressure or distraction, can lead to mistakes or slips. "Complex automaticity" would involve decisions which require knowledge level processing in order to be made correctly, but which would, if the same decision was made frequently enough, encourage the decision maker to become complacent and not use this knowledge level.

Another reason why the results do not uphold the automaticity hypothesis was that the participants were not actively controlling the process for a significant amount of time. This is shown by the purity and flow-rate levels. The results indicate that most participants did not attempt to maximise the flow rate at the required level, that is, stabilise the purity at a low flow rate. This means that they had more time to wait for an event to occur. This compounded the problem rather than caused it, since if they were actively controlling the process this would have brought a significant amount of distraction to the task, but would not have changed the complexity of the decision.

The event-handling used in the experiment dealt with "simple automaticity", that is, skill/rule level decisions (yes/no decisions) which are simple enough not to warrant any real consideration of consequences under any conditions, but which, under conditions of time pressure or distraction, can lead to mistakes or slips. "Complex automaticity" would involve decisions which require knowledge level processing in order to be correctly made, but which would, if the same decision was made frequently enough, encourage the decision maker to become complacent and not use this knowledge level. This type of decision would be typified by fault diagnosis, information seeking or interpretation, or event/situation-following decisions (where close attention must be paid to a changing situation). Therefore, rather than automaticity implying simple decisions, it should imply complex decisions which have become routine and thus lowered to a skill/rule level, that is, the decision maker takes the wrong path in the IDM.

Discussion

The IDM appears to provide a useful mechanism for characterising decision making processes although the question of automaticity requires attention. To validate the model further, it would be advantageous to address other error types within the taxonomy. It is not immediately clear how one would seek to address B, since the nature of a participant's mental simulations are likely to be impenetrable (maybe even to the participant him/herself). However, the two remaining error types, A and D, are open to enquiry, and can be tested within the same process control task.

The IDM highlights areas where decision support could intervene to aid the crew. For example, an effective intervention point for decision support could be to provide predictive information and feedback on the effects or consequences of crew actions. This could also help to clarify or even restore crew situation awareness. Such a system would essentially be a warning system, which gathers information on the aircraft and its present environment, so that it can provide the crew with an accurate picture of the situation.

It is suggested that at the moment, we have an inadequate understanding of decision making in relation to complex systems that will allow us to do this. As Moray (1999) pointed out, we know a lot about human information processing, but little research has been carried out on locating the optimal level of human-machine co-operation. An understanding is required of how information can be provided at the interface, which will lead to sound decisions being made. The issue concerns how we can design an interface which lets the crew know that the action path they are currently undertaking does not match their current mental model of the world, without flashing everything as a warning. With this information, decision support systems could be produced that will help the crew make better decisions and allow them to react appropriately to information that does not fit.

In order to see the maturity of this opportunity, many issues still need to be resolved. On the one hand, there are many merits attached to the development and implementation of automated systems, while on the other, there are a number of unresolved issues relating to highly automated systems. The ultimate way forward is undoubtedly to strike a balance somewhere between the two extremes of "no automation" through to "total automation". There is a need to use the automation and capabilities of a system to present information, which facilitates the crews' cognitive process and thinking, supports their situation awarenss and understanding, and assists in decision making and implementation.

Acknowledgements

Thanks are extended to Alison Starr of Smiths Aerospace for her advice on the avionics content of the research we have carried out together over the last 20 years. Also, thanks to British Airways for their help and feedback on this research, and to Doug Donnelly for initiating the work on the IDM model. Finally, the DTI (Department of Trade and Industry) under their Aeronautics Research Programme, is acknowledged for funding the experimental programme currently taking place.

References

Donnelly, D. M., Noyes, J. M. and Johnson, D. M. (1998). "Development of an integrated decision making model for avionics application". In M. Hanson (ed.) *Contemporary Ergonomics* (pp. 424-428). London: Taylor & Francis.

Endsley, M. (1995). "Towards a theory of Situation Awareness in dynamic systems". *Human Factors*, 37(1), 32-64.

Green, R. (1990). "Human error on the flight deck". *Philosophical Transactions of the Royal Society of London Series B*, *327*, 503-512.

Jensen, R. (1995). *Pilot judgment and crew resource management.* Cambridge: Avebury Aviation.

Kaber, D. B. and Endsley, M. R. (2004). "The effects of level of automation and adaptive automation on human performance, situation awareness and workload in a dynamic control task". *Theoretical Issues in Ergonomics Science*, 5, 113-153.

Klein, G. (1993). "A recognition-primed decision model of rapid decision making". In G. Klein, J. Orasanu, R. Calderwood and C. Zsambok (eds), *Decision making in action: Models and methods*. Norwood, NJ: Ablex.

Moray, N. (1999). "Human operators and automation". In *Proceedings of People In Control conference*, p.8. London: IEE.

Moray, N. (2001). "Humans and machines: Allocation of function". In J. M. Noyes and M. L. Bransby (eds) *People In Control* (pp. 101-115). London: IEE.

National Transportation Safety Board (1986). *Aircraft accident report – China Airlines 747-SP, N4522V, 300 Nautical miles northwest of San Francisco, 19th Feburary 1985*. Report No. NTSB/AA-86/03. Washington: NTSB.

Rasmussen, J. (1993). "Deciding and doing: Decision making in natural contexts". In G. Klein, J. Orasanu, R. Calderwood and C. Zsambok (eds), *Decision making in action: Models and methods*. Norwood, NJ: Ablex.

Skitka, L. J., Mosier, K. L. and Burdick, M. (1999). "Does automation bias decision making?" *International Journal of Human-Computer Studies*, 51, 991-1006.

Chapter 8

The Flightdeck of the Future: Field Studies in Datalink and Freeflight

Gemma Cox, Sarah Sharples, Alex Stedmon, John Wilson and Tracey Milne

Introduction

The national airspace system involves certain key components including: air traffic control personnel, airline management, and pilots and their aircraft systems. These main components differ with respect to their goals, and the information available to them (Wickens, Mavor, Parasuraman and McGee, 1998). An air traffic controller's overall goal is the safety of all aircraft in the system whereas airline management, although concerned with safety, have a more vested interest in expediency and efficiency, which is driven ultimately by profit. The pilot's interests are more local, concerned primarily with the safety and expediency of their aircraft. Each component maintains different information about the airspace, for example airline management are likely to possess the greatest information about global weather patterns in order to compute the most efficient flight plans.

Air traffic controllers, conversely, may have less precise information about current weather conditions, although more information regarding global traffic patterns. Thus across these key components there is an uneven distribution of information scope and detail. Despite this conflict of interests the challenge facing all the key components of the airspace system is undoubtedly the relentless increase in the number of aircraft that are expected to occupy the world's airspace in the future, and how to accommodate these extra aircraft so that they can operate economically, efficiently and safely.

New initiatives

The Flightdeck and Air Traffic Control Collaboration Evaluation (FACE) project, funded by the UK Engineering and Physical Sciences Research Council (EPSRC), ran from November 2002 until January 2006. It took a systems perspective to investigate the human factors requirements for the flightdeck-air traffic control (FD-ATC) system of the future in relation to two specific elements: datalink and freeflight.

Datalink

Datalink is designed to provide electronic exchange of information between the ATCO and the flight deck. Providing visual rather than aural information has the potential to reduce misunderstandings that arise from the current radiotelephony (R/T) communications by providing information that is permanently visible on a display in the form in which it was sent. The message transfer system can be used for many routine operations such as flight level requests and clearances, route heading clearances and requests and speed clearances. Datalink offers an alternative communication medium and so reduces the congestion of the current R/T system. There are many advantages associated with datalink including the potential to resolve misunderstandings arising from failures of the R/T such as instances of stuck microphones and blocking of frequencies by simultaneous transmissions. Air/ground datalink not only offers the benefit of reducing communication workload of pilots and ATCOs and allowing them to concentrate on other essential tasks, it also contributes to ensure higher safety levels in air transport (EATCHIP, 1999).

There are, however, issues relating to the introduction of datalink. Datalink has the potential to reduce the amount of information available to flight crews, often relayed via non verbal cues in verbal communication. One major issue concerning datalink is described as the party line effect. In conventional R/T communication multiple aircraft on the same frequency are able to "listen in" to a great deal of information about weather conditions and traffic density (Pritchett and Hansman, 1993) that would not be available with the datalink system. In addition to this, the absence of verbal interaction between controller and ATCO may make it difficult to establish an effective team relationship under emergency conditions. In order to assess the viability of datalink it is necessary to examine the relevant human factors issues.

Freeflight

Freeflight is an American concept, put forward by the US Federal Aviation Authority (FAA). The concept of freeflight involves the transfer of responsibility from the ATCO to the pilot for determining flight paths of an aircraft. The role of the ATCO changes from one of direct manipulation to one of passive monitoring; the ATCO acts as a system manager monitoring the position of aircraft in their sector and intervening only when needed to resolve potential conflicts.

Freeflight is the idea that pilots may take greater control over the planning and execution of their own flights. For example, if a pilot knows that there are better winds at another altitude, they can go ahead and seek that altitude without clearing it with air traffic control or flight dispatch. The potential gains of freeflight are momentous, although in order to implement the concept of freeflight great changes would need to be undertaken to revolutionise the present air management system. These changes have implications that must be investigated thoroughly in order to assess the viability of freeflight.

FD-ATC distributed cognitive system

The FACE project employed a distributed cognition approach. A distributed cognitive system is one where work is carried out by interacting systems of people and computers distributed over space and time, with constant change in the nature of the relationships between the parties and in the type of work being carried out (Wilson, 2000; Wilson et al., 2002). In studying such systems the investigators are as much concerned with social and organisational factors as with the more traditional view of cognitive factors in human-computer interaction. Using a distributed cognition approach the representation of knowledge is examined as well as the propagation of knowledge through the system. As such, the external representations of knowledge can be used to assess what the internal representations (or the cognition of the individual actors) may be (Fields et al., 1998). A distributed cognition approach provides a basis for describing the collaborative activity of the FD-ATC system. Within the FD-ATC system a distributed, collaborative, decision making network exists whereby the goals of safety and efficiency are mutual but the preferred tactics/ procedures used by each part of the team may be different (Stedmon et al., 2003).

A distributed cognition analysis examines both the work and the division of labour required to co-ordinate the activities of individual agents within a single framework (Hollnagel, 2002). For example, during a typical flight, a pilot will communicate with other members of flightcrew and different ATCOs; will receive information from flightdeck instruments and displays; and, through "eaves-dropping" R/T communications between other aircraft and ATCOs, may develop an awareness of activities occurring nearby. These sources of information contribute to both pilot and ATCO attention demands, mental workload and situation awareness (SA), and may affect subsequent communications and behaviour. Figure 8.1 represents the current FD-ATC system. Pilots have access to flightdeck technology and ATCOs have access to ATC technology, which is communicated between each party via R/T communications.

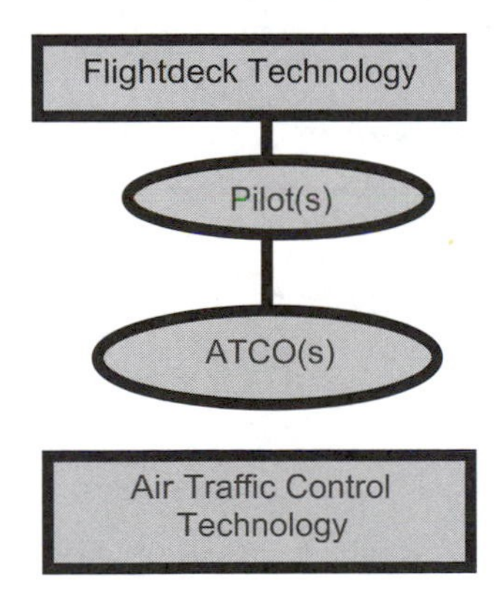

Figure 8.1 Representation of the currect FD-ATC system

Figure 8.2 represents the FD-ATC system with the introduction of datalink technology. Datalink will enable the ATCOs to have direct access to flightdeck information and, theoretically, for pilots to have access to ATC information, without the need to communicate such information directly via R/T.

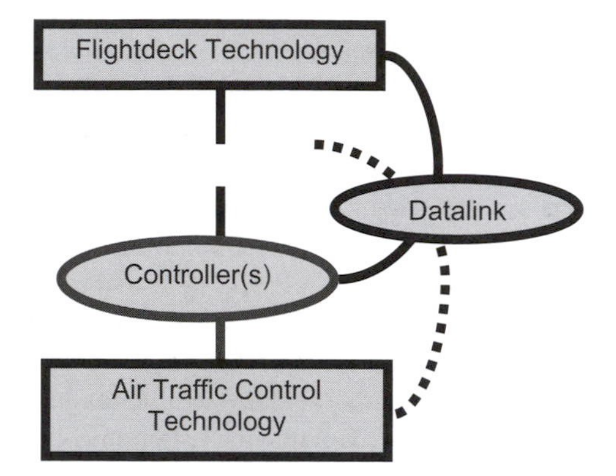

Figure 8.2 Representation of the FD-ATC system with the introduction of datalink

Figure 8.3 represents the FD-ATC system with regard to datalink and freeflight. With the introduction of freeflight technology, pilots will have direct access to the ATC information, allowing more responsibility for navigation.

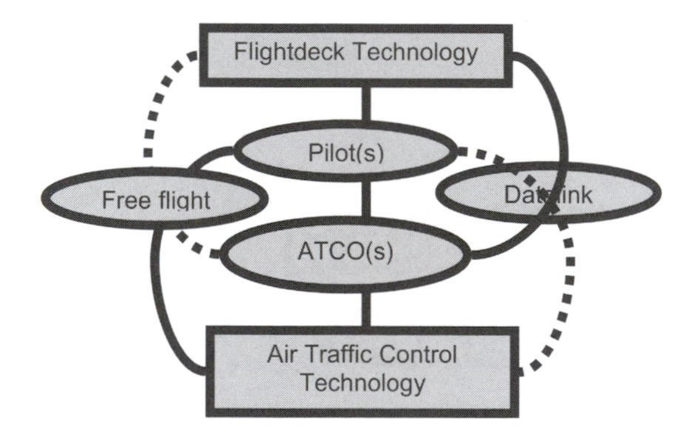

Figure 8.3 Representation of the FD-ATC system with the introduction of datalink and freeflight

Building a picture of the distributed cognitive system using field studies

Distributed cognition is used to identify information processing activity through the inputs, outputs, processes and representations in a particular task system (Perry, 1998). The framework provided by distributed cognition allows an effective and practical strategy for managing the process of data collection.

A blend of both qualitative and quantitative methods was utilised within the FACE project. Quantitative methods involve causal determination, prediction and the generalisation of findings. Qualitative methods allow researchers to understand a situation through preserving chronological flow, identifying links between events and consequences, and providing rich explanations of processes (Miles and Huberman, 1994). A number of field visits were made to the Swanwick new en-route centre (UK), the Prestwick en-route centre (UK) as well as the Airbus simulator at BAE Systems in Bristol (UK) where valuable observation work and interviews with subject matter experts (SMEs) were conducted, culminating in the development and use of an observation tool (Cox et al., forthcoming).

The application of the distributed cognition framework gave structure to the observational work and helped guide the design of the observation tool which focused on the whole system and the potential impact of the introduction of datalink and freeflight technologies. Furthermore, fundamental issues were identified which were then investigated in laboratory-based experiments.

Identification of key themes

Reviews of the literature identified certain key themes to be investigated further.

Misunderstandings in R/T communications

These are cited as the third most common cause of aviation incidents (Cox et al., forthcoming). An investigation into pilot error when copying ATC clearances found a very high number of a few errors of a particular type, for example, altitudes and speeds (Cushing et al., 1994). Pilot expectancy of a clearance will presumably have a significant impact on how accurately the clearance is received. The sensory register only holds auditory information for between three to five seconds before focused information is passed on to working memory (WM). Expectancy will therefore have a considerable effect on the information that is passed on to WM as there is very little time to compute information from the sensory register.

With the introduction of datalink, information from the flightdeck could be automatically transferred to the ATCO, reducing the dependence on R/T communication between the pilot and ATCO. Information from flightdeck technology such as the Traffic Collision Avoidance System (TCAS) would be directly available to ATCOs and this may be effective in increasing their SA.

Importance of "party line" information

An implication of datalink involves the reduction in attention requirements regarding the necessity to listen continuously to the "party line" and identify call signs. Although contrary to this it is proposed that monitoring of the "party line" has some advantages, such as giving information about the current traffic situation, weather conditions, and so on, and loss of this information might ultimately affect the pilot's situation awareness (Pritchett and Hansman, 1993).

Working memory

When reaching WM limits, mental workload increases and performance decreases. WM limits the capacity of the human processing system (Rantanen and Kokayeff, 2002). "Losing the (traffic) picture" is one of the most important concerns of ATCOs and can be attributed to the limited capacity and/or the proneness of WM to interferences.

Information, once recognised, either leads directly to a response (automatic processing) or goes to WM for further processing. Only a very limited amount of information can be brought from the sensory register to WM (Baddeley and Hitch, 1974; Anderson, 1995). The sensory register holds visual information for approximately one second, and auditory information for approximately three to five seconds. During this time attention acts on the sensory register and the subset of information upon which attention is focused is brought into WM for further processing.

Datalink has the potential to offer the "permanence" of information in a way that buffers the vulnerabilities of WM. This would allow ATCOs to devote their cognitive resources to other demanding cognitive tasks, for example, solving conflicts and so on. An investigation into the effectiveness of three different datalink interfaces, auditory, visual and redundant (auditory and visual) found that, whilst the change from auditory to a visual display has benefits in terms of load on working memory, there are other implications for how this information is processed by both the ATCO and the pilot (Best, 1995). The findings indicate that the visual display format fares best, the auditory format fares worst, and the redundant format is intermediate. The study also employed eye-tracking equipment to measure the allocation of visual attention. Greater allocation of visual resources to the instrument panel and the outside world leads to better flight path tracking and traffic detection performance (Best, 1995). The study reported that the visual interface supported the greatest allocation of visual resources to the outside world and the instrument panel, consequently "head down time" was significantly reduced with the visual datalink interface. The results of this study indicate that with the use of a visual datalink interface, allocation of attention to the interface is reduced enabling a greater proportion of "head up time". This in turn leads to better flight path tracking and traffic detection performance.

Impact of technology

As noted above, datalink by definition should reduce the load on WM and decrease workload, seemingly solving many of the human factors issues associated with verbal communication via R/T. This change from speech to text-based communication between the air and the ground may itself have implications and raise new human factors issues.

The effect of datalink communications on demands, workload and situation awareness was investigated (Helleberg and Wickens, 2001). The results indicated that pilots' ratings of the usability of the system were positive, although acceptance of datalink depended on flight phase. Datalink was rated positively for use in the cruise phase of flight, and negatively for the approach and take-off phase where the implementation of datalink led to higher pilot workload. There was evidence that datalink led to more complex activities within the cockpit, and the most disadvantageous combination of experimental factors was high workload in combination with datalink communication. The issue of transaction times was raised, although the possible increase in transaction times was offset by the reduction and simplification in air-ground communication in terms of communication acts.

Intra-crew communications

With the introduction of datalink, information from ATCOs may not be available to other members of the flightdeck. Loss of intra-crew communication may adversely affect the joint cognitive system as the shared mental models of the crew members may change or even break down. Without the existence of shared mental models, team members lose the common knowledge base necessary to form accurate and similar expectations. Such shared expectations are essential if team members are to co-ordinate their activities effectively (Muller and Geisa, 2002).

Workload

The concept of workload is most easily understood in terms of ratio of time required to perform tasks to time available (Rouse, Cannon-Bowers and Salas, 1992). In terms of ATC, the controller's workload is related to the capacity to manage traffic, the more traffic that has to be handled, the higher the workload. Under routine conditions, controller workload may be reduced with the introduction of freeflight (Wickens, Gordon and Liu, 1998), but the change in the controller's role (from active controller to passive monitor) may ultimately result in a higher controller workload. In addition to this, the loss of airspace structure and the increase in traffic complexity with the introduction of freeflight (Hillburn et al., 1997) will impose greater cognitive workload in trying to predict traffic behaviour to maintain adequate SA. This increase in controller workload resulting from the decreased structure of the airspace was observed in a number of simulation experiments (Wyndemere, 1996).

Conclusions

The overall aim of FACE was to understand better the quality of collaboration within aviation in the light of two potential information technology developments which will affect FD-ATC in the future: possible changes from voice to datalink communications, and shifts in the delegation of tasks, including freeflight. It is likely that there are positive and negative implications of each for human factors. The use of a distributed cognition approach and the subsequent design of an observation tool based on this approach provided a framework to guide field data collection with particular reference to the people involved in the system, their relationships to each other, the different technologies used, and the different environments within which the whole cognitive system functions. The FACE project investigated the FD-ATC system in terms of collaboration and how this is currently managed. This allowed predictions to be made about potential problems/issues with regard to the introduction of new technology and, more specifically, identification of where these new technologies would be useful, as well as where their introduction might potentially cause more of a disruption to current working practices.

Acknowledgements

The work presented in this paper is supported by the Engineering and Physical Sciences Research Council, Project GR/R86898/01: Flightdeck and Air Traffic Control Evaluation.

References

Anderson, J. R. (1995). *Cognitive psychology* (4th ed.). New York: W.H. Freeman.

Baddeley, A. D. and Hitch, G. J. (1974). "Working memory". In G. H. Bower (ed.), *The psychology of learning and motivation, Vol. VIII* (pp. 47-90). New York: Academic Press.

Best, J. B. (1995). *Cognitive psychology* (4th ed.). St. Paul, MN: West Publishing.

Cox, G., Sharples, S. C., Patel, H., Cole, H. and Stedmon, A. W. (forthcoming). "Methods for examining human factors issues of complex collaborative systems". *Special Issue of Applied Ergonomics*.

Cushing S. (1994). *Fatal words: Communication clashes and aircraft crashes*. Chicago: University of Chicago Press.

EATCHIP IIIb Simulation report/Ref:EEC. (1999). "Air-Ground Data Communications Projects". EUROCONTROL.

Endsley, M. R., Mogford, R., Allendoerfer, K. and Stein, E. (1997). *Effect of Free flight Conditions on Controller Performance, Workload, and Situation Awareness: A Preliminary Investigation of Changes in Locus of Control Using Existing Technologies*. Lubbock: Texas Tech University.

Fields, R. E., Wright, P., Marti, P. and Palmonari, M. (1998). "Air traffic control as a distributed cognitive system: A study of external representations". In *Proceedings of the 9th European Conference on Cognitive Ergonomics*, T. R. G. Green, L. Bannon, C. P. Warren and J. Buckley (eds). European Association of Cognitive Ergonomics (EACE), Le Chesnay, France.

Helleberg, J. and Wickens, C. D. (2001). "Effects of data link modality on pilot attention and communication effectiveness". In *Proceedings of the 11th International Symposium on Aviation Psychology* (6 pp). Columbus, Ohio.

Hilburn, B. G., Baker, M. W. P., Pakela, W. D. and Parasuraman, R. (1997). "The effect of freeflight on air traffic controller mental workload, monitoring, and system performance". In *Proceedings of the 10th International CEAS Conference on Free Flight*, Amsterdam.

Hollnagel, E. (2002). "Cognition as control: A pragmatic approach to the modelling of joint cognitive systems". Available online: http://www.ida.liu.se/~eriho/

Kirwan, B. and Rothaug, J. (2001). "Finding Ways to Fit the Automation to the Air Traffic Controller". In M. Hanson (ed.), *Contemporary Ergonomics 2001*. London: Taylor & Francis Ltd.

Miles, M. B. and Huberman, A. M. (1994). *Qualitative data analysis* (2nd ed.). Thousand Oaks: SAGE Publications, Inc.

Muller, T. and Geisa H. G. (2002). "Effects of airborne datalink communication on demands, workload and situation awareness". *Cognition, Technology & Work*, 4(4), 211-228.

Perry, M. (1998). "Process, representation and taskworld: Distributed cognition and the organization of information". In *Proceedings of ISIC'98*. London: Taylor Graham.

Pritchett, A. and Hansman, R .J. (1993). "Preliminary analysis of pilot ratings of 'party line' information importance". In *Proceedings of the 7th International Symposium on Aviation Psychology*. R. S. Jensen and D. Neumeister (eds), Volume 1 (pp. 360-366).

Rantanen, E. M. and Kokayeff, N. K. (2002). "Pilot error in copying air traffic control clearances". In *Proceedings of the 46th Annual Meeting of the Human Factors and Ergonomics Society*, Baltimore, Maryland (pp. 145-149).

Rouse, W. B., Cannon-Bowers, J. A. and Salas, E. (1992). "The role of mental models in team performance in complex systems". *IEEE Transactions on Systems, Man, and Cybernetics*, 22 (6), 1296-1308.

Stedmon, A. W., Nichols, S. C., Cox. G., Neale, H., Jackson, S., Wilson, J. R. and Milne, T. J. (2003). "Framing the flightdeck of the future: Human factors issues in free flight and datalink". In *HCI International '03. Proceedings of the 10th International Conference on Human-Computer Interaction*. Lawrence Erlbaum Associates.

Wickens, C.D., Gordon, S. E. and Liu, Y. (1998). *An introduction to human factors engineering*. New York: Addison-Wesley Longman Inc.

Wickens, C. D., Mavor, A. S., Parasuraman, R. and McGee, J. P. (eds) (1998). *The future of air traffic control: Human operators and automation*. National Research Council. Washington, D.C.: National Academy Press.

Wilson, J. R. (2000). "Fundamentals of ergonomics". *Applied Ergonomics*, 31, 557-567.

Wilson, J. R., Cordiner, L., Nichols, S., Norton, L., Bristol, N., Clarke, T. and Roberts, S. (2002). "On the right track: systematic implementation of ergonomics in railway network control". *Cognition, Technology & Work*, 3(4), 238-253.

Wyndemere, J. (1996). *An evaluation of air traffic control complexity*. Final report, contract number NAS 2-14284. Boulder, CO: Wyndemere.

Chapter 9

The Flightdeck of the Future: Perceived Urgency of Speech and Text

Alex Stedmon, Sarah Sharples, Edward Nicholson,
Gemma Cox and John Wilson

The flightdeck of the future

Air Traffic Management (ATM) represents all the characteristics of a complex organisation (Siemieniuch and Sinclair, 2001) and with aircraft levels set to double in the next 15 years, some degree of automation will be needed to enable safe increases in air traffic capacity (Kirwan and Rothaug, 2001). Most flightdeck automation has taken place within the immediate cockpit environment with a dramatic increase in the number of displays/systems available to pilots over the last 50 years (Hancock, 1996). However, the modern flightdeck system encompasses more than pilots and their immediate displays, it involves the complex integration of Air Traffic Controllers (ATCOs), ATM procedures, and ground crew, as well as auxillary agencies, airline companies and service staff. In many ways the flightdeck system represents a working team characterised by trust in the system, functionality of team members, communication within the team, and where authority should be invested in the team (Taylor and Selcon, 1990).

Within the flightdeck-air traffic control (FD-ATC) system, a distributed, collaborative, decision making network exists whereby the goals of safety and efficiency are mutual but the preferred tactics/procedures used by each part of the team may be different (Stedmon et al., 2003). From this perspective, a joint cognitive system emerges incorporating a number of operators and a number of systems (Hollnagel, 2002). For example, during a typical flight, pilots will communicate with other members of flightcrew and different ATCOs; receive information from flightdeck instruments and displays; and may develop an awareness of other activities occurring in nearby airspace through "eavesdropping" radio communications between other aircraft and ATCOs. These sources of information contribute to both the pilot and ATCO attention demands, mental workload and situation awareness (SA) and may affect subsequent communications and/or behaviour.

The impact of automation on the FD-ATC system has a number of potential impacts on new error forms, potential skill-set changes, trust in the system, changed team roles and recovery from system failure (Kirwan and Rothaug, 2001). Given the safety critical nature of ATM, significant effort is being made to harmonise the automation of ATC roles. Experience in other related domains such as cockpit automation has shown that failure to take early account of human factors can lead to "start-up" problems or in extreme cases "automation-assisted accidents" (Kirwan and Rothaug, 2001). Rather than following technical driven approaches that prescribe how safety critical tasks may be automated, another approach is to consider function re-allocation amongst existing components of a system – a socio-technical perspective that combines both macroscopic (overall air-ground system) and microscopic (local practices) entities (Rognin, Grimaud, Hoffman and Zeghal, 2001). It is crucial, therefore, to understand how these different entities co-exist, and interact, and how automation might impact upon the behaviour of the system as a whole (National Research Council, 1997).

In any domain, it is crucial that operators remain in the control loop and aware of the overall situation at all times (Weiner and Curry, 1980) and that systems remain "transparent" so that operators understand what is happening (Norman, 1990). It has been suggested that errors in future state prediction are linked to failures in controller SA, with failure in state awareness forming the largest category of errors in aviation accidents (Jones and Endsley, 1996). It is the human controller's skill in identifying and addressing potential future problems before they occur which makes their predictive ability one of the most significant aspects of the integrity of the total system (Forrest and Lamoureux, 2000). With any degree of automation, the FD-ATC system should be reasonably intuitive and not place excessive (or insufficient) demands on ATCOs and pilots. It should also be a system where the human operator has enough knowledge of the system and appropriate skill sets to optimise the ATM system performance and where current levels of error detection and recovery are maintained (Kirwan and Rothaug, 2001).

The Flightdeck and Air Traffic Control Collaboration Evaluation (FACE) project ran from November 2002 to January 2006 and was funded by the UK Engineering and Physical Sciences Research Council (EPSRC). The project was led by the Human Factors Research Group at the University of Nottingham, with partners in the Centre for Human Sciences at QinetiQ. Expertise and field access were provided by National Air Traffic Services (Bournemouth and Prestwick) and flight consultants. The FACE project took a joint cognitive systems approach to investigate the human factors requirements for the flightdeck of the future. A key component of the project was the potential move away from speech-based radiotelephony (R/T) communications towards text-based datalink communications.

Datalink

Aircraft safety during flight is highly dependent on information exchanges, via R/T, between ATCOs and pilots (Navarro and Sikorski, 1999). Datalink is designed to provide an electronic exchange of information between the ATCOs and pilots where the message transfer can be used for operations such as flight level requests and clearances, route heading clearances and requests and speed clearances. The development of datalink technology has focused on reducing the burden placed on R/T channels and enhancing the overall effectiveness of the communications, surveillance and navigation network (Harris and Lamoureux, 2000).

It is estimated that 37 per cent of current communication failures could be prevented if datalink replaced all standard verbal controller-pilot communications, and if additional systems were devised to check that pilot understanding matched a controller message, this would provide an additional 30 per cent improvement (Gibson, Megaw and Donohoe, 2001). Datalink offers significant benefits in terms of increased consistency between controller messages and pilot understanding of those messages (and vice versa). Problems associated with voice transmissions (such as background noise, channel distortions, phraseology, pronunciation problems, foreign and regional accents) would not exist with datalink and might lead to a greater understanding of information and reduced memory load (Rebello, 2001).

Whilst there are potential advantages associated with datalink, it also has the potential to reduce the amount of information available to flight crews, often relayed via non-verbal cues, and impact on SA through the loss of the "party line" effect. This is where multiple aircraft on the same R/T frequency are able to "eavesdrop" on information about weather conditions and traffic density that would not be available with a datalink system. In addition to this absence of verbal interaction between pilots and ATCOs, datalink could be operationally unacceptable due to high task densities and small task completion windows (Reynolds and Neumeier, 1991); increased workload to unacceptable levels during aircraft departure and arrival (Kerns, 1991); increased task complexity due to resending failed messages or keeping notice of responses (Harris and Lamoureux, 2000).

For these reasons datalink could make it difficult to establish an effective team relationship under emergency conditions where the immediacy of speech communication would be lost. Care needs to be taken, therefore, in the investigation and implementation of datalink. Information must be presented in the right form and an appropriate balance between direct voice and datalink communication must be established. Pilots and ATCOs may wish to revert back to R/T during take-off and approach activities when tasks are constrained and demanding, since datalink may be too slow or complex to maintain control in highly tactical airspace (Shingledecker, 1992).

New procedures with datalink will completely change the operational image of the FD-ATC system (Rebello, 2001). It is not clear which communication errors and problems might be alleviated or arise; and the risks of crew exclusion from communications, workload modification, transaction time modification, interference between simultaneous communications and flying actions, and increasing task and visual channel overload are not adequately understood (Navarro and Sikorski, 1999).

The change from voice communications to a text-based system also raises specific issues through the immediate loss of paralinguistic information, such as identity/nationality, emotion and urgency within the voice. A text-based system may therefore have to include additional information to supplement these subtle speech-based cues (such as the colour, size, and flash rate of text, and the choice of words themselves) and this could have important implications for the authority and legitimacy of communications. For example, if an aircraft was hijacked, suffered a system failure or passenger crisis, and the communications between ATC and the aircraft were purely text-based, the ATC controller could be less likely to quickly recognise that something was wrong or judge the severity of the situation by the pilot's voice.

Perceived urgency of speech and text

The FACE project developed two integrated programmes of data collection and analysis: an experimental programme run at the University of Nottingham, with support from QinetiQ, and field research at Air Traffic Control Centres. Of particular importance was the issue of perceived urgency of speech and text communications. With systems delivering information in the visual modality it is important to address the circumstances under which operators could miss critical information, or become habituated to visual stimuli (Thorley, Hellier and Edworthy, 2001) and the impact this might have on the patterns of interaction within the flightdeck system. Consideration also needs to be given to the wider implications of SA in datalink communications since the impact of any task re-distribution could be expected to contribute to safety in enabling controllers and pilots to maintain an up-to-date picture of the relevant situation, as well as a shared cognitive environment. Controllers and operators should, therefore, share a consistent representation of the on-going delegation (Rognin, Grimaud, Hoffman and Zeghal, 2001). The findings are discussed in relation to perceived urgency, workload, SA, attention and overall task performance.

Method

Participants

Forty-eight male and female students from the University of Nottingham were recruited for the study. Ages ranged from 19 to 42 years (mean age = 23years). All participants spoke English as their first language and had normal, or corrected to normal, vision.

Apparatus

Two computers were used to run the study (as shown in Figure 9.1). One computer ran a tracking task in which participants were required to track a target using a joystick, whilst the other computer ran a perceived urgency task using Superlab software. Urgency rating and performance data were recorded by computer software, whilst NASA-TLX questionnaire and SAGAT-style memory task data were collected after each trial.

Figure 9.1 Primary and secondary task computers

Design

A repeated measures, within-subjects, design was used. The independent variables were mode of command presentation (text or speech), content of command (urgent or non-urgent messages) and display of command (stressed or un-stressed text or speech). A 2x2x2 design with eight experimental conditions was generated. Eight commands (see Table 9.1) were presented twice in each trial and the same commands were used in both the speech and text modes. To minimise any order, practice or learning effects, the stimuli were randomly presented and all conditions were counterbalanced. Dependent variable measures were taken for perceived urgency ratings, workload, SA, number of commands responded to, tracking task accuracy, and response time to rate urgency.

Table 9.1 Urgent and non-urgent communications

Urgency	Communications
Urgent	"attention all aircraft in the vicinity of gatwick, emergency descent in progress, stay clear"
	"all stations, this is SRE 670, stop transmitting – mayday"
	"avoiding action, climb immediately to flight level 260"
	"dangerous winds immediately ahead, seek new route"
Non-Urgent	"thank you, for further contact, Padua 120 decimal 7.2 bye bye"
	"expect departure at 14.50, start up at own discretion"
	"no traffic delay expected due to weather"
	"taxi to holding point"

The commands were derived from typical ATCO communications which were representative of real communications but were also readily understood and differentiated by novice participants. When the commands were piloted, they were tested for the number of words and syllables in each of the urgent and non-urgent communications. No significant effects were observed ($p>0.05$), illustrating that both sets of communications had a similar number of total words and syllables and any differences observed could be associated to the content of the commands.

Procedure

Participants completed trials with commands presented as speech and with commands presented as text (counterbalanced between-subjects). Participants conducted two simultaneous tasks:

- A primary tracking task was performed where participants had to track a randomly moving target using a joystick to control a cursor on the screen
- A secondary urgency task was performed where participants had to rate the urgency of each communication they read or heard (depending on the experimental condition).

Participants were instructed to rate the urgency of the communications according to a 7-point scale (where 1=low urgency and 7=high urgency). The urgency ratings were collected via a computer keyboard. Whilst attending to the primary tracking task, participants rated the urgency of the commands. After each trial a workload questionnaire and SA test were administered. At the end of the experiment, participants were thanked and paid for their time.

Results

The data were tested for normality, equality of variance, and were found to meet the assumptions for parametric analysis.

Perceived urgency

This was rated on a 7-point scale (from 1 = non-urgent to 7 = urgent). The ratings were analysed using a 2x2x2 repeated measures ANOVA. A significant main effect for mode was observed [$F(1,71)=4.748$, $p<0.05$ (2-tailed)], illustrating that speech (mean urgency = 3.93) was perceived as less urgent than text (mean urgency = 4.10). A significant main effect for content was also observed [$F(1,71)=1096.13$, $p<0.001$ (2-tailed)], as urgent communications (mean=6.04) were rated as more urgent than non-urgent communications (mean=1.99), which in addition to the pilot study, further verified the commands used. No other significant effects were observed ($p>0.05$).

Workload

This was assessed using NASA-TLX questionnaires, rated along six 100-point scales. The scores along the different scales were combined into an overall score as well as the individual scales. Paired-samples T-tests were conducted on the overall NASA-TLX scores and individual scales. A significant effect was observed between speech and text [t(47)=3.585, p<0.01 (2-tailed)], illustrating that participants found speech (mean=45.19) less effort than text (mean=50.52). Significant effects were observed between speech and text on the individual scales of mental demand [t(47)=4.41, p<0.001 (2-tailed)] and temporal demand [t(47)=2.61, p<0.05 (2-tailed)], illustrating that participants rated mental and temporal demand higher when referring to text commands (mental=67.02; temporal=46.38) than speech commands (mental=55.90; temporal=40.23). No other significant effects were observed (p>0.05).

Situation awareness

This was assessed via a SAGAT-style memory task which participants completed at the end of each trial. Data were recorded for components of three shapes that appeared on the primary task display during each trial (shape, colour and position). These component elements also combined to provide an overall score. Paired-samples T-tests were conducted on the total recalled and individual components of shape, colour and position data. A significant effect was observed between speech and text [t(47)=-2.27, p<0.05 (2-tailed)], illustrating that participants remembered more components with speech (mean=5.02) than text (mean=4.38). When the individual components were analysed a significant effect was observed for colour [t(47)=-2.35, p<0.05 (2-tailed)], illustrating that participants remembered more colour detail with speech (mean=2.54) than text (mean=2.17). No other significant effects were observed (p>0.05).

Number of communications

These were analysed from the Superlab software where it was possible to calculate the number of communications that participants responded to. These were analysed using paired-samples T-tests. A significant effect was observed between speech and text [t(47)=2.79, p<0.01 (2-tailed)], illustrating that participants missed more speech communications (mean=7.52) than text communications (mean=7.83).

Response times (RTs)

These were also analysed using the Superlab software as it was possible to calculate the time (in seconds) that participants took to respond to each command. RTs were analysed using a 2x2x2 repeated measures ANOVA. A significant main effect for mode was observed [F(1,71)=627.57, p<0.001 (2-tailed)], illustrating that speech (mean=1.59secs) was responded to more quickly than text (mean=4.52secs). No other significant effects were observed ($p>0.05$).

Tracking task

Tracking task accuracy was calculated as a pixel index along the hypotenuse of the x-y deviation from the cursor and the moving target. The Superlab software recorded the deviation every 0.025secs. Tracking task data were analysed by conducting paired-samples T-tests on the mean deviation scores. A significant effect was observed between speech and text [$t(47)=5.14$, $p<0.001$ (2-tailed)]. Participants performed better at the tracking task with text communications (mean deviation=25.06 pixels) than with speech communications (mean deviation=28.95 pixels). A significant effect was also observed between the first and second trials [$t(47)=-3.04$, $p<0.01$ (2-tailed)], illustrating that participants were more accurate the first time (mean deviation=25.69 pixels) than the second time they did the task (mean deviation=28.32pixels). No other significant effects were observed ($p>0.05$).

Discussion

The overall results are summarised in Table 9.2.

Table 9.2 Summary of findings

Text	Speech
more urgent	less urgent
increased workload	lower workload
more attended to	less attended to
SA decreased	improved SA
better tracking task performance	decreased tracking task performance
slower response times	quicker response times

From the findings text commands were considered more urgent than the speech-based commands, possibly because they provided a constant visual cue. That the urgent commands were rated as more urgent than the non-urgent commands supports the validity of the commands used in the trials. Participants felt the text-based commands created a higher overall workload, especially in the mental and temporal dimensions. This is because it probably took more effort to read the commands rather than listen to them whilst simultaneously performing the tracking task. Although participants felt that the text-based commands created higher workload, they attended to more of them. This in itself could explain the higher workload as participants missed speech-based commands possibly because once they were spoken they were no longer available to refer to. It was possible that SA decreased when participants read text because they had to look away to attend to the commands and may have missed information on the primary task display. In addition, participants performed worse in the tracking task in the speech condition and better in the text condition. However, participants took longer to respond to the text-based commands, perhaps due to the extra processing time it took to read them. Furthermore, the second time participants conducted the task, tracking task accuracy suffered. This may have been due to cognitive or physical fatigue.

From the findings it is clear that both speech and text commands in expected or unexpected situations have their relative merits. It is likely that in routine, low workload communications, such as a request for change in height as stated on the flight-plan, the use of datalink could avoid errors that may occur due to mishearing, low radio quality, or perceptual confusion between similar flight numbers. However, for non-routine situations, such as a pilot running low on fuel, the potential impact of datalink could be more critical. A text-based mode of communication could mask urgency that would be evident in a spoken communication, take longer to process and impact on primary task performance. Furthermore, the impact of the loss of radio communications and paralinguistic information on pilot "eavesdropping" other pilot-ATCO communications, and using the information to maintain SA, cannot be ignored.

Fundamentally, information must be presented in the right form; an appropriate balance between direct voice and datalink communication must be established; and the content of the data transmitted to and from the aircraft needs to be considered in an integrated and holistic manner rather than examining each requirement separately (Stedmon et al., 2003).

However, the handling of datalink needs careful study, information must be presented in the right form, and an appropriate balance between direct voice and datalink communication must be established. Options for datalink protocols and communications media require further research to understand which provide solutions capable of meeting long-term bandwidth, signal integrity and cost requirements. The total content of the data transmitted to and from the aircraft needs to be considered in an integrated way rather than examining each requirement separately, and research is needed to investigate possibilities for information interchange, including FD-ATC dialogues and the balance between datalink and voice communications.

Conclusions

The direct impact of the FACE project has been to inform further the ATM community about potential human factors implications of the introduction of new flightdeck technology and present the benefits of taking a distributed cognition perspective in understanding the FD-ATC "system". A flexible flightdeck system is a strategic target, with increased freedom for management shared between service providers on the ground and in the air. With so much automation predicted for future ATM, concern has been expressed about the changes in user roles of the future. Human factors should have its own vision of future ATM roles and such a vision must be based on system goals of operability and safety (Kirwan and Rothaug, 2001). With the introduction of datalink, pilots and ATCOs would manage and optimise the smooth flow of traffic via the use and supervision of automation tools. As such, a better understanding is needed of the relationship between the pilot, ATCO and the content of the data transmitted within the flightdeck "system" in an integrated and holistic manner (Stedmon et al., 2003). It is crucial, therefore, when considering these aspects to establish the extent to which datalink, along with different levels of task delegation, supports or detracts from safe operations in the flightdeck system of the future.

Acknowledgement

The work presented in this paper is supported by the Engineering and Physical Sciences Research Council, Project GR/R86898/01: Flightdeck and Air Traffic Control Evaluation.

References

Forrest, D. and Lamoureux, T. (2000). "Future system state prediction by novice and expert air traffic controllers". In M. Hanson (ed.), Contemporary Ergonomics 2000. London: Taylor & Francis Ltd.

Gibson, H., Megaw, T. and Donohoe, L. (2001). "Failures in pilot-controller communications and their implications for datalink". In D. Harris (ed.), Engineering Psychology and Cognitive Ergonomics – Vol. 5. London: Ashgate.

Hancock, P. A. (1996). "On convergent technological evolution". Ergonomics in Design, 4(1), 22-29.

Harris, S. and Lamoureux, T. (2000). "The future implementation of datalink technology: The controller-pilot perspective". In M. Hanson (ed.), Contemporary Ergonomics 2000. London: Taylor & Francis Ltd.

Hollnagel, E. (2002). "Cognition as control: A pragmatic approach to the modelling of joint cognitive systems". Available online: http://www.ida.liu.se/~eriho/

Jones, D. G. and Endsley, M. R. (1996). "Sources of situational awareness errors in aviation". Aviation, Space, and Environmental Medicine, 67(6), 507-512.

Kerns, K. (1991). "Datalink Communication between controllers and pilots: A review and synthesis of the simulation literature". International Journal of Aviation Psychology, 1(3), 181-204.

Kirwan, B. and Rothaug, J. (2001). "Finding ways to fit the automation to the Air Traffic Controller". In M. Hanson (ed.), Contemporary Ergonomics 2001. London: Taylor & Francis Ltd.

National Research Council (1997). Flight to the future: Human Factors in Air Traffic Control. Washington DC: National Academy Press.

Navarro, C. and Sikorski, S. (1999). "Datalink communication in flightdeck operations: A synthesis of recent studies". International Journal of Aviation Psychology, 9, 361-376.

Norman, D. A. (1990). "The 'problem' with automation: Inappropriate feedback and interaction, not 'over-automation'". Philosophical Transactions of the Royal Society of London. Series B, Biological Sciences, 585-593.

Rebello, L. H. B. (2001). "The control-system interface in Air Traffic Control: An ergonomic approach". In D. Harris (ed.), Engineering Psychology and Cognitive Ergonomics – Vol. 5. London: Ashgate.

Reynolds, M. C. and Neumeier, M. E. (1991). "Mode-S datalink pilot-system interface: A blessing in de skies or beast of a burden?" Proceedings of 6th International Symposium on Aviation Psychology, 1, 154-159.

Rognin, L., Grimaud, I., Hoffman, E. and Zeghal, K. (2001). "Implementing changes in controller-pilot tasks distribution: The introduction of limited delegation of separation assurance". In Proceedings of HESSD-01 (4th International Workshop on Human Error, Safety and Systems Development), Linkoping, Sweden.

Shingledecker, C. A. (1992). "Controller evaluations of ATC data link services". Society of Automotive Engineers Technical Paper Series. Report number 922027.

Siemieniuch, C. E. and Sinclair, M. A. (2001). "The process owner – A role to overcome problems of manufacturing complexity and organisational learning". In M. Hanson (ed.), Contemporary Ergonomics 2001. London: Taylor & Francis Ltd.

Stedmon, A. W., Nichols, S. C., Cox. G., Neale, H., Jackson, S., Wilson, J. R. and Milne, T. J. (2003). "Framing the flightdeck of the future: Human factors issues in free flight and datalink". In Proceedings of the 10th International Conference on Human-Computer Interaction. Lawrence Erlbaum Associates.

Taylor, R. M. and Selcon, S. J. (1990). "Psychological principles of human-electronic crew teamwork". In T. J. Emerson, M. Reinecke, J. M. Reising and R. M. Taylor (eds), The Human Electronic Crew: Is the Team Maturing? Proceedings of the 2nd Joint GAF/USAF/ RAF Workshop. RAF Institute of Aviation Medicine, PD-DR-P5.

Thorley, P., Hellier, E. and Edworthy, J. (2001). "Habituation effects in visual warnings". In M. Hanson (ed.), Contemporary Ergonomics 2001. London: Taylor & Francis Ltd.

Weiner, E. L. and Curry, R. E. (1980). "Flight-deck automation: Promises and problems". Ergonomics, 23, 955-1011.

Chapter 10

Operator Interface Research Testbed for Supervisory Control of Multiple Unmanned Aerial Vehicles (UAVs)

Gloria Calhoun, Mark Draper and Heath Ruff

Introduction

The US Air Force Research Laboratory (AFRL) Human Effectiveness Directorate supports research addressing human factors challenges associated with Unmanned Aerial Vehicle (UAV) control. Earlier research has focused on teleoperated UAV control (Draper, Calhoun, Ruff, Fontejon and Guilfoos, 2003). However, advances in technology will enable UAVs to operate more autonomously, requiring little direct operator control. As a result, a single operator will likely be expected to monitor and control multiple semi-autonomous UAVs. Innovative methods will be required to keep the operator "in the loop" for optimal situation awareness, workload, and decision making (Bonner, Taylor, Fletcher and Miller, 2000; Parasuraman, Sheridan and Wickens, 2000). One method that may enhance supervisory control is multiple levels-of-automation (LOA), whereby each level specifies the degree to which a task is automated. Thus, automation can vary across a continuum of levels, from the lowest level of fully manual performance to the highest level of full automation. Use of higher LOA might allow for more vehicles to be controlled by a single supervisor. However, these high LOA tend to remove the operator from the task at hand and can lead to poorer performance during contingencies and automation failures. In contrast, an intermediate LOA that involves both the operator and the automation system in operations may preclude multi-UAV control due to increased operator task requirements. However, it has been hypothesised that an intermediate LOA can improve performance and situation awareness, even as system complexity increases and automation fails. Some research supports this hypothesis (for example, Ruff, Narayanan and Draper, 2002) and other results (for example, Endsley and Kaber, 1999) suggest that there are factors that can have an impact on the benefit of an LOA (for example, whether the task involves option selection versus higher-level cognition). Such results demonstrate the need for more research comparing LOA in different task environments. To evaluate candidate LOA schemes for single operator supervision of multiple UAVs, a relevant synthetic task environment simulation testbed is needed.

The Multi-modal Immersive and Intelligent Interface for Remote Operation (MIIIRO) was developed to provide a generic test environment in which a single operator supervises multiple UAVs (Tso, Tharp, Tai, Draper, Calhoun and Ruff, 2003). The present chapter will briefly describe this prototype multi-UAV synthetic task environment, as well as the results from two evaluations using MIIIRO as the apparatus. Finally, some recommendations are made for testbed design refinements to better support the evaluation of supervisory control of multiple UAVs.

MIIIRO multi-UAV synthetic task environment

The basic MIIIRO testbed configuration (Figure 10.1) consists of two monitors, a keyboard, and mouse (Tso et al., 2003). One monitor (Figure 10.2) presents the Tactical Situation Display showing the colour coded UAV routes, suggested route re-plans, waypoints, targets, threat rings, and any unidentified aircraft. As each simulated UAV passes a target, its camera takes images and these appear in the queue at the bottom of the Image Management Display (Figure 10.3). The image in the top row of the queue is displayed in the top window. Suspected hostile targets within the image are highlighted with red square outlines generated from a simulated automatic target recogniser cueing (ATC) system.

Figure 10.1 Multi-UAV supervisory control MIIIRO test environment

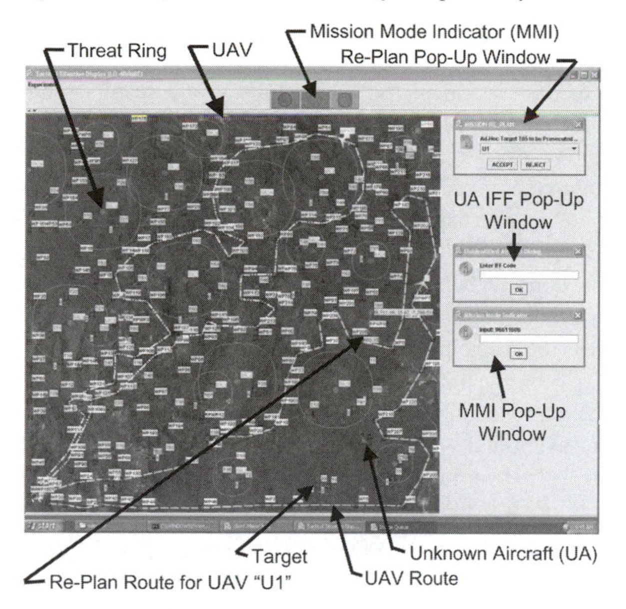

Figure 10.2 Example MIIIRO tactical situation display format

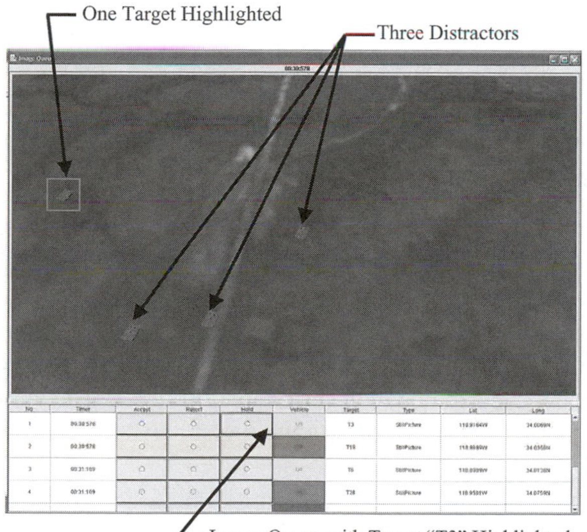

Figure 10.3 Example MIIIRO image management display format

MIIIRO evaluations: experimental design

A more complete description of the methodology and results for the two AFRL evaluations conducted with MIIIRO are available (Ruff, Calhoun, Draper, Fontejon and Guilfoos, 2004). Both evaluations examined LOA as a within-subject variable (blocked and counterbalanced). In one of the two levels, management-by-consent, the participant had to agree explicitly to suggested actions before they occurred. The automation proposed route re-plans and target identifications within each image, but required operator consent before acting. In the other LOA level, management-by-exception, the system automatically implemented suggested actions after a preset time period, unless the participant objected. This preset time period differed between the two studies. In Study 1, participants had 40 seconds and 15 seconds, respectively, to respond to image prosecutions and suggested route re-plans. In Study 2, two time limit levels were examined within each LOA: "long" (what was used in Study 1; 40/15) and "short" (15/10 for images and re-plans, respectively).

Automation reliability for the image prosecutions and route re-plans was also manipulated in both studies as a between-subjects variable (low: 75 per cent and high: 100 per cent). The number of UAVs (two versus four) was a within-subjects variable in Study 1; four UAVs were examined in all trials for Study 2. There were 16 participants in each study.

Participants were required to respond to several types of events in each of the16-minute missions (one with each combination of independent variables). The events, listed in order of priority, were:

- Unidentified Aircraft (two per mission). Response: click on any red airplane icon that randomly and temporarily appeared on the Tactical Situation Display and enter a code in a pop-up window. This task emulated having a highly unexpected, non-routine, high-priority event occur during a mission.
- Route Re-plans (16 per mission). Response: inspect alternate routes suggested by the automation in response to *ad hoc* targets and threats on the Tactical Situation Display. These were displayed as a star (target) or a new orange ring (threat), along with an automatically generated route-re-plan proposal. Make an accept/reject decision in a pop-up window, based on (Study 1) whether the re-plan crossed another threat or another UAV's route or (Study 2) the route's relative safety, in addition to the mathematical relationship of three displayed values (added to increase the cognitive complexity of the decision making process).
- Image Prosecution [per mission: 34 (two UAVs), 66 (four UAVs)]. Response: view the top window of the Image Management Display and verify that red boxes were only around targets (versus distracters). If there were errors, red boxes could be added or deleted with mouse clicks. Then, make accept/reject decision.
- Mission Mode Indicator. Response: whenever a green light changed to yellow or red on a status panel, click on the panel and make an entry in a pop-up

window. This was an abstract secondary monitoring task representing the various contingency management panels that will likely exist in future control stations. The number of events increased as the number of UAVs under supervisory control increased [per mission: 16 (two UAVs), 32 (four UAVs)], reflecting the likely increase in status panel monitoring required.

MIIIRO evaluations: results

Data recorded included time and accuracy in responses to the four tasks described above: 1) unidentified aircraft; 2) proposed re-plans; 3) image prosecutions; and 4) changes in mission mode indicator. Subjective ratings were also obtained with regard to workload, situation awareness, and trust in decisions made.

The number of UAVs supervised (two versus four) was only examined in Study 1. In general, performance was better and the subjective ratings were more favourable when only two UAVs were being supervised. Average task completions times were faster with two UAVs than four for the image prosecutions, route re-plans, and monitoring tasks and less time was spent in threat zones. The subjective ratings indicated that participants viewed the four UAV condition as higher workload, more difficult, and less trustworthy.

Two automation reliability levels were examined in the image prosecution and route re-plan task data. In Study 1, fewer images were prosecuted and more errors were made in the low reliability level compared to the high level. The subjective data also indicated that the participants had less trust when reliability was low. Only one measure showed a significant effect of reliability in Study 2. The percentage of images correctly prosecuted was less in the low reliability level compared to the high.

Results with regard to the LOA were more interesting. In Study 1, performance between the two LOA varied little and did not show a consistent trend across measures. The design dictated that trials with the management-by-consent automation never timed out. With management-by-exception, participants typically responded (manually) rather than let the action automatically occur. In fact, image prosecution time averaged 12 seconds for both LOAs, much shorter than the criterion time limit employed (40 seconds). Thus, the results pertaining to LOA were questionable, as the automation was not utilised as designed. Rather the results suggested that the time criterions employed in the LOAs should be shortened significantly, to determine whether automation is a benefit in this simulated task environment. Study 2 was conducted to evaluate this change, evaluating both the time limits used in Study 1 (40 and 15 seconds for image prosecution and route re-plans, respectively) and shorter time limits (15 and 10 seconds, respectively).

For Study 2, there were no significant differences in the performance and subjective measures in regards to LOA, except as a function of the Time Limit variable. With these LOA, participants' ratings indicated that the shorter limit was higher workload (Figure 10.4) and more difficult.

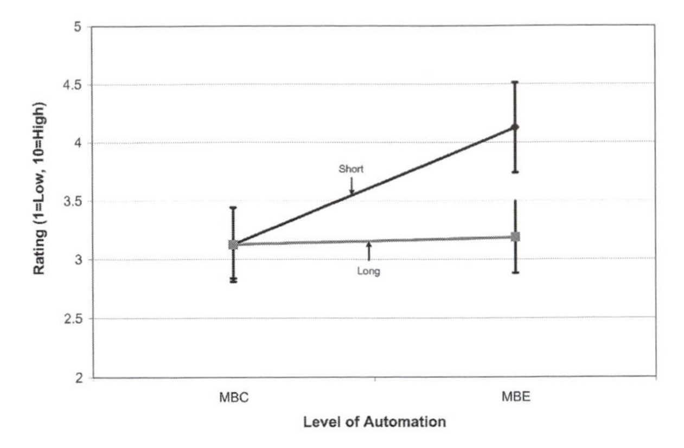

Figure 10.4 Average Modified Cooper-Harper Rating for workload for each LOA (management-by-consent and management-by-exception) and time limit (short and long), with standard error of the mean

The participants' ratings may reflect the fact that their average time to complete image prosecutions was faster with the shorter time limit in management-by-exception than that for the other three combinations of LOA and time limit (Figure 10.5). These findings may be related to the participants' ratings of less confidence with the shorter time limits and the nature of the LOA. In management-by exception, if the participant did not respond to images before the time limit, they were automatically prosecuted. The fact that an erroneous action could occur, and more likely with the shorter time limit, may have pressured participants to respond faster and view it as high workload. Thus, although management-by-exception was hypothesised to be a workload reducer, it actually appeared to add to perceived workload.

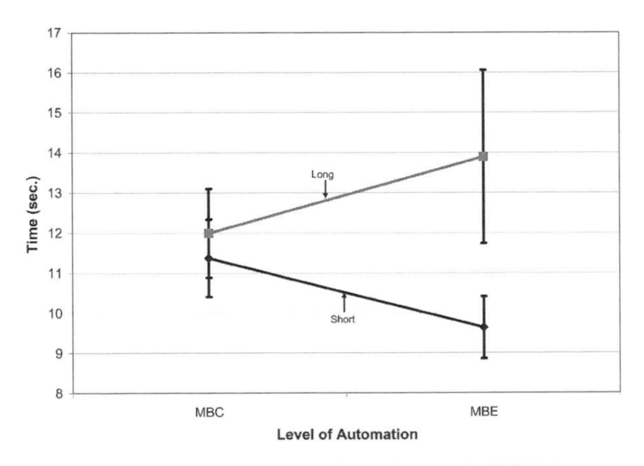

Figure 10.5 Average image prosecution time for each LOA (management-by-consent and management-by-exception) and time limit (short and long), with standard error of the mean

The frequency in which the automation was utilised, as opposed to the participant making a manual response before the time limit, was also a function of the length of the time limit. As was mentioned earlier, the automation was rarely used in Study 1; participants responded manually before the time limit. The longer time limit in Study 2 was the same limit used in Study 1 and the results were similar between the two studies (for example, the automation was utilised in only 1 per cent of the image prosecution trials). However, for the shorter time limit in Study 2, the automation was exercised in 12 per cent of the image prosecution trials. Even though the percentage of trials was higher, the automation was still rarely utilised and the low percentage suggests that the participants were not intentionally letting the automation exercise. Rather, they just were not quick enough to respond manually, in some instance, with the shorter limit. In fact, in Study 2, most re-plans (Figure 10.6) and image prosecution tasks (Figure 10.7) were completed manually, in less time (averages were 7.2 and 11.7 seconds, respectively) than the available shorter time limits (10 and 15 seconds).

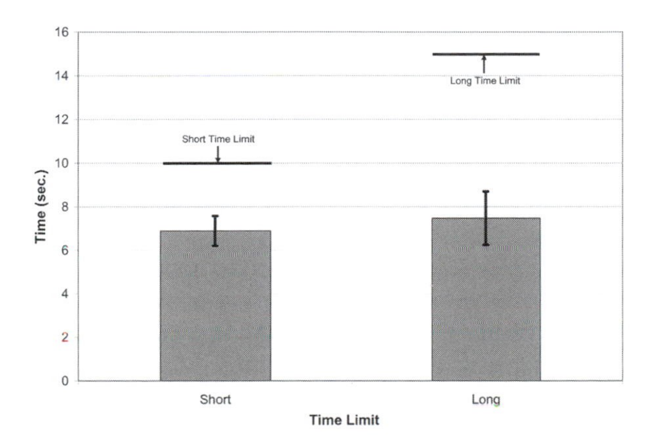

Figure 10.6 Average re-plan task time with each time limit (short and long), with standard error of the mean

MIIIRO evaluations: conclusions

The results from these evaluations with the MIIIRO multi-UAV task environment showed that for the experimental paradigm employed, participants preferred to respond manually, rather than rely on the automation. In effect, the participants rushed to manually respond fast enough to avoid having the system automatically respond. This conclusion is supported by the rarity of automated actions, together with the increased workload and decreased re-plan and image prosecution times, and lower confidence ratings with the shorter time limit.

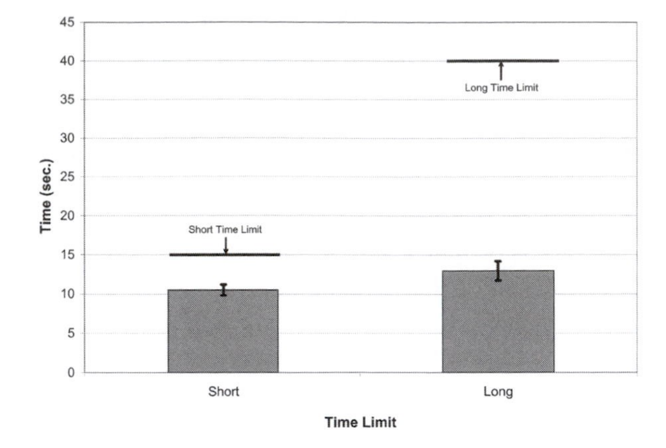

Figure 10.7 Average image prosecution time with each time limit (short and long), with standard error of the mean

What is a more interesting question is why did the participants not rely on the automation? Was it the manner in which the LOA were implemented in the experiments? Were the tasks that were automated not complex enough to warrant human reliance on automation? Is there a better design for an intermediate LOA? Do the results reflect an innate preference for humans to be engaged in important tasks? Or was the workload too low or the task environment too boring, resulting in a preference for manual control? Would increases in cognitive demands make it difficult for operators to respond fast enough and beat having the action automated? Should experimental trials be longer such that vigilance effects are more likely to occur? At the very least, the results from the MIIIRO experiments illustrate the complex relationship between LOA, time limits, and perception of difficulty, workload and confidence.

Requirements recommendations for multi-UAV research testbed

MIIIRO has been shown to be a valuable testbed for uncovering important issues in multi-UAV supervisory control. Additionally, it has proven to be an effective testbed for exploring information presentation concepts (Nelson, Lefebvre and Andre, 2004). However, it is evident from the evaluations described above that additional testbed capability is needed to explore fully the role of LOA in multi-UAV supervisory control.

One issue raised by these evaluations is the importance of representing realistic operator workload. By increasing the number of events per mission, a task environment can be created such that the operator is *so* busy that there is more reliance on the automation. However, if such a task environment is unrealistic, it would not provide a useful assessment of LOA for multi-UAV supervisory control. Actually, the key difficulty is determining what *is* a realistic task environment, given that there are no operational systems to examine. Although some insights can be gained by examining research focusing on missile (for example, Cummings and Guerlain, 2004) and air traffic control (for example, Lamoureux, 1999), these domains may not make comparable demands on operator resources that are envisioned for complex multi-UAV missions. More in-depth analysis of current UAV operations, as well as interviews with subject-matter-experts (SME), may help formulate specific multi-UAV system control station requirements. Further development of a multi-UAV research testbed should incorporate this knowledge as well as feature flexible software architecture such that changes can easily be made corresponding to future technological advances and refinements in UAV concept of operations.

Several testbed design requirements can be culled from our work with the MIIIRO prototype system. First, recommendations can be made with regards to the synthetic task environment to be created. For instance, the testbed needs to support the three major visually distributed subtasks for UAV operations (Dixon, Wickens and Chang, 2005): 1) mission completion/UAV navigation; 2) monitoring of on-board system health; and 3) surveillance through a camera image and image manipulation/identification. Even though many sub-functions will be automated (for example, navigation and health monitoring) for single operator control of multiple UAVs, consideration is needed on information/task requirements for operator supervision of all automated systems. As automated system operation can be unreliable, it is essential to have experimenter control of the reliability of each automated system in the testbed. Also, it is important to identify the display requirements to provide the operator with adequate information to compensate for the inherent unreliability of automated system operation. Since the automated systems can also fail, the testbed needs to include this possibility in the simulation, so that operator intervention during failures can be examined. Information requirements to help the operator rapidly acquire situation awareness relevant to the failure also need to be defined.

Reflection of the envisioned UAV concept of operations in the testbed will help ensure that the displays and tasks represent the complexity of future task environments. For instance, what are the anticipated information requirements in a net-centric warfare environment? Since it is likely that the UAV supervisor will be a member of a collaborative teaming environment, how will the anticipated communication/task sharing between other personnel/systems in heterogeneous operations be represented? The testbed should also support lengthy missions to reflect anticipated operations and stimulate anticipated vigilance effects. The key goal, in this regard, is to have the task environment provided by the testbed approximate the anticipated cognitive/attentional demands and level of decision making that an operator supervising multiple UAVs will have to face.

Particular attention needs to be devoted to designing the testbed architecture to support more finely grained LOA, rather than a unitary LOA that applies to the entire task environment (Parasuraman, et al., 2000). The testbed architecture should support a variety of LOA schemes for experimentation, as it is still unknown how best to balance tasks between operator and automation. One possibility is having the LOA not only task/contingency specific, but also adaptive during the mission, based on experimenter defined criteria. Another is to have a global LOA set that, in turn, defines LOA for each individual task (with the possibility that LOA will differ across tasks). In determining the LOA scheme, the operator's trust in the automation and the cost/benefit of the automated function in the implementation of the LOA needs to be considered (Proud and Hart, 2005).

Besides supporting a variety of LOA schemes, the testbed design needs to be flexible in interaction protocols between the operator and the automated systems such that a variety of methods to delegate commands and stipulate intent can be explored (Goldman, Miller, Wu, Funk and Meisner, 2005). In fact, LOA management is a design issue in itself, as it should be transparent to the operator and not impose additional workload. Some example research questions include: how the operator should be informed on what LOA are in effect such that it is intuitive at all times how the task loading is divided between the operator and the automation; for an adaptable automation system, how should the operator best change/manage LOA within individual UAVs and across multiple UAVs; how many different LOA assignments are manageable within/across UAVs; how best to ensure the operator maintains current LOA awareness, and so on.

Finally, to support meaningful evaluations, the testbed needs to meet the needs of the researcher, providing control over the setting of manipulated variables as well as the number, order and timing of events that occur in each test mission. Moreover, there should be a method to generate multiple missions that are relatively equal in complexity/difficulty for use in conducting a series of trials. In order to judge the adequacy of candidate LOA schemes and the effect of other manipulated variables, the testbed should support real time recording of multiple measures of performance and situation awareness for each task event occurring in the mission.

As illustrated by the number of issues raised above, a great deal of research is needed to determine the optimal coupling of humans and autonomy in single operator control of multiple UAVs. Designing a testbed that supports all of these aspects is probably not feasible. However, the more these issues can be examined in operationally relevant multi-UAV simulation testbeds, the more likely it is that the evaluation results will be generalisable to future UAV supervisory control applications.

References

Bonner, M., Taylor, R., Fletcher, K. and Miller, C. (2000). "Adaptive automation and decision aiding in the military fast jet domain". In *Proceedings of the Human Performance Situational Awareness and Automation Conference*, 154-159.

Cummings, M. L. and Guerlain, S. (2004). "Human performance issues in supervisory control of autonomous airborne vehicles". In *Proceedings of AUVSI's Unmanned Systems North American 2004 Symposium*, 1-15.

Dixon, S. R., Wickens, C. D. and Chang, D. (2005). "Mission control of multiple unmanned aerial vehicles: a workload analysis". *Human Factors*, 47(3), 479-487.

Draper, M. H., Calhoun, G. L., Ruff, H. A., Fontejon, J. V. and Guilfoos, B. (2003). "Multi-sensory interface concepts and advanced visualization techniques for UAV systems". In *Proceedings of the Association for Unmanned Vehicle Systems International (AUVSI) Meeting*, 1-15.

Endsley, M. R. and Kaber, D. B. (1999). "Level of automation effects on performance, situation awareness and workload in a dynamic control task". *Ergonomics*, 42(3), 462-492.

Goldman, R., Miller, C., Wu, P., Funk, H. and Meisner, J. (2005). "Optimizing to satisfice: Using optimization to guide users". In *Proceedings of the American Helicopter Society's International Specialist Meeting on Unmanned Aerial Vehicles*, January 18-20, Chandler, AZ.

Lamoureux, T. (1999). "The influence of aircraft proximity data on the subjective mental workload of controllers in the air traffic control task". *Ergonomics*, 42, 1482-1491.

Nelson, J. T., Lefebvre, A. T. and Andre, T. S. (2004). "Managing multiple uninhabited aerial vehicles: Changes in number of vehicles and type of target symbology". In *Proceedings of the Interservice/Industry Training, Simulation, and Education Conference*, USA, 1213-1218.

Parasuraman, R., Sheridan, T. B. and Wickens, C. D. (2000). "A model for types and levels of human interaction with automation". *IEEE Transactions on Systems, Man, and Cybernetics – Part A: Systems and Humans*, 30(3), 286-297.

Proud, R. W. and Hart, J. J. (2005). *FLOAAT, a tool for determining levels of autonomy and automation, applied to human-rated space systems*. Arlington, VA: AIAA.

Ruff, H. A., Calhoun, G. L., Draper, M. H., Fontejon, J. V. and Guilfoos, G. J. (2004). "Exploring automation issues in supervisory control of multiple UAVs". In *Proceedings of the Human Performance, Situation Awareness, and Automation Technology Conference*, 218-222.

Ruff, H. A., Narayanan, S. and Draper, M. H. (2002). "Human interaction with levels of automation and decision-aid fidelity in the supervisory control of multiple simulated unmanned air vehicles". *Presence: Teleoperators and Virtual Environments*, 11(4), 335-351.

Tso, K. S., Tharp, G. K., Tai, A. T., Draper, M. H., Calhoun, G. L. and Ruff, H. A. (2003). "A human factors testbed for command and control of unmanned air vehicles". *Digital Avionics Systems 22th Conference*, 8.C.1. 1-12.

Chapter 11

Virtual Environments for Military Decision Making

John Kostaras and Georgios Detsis

Introduction

Modern armies are currently in the process of changing their paper-based work to computer-assisted systems. This is a result of the increasing need to deal with large amounts of information, which must be processed with speed, accuracy and mental load in mind. Military intelligence is an area where information load to intelligence officers has increased radically in recent years due to advancements in the technology of sensors and data gathering techniques.

The usual interaction between the computer and the human is based on the two dimensional (2D) screen or presentation metaphor. Modern battle management systems and C^{4}IS (Command, Control, Computers, Communications and Information Support) present information in 2D (either graphical or text) form. This form of human-computer interaction (HCI) has severe limitations in the sense that it does not take advantage of the processing capacity of the human visual system. Virtual reality (VR) and/or virtual environment (VE) techniques allow for humans to process vast amounts of information in a rapid way and are considered in this chapter for military intelligence decision making.

A VE is a computer-generated space, where the decision objects can be observed and/or manipulated. The user can either, for example, by use of a head-mounted display be given an impression of being in the environment (immersive VR), or alternatively look at the VE from outside, so that the computer display acts as a window (non-immersive VR or desktop VR).

This chapter reports on the results of the EUCLID RTP 6.14 MARVEL (Military Applications for VE Techniques in Logistics and Intelligence) research project with the objective of demonstrating the feasibility and use of virtual environments in military decision making. The main purpose is to show how the decision maker's situation awareness (Endsley, 1995), and hence their performance, can be improved by these VE techniques. The research has focused on the intelligence army staff at battalion level and below.

VE design approach

A user-centred approach has been adopted in the development of the VE. This involved task analysis, design reviews with military experts and evaluation with intelligence officers.

From the Hierarchical Task Analysis (HTA) and task decomposition (Kirwan and Ainsworth, 1992; Perrie et al., 2000), a number of pre-battle and during-battle intelligence tasks have been identified and a number of them have been selected which are likely to benefit from the application of virtual environments. These tasks have been grouped, according to their relevance, to four categories: visualisation, interaction, group-working and decision support. These tasks have also been grouped as pre- or during-battle tasks.

For the VE design, prototypes of the virtual environment were built and evaluated with intelligence officers in workshops. Hence, an incremental approach has been followed for the resulted demonstrator software that has been developed after taking into consideration the results of the evaluations from the intelligence officers.

A typical military scenario has been used to evaluate and compare a 2D battle management system with the VE demonstrator. The intelligence officers were asked to perform certain intelligence tasks on the 2D and the 3D (three dimensional) system. After completion of each task, the officers completed a questionnaire in which they rated the difference between the virtual environment demonstrator and their normal 2D manner of working for speed, accuracy and mental workload.

VE architecture

The demonstrator's architecture is shown in Figure 11.1. The user(s) view a 3D world in VE display device(s) and can interact with it via VE input device(s). A viewer that contains a 3D scene built up out of prepared 2D and 3D models and a terrain/environment model drives the display and input devices. The 3D scene is a dynamic data structure that contains all the information that the VE application is going to show to the user. The 3D models describe the classes of visible objects of the 3D scene. The terrain/environment model describes the landscape of the operations environment in 3D view. A Computer Generated Forces (CGF) tool is used in order to simulate and command the forces according to the scenario in order to have a more dynamic and realistic demonstration.

MARVEL demonstrator

The VE has been designed on the basis of the results of the task analysis (Kirwan and Ainsworth, 1992; Perrie et al., 2000), results found in "human factors" experiments on the effect of VEs (Shallman et al., 2001; Wickens, Thomas and Young, 2000; Woods, 1984) and results from the evaluation workshops performed with intelligence officers.

The produced demonstrator contains a number of functionalities for each of the groups of tasks mentioned above, that is, for visualisation, interaction, group-working and decision support.

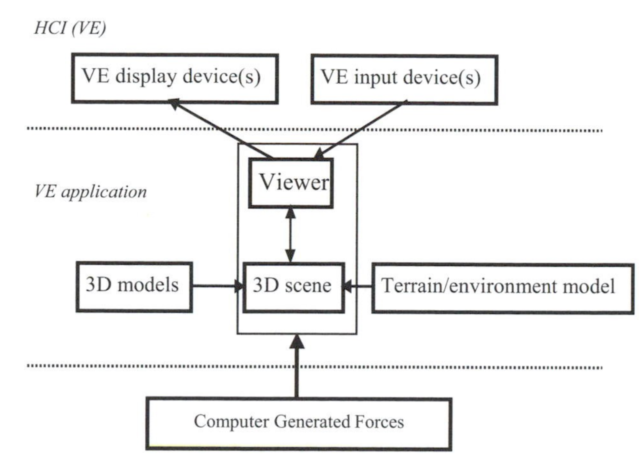

Figure 11.1 MARVEL demonstrator architecture

The demonstrator corresponds to a command and control system at battalion level, to be used in a mechanised infantry or armoured tank battalion, widely known as a Battle Management System. The examined functionality is limited to the roles of specific command levels within the battalion.

Figure 11.2 MARVEL VE demonstrator

The application interface adheres to a pretty standard layout for 3D modelling/editing software (see Figure 11.2). A manipulator's toolbar is placed on the topmost part of the viewport, while an element management interface is implemented via an operations tool panel on the right side. The largest part of the viewport is covered by the main 3D view and is complemented by an information status bar in the bottom.

Visualisation techniques

The first task a military intelligence officer has to deal with is mind setting. Mind setting involves S2 (intelligence officers) focusing their thoughts on general and specific entities that can be expected to be encountered during the forthcoming battle. These entities can be: type of battle to be fought, environmental obstacles, enemy units, non-combatants, and own forces.

To deal with this task, a number of visualisation techniques have been implemented. The user should be able to manipulate the terrain in any possible way in order to gain an as better knowledge of it as possible. Both immersive and exocentric views of the VE are available to the user. Additionally, a 2D minimap or overview map to facilitate navigation in both views is always visible. More than one view is simultaneously visible because exocentric views do not offer enough detail while egocentric views do not provide enough overview (Woods, 1984).

Previous research has shown that an immersed view can lead to cognitive tunneling or "out-of-sight-out-of-mind" bias (Wickens, Thomas and Young, 2000; Thomas, Wickens and Merlo, 1999; Thomas and Wickens, 2001). Cognitive tunnelling means that humans tend to decide on the basis of visible information only. This is the retrievability/accessibility bias in which easily retrievable information is given more weight while information that is "out-of-sight" and hence less easily-retrievable is given less weight, that is, it is "out-of-mind". For the counting of enemy units an exocentric view produced better performance (60 per cent correct for exocentric view and 40 per cent correct for the immersed display). This was accompanied by a high degree of confidence for both the immersive and exocentric views, indicating a cognitive tunnelling effect.

In a follow-up experiment (Thomas and Wickens, 2000) to investigate the potential cause of this "display-induced" cognitive tunnelling, participants in both immersed views seemed to fail in obtaining information from the 2D inset map, suggesting that integration information across the two views was not accurate.

To cope with cognitive tunnelling, an exocentric VE is always full screen coupled with a 2D minimap for fast navigation. The user, however, can switch between immersed and exocentric views at will. To avoid the user being disoriented when switching abruptly between the two views, a smooth continuous transition has been implemented.

Care should be taken so that the user does not "get lost" with the multiple views. A possible difficulty exists in relating the information of the immersed view to corresponding information on the navigation map. By using a wedge on the 2D minimap, the visual momentum (Woods, 1984) between the two views is improved, because the user can see how the displays relate to each other. Another helpful "widget" is the integration of a compass for north indication as a landmark on the terrain (see Figure 11.3). The compass always remains centred on screen.

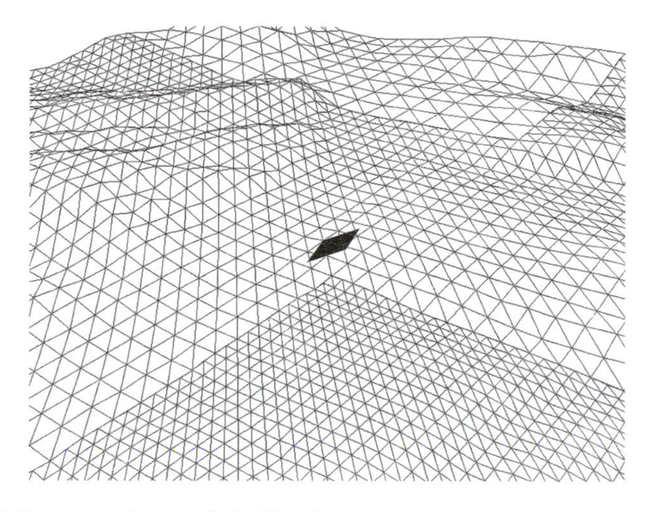

Figure 11.3 Compass for north indication

Finally, intelligence officers gain a better understanding of the terrain by combining, for example, a topographical map to an aerial or satellite photograph. For example, they can determine whether a certain road can be passable for tanks or not. The demonstrator contains a layer control tool that allows for different overlays to be added on top of the terrain as textures.

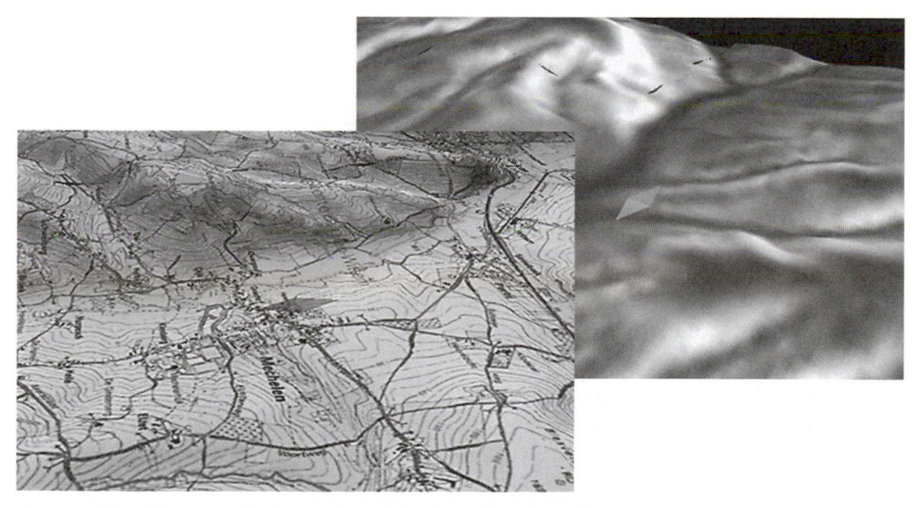

Figure 11.4 Layers that can be added on the terrain

Hence, the intelligence officer can choose between a topographical or aerial photo, or even a steepness visualisation map (Figure 11.4) to add to the terrain and/or remove it later. Other options include iso-contours and grid overlays to be added on top of the above maps.

Interaction

An intelligence military officer should be able to interact with the VE in order to produce mission plans. These mission plans allow, for example, for S2 and S3 officers to wargame. These mission plans contain friend and enemy military forces as well as various "drawings" that allow them to accomplish tasks such as: identify possible friend or enemy courses of action (COA), named areas of interest or tactically important areas, create objectives, fire and observation coverage, strong lines of defence, mobility corridors, and so on.

The demonstrator allows the user to add NATO standard military symbols on the terrain that represent friend, enemy, neutral or unknown military echelons (companies, battalions, and so on). Additionally, the user can handle deployment elements such as arrows, polygons, lines, and so on to draw courses of action, objectives, and to mark names and areas of interest, and so on. Collision detection and terrain following have been implemented to allow for easy and accurate manipulation of elements and symbols.

A 2D symbol representation of units and deployment elements is preferred to 3D icons. Realistic icon recognition is poor because it is difficult for users to discriminate between the subtle visual differences of military platforms (Shallman et al., 2001; Shallman et al., 2000). Users perform better with a distinctive and, sometimes unrealistic, 2D depiction than with a 3D accurate icon.

The 2D representations of symbols are added as billboards or signposts on the 3D terrain that are always oriented towards the position of the camera, hence always facing (view-aligned to) the user (Thomas and Wickens, 2001).

Group-working

During the pre-battle phase, military officers have to determine positions of reconnaissance units, what the enemy courses of action might be, impacts of air or logistics support, and so on. This results in S2 and S3 officers being able to wargame, that is, to try different possible scenarios of the forthcoming battle. The officers then discuss the results of the scenarios.

The equivalent to group-working in VR is Collaborative Virtual Environment (CVE). Collaborative environments are on-line multi-user workspaces, equipped with multi-modal communications and collaborative work interfaces. A Collaborative (or multi-user or shared) Virtual Environment (CVE) or Shared Virtual Space (Disz et al., 1997) is a computer-mediated tool that actively supports human-human communication in addition to human-machine communication and which uses a virtual environment as the user interface. Such environments use strong spatial metaphors for navigation, communication, and interaction scoping and object manipulation.

A user can use a tool to align the camera with the view of an information-gathering unit on the terrain, thus allowing him/her to have a first-person view of the information-gathering unit's field of view. This allows for the determination of the best positions in the terrain for the location of reconnaissance units.

A number of tools for wargaming have also been developed. One is a tool that allows for real time interoperability between two (or more) instances of the demonstrator. In this way, two officers can choose either "friend" or "foe" and deploy their units on the terrain from different instances of the application but both viewing the same VE. An opponent unit becomes visible only when there is line of sight to it by an own unit. Hence, the two opponents do not have the full picture of others' units and positions, allowing them to wargame in a similar way to the popular naval battle game.

A scenario recorder and player component allows the creation of scenarios by recording S2 and S3 officers while wargaming. The recordings are saved as text files and can be reproduced later for evaluation and discussion or even for training new officers' strategic tactics. A voice component allows for real time audio communication in order to enhance group-working.

Decision support

A line of sight visibility tool has been implemented. Visibility of 360° for its field of view is shown for the selected unit. This tool helps to decide on the best location of reconnaissance units on the terrain.

A history tool has also been implemented. The history tool allows for the visibility of Virtual Prints (ViPs) (Mouzouris et al., 2003) of real time armoured vehicles' tracks on the terrain; that is visualisation of a continuous 2D line representation of the path followed by a unit in the virtual space during time. The demonstrator is fed by a CGF application with unit data positions that are moving in near real time, thus allowing the visualisation of their position history.

Results

Figures 11.5 to 11.7 show the evaluation results for the visualisation and interaction, group-working and decision support phases for speed, accuracy and mental workload.

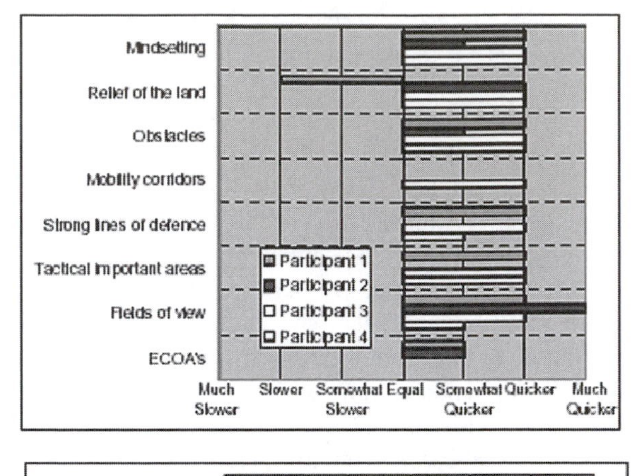

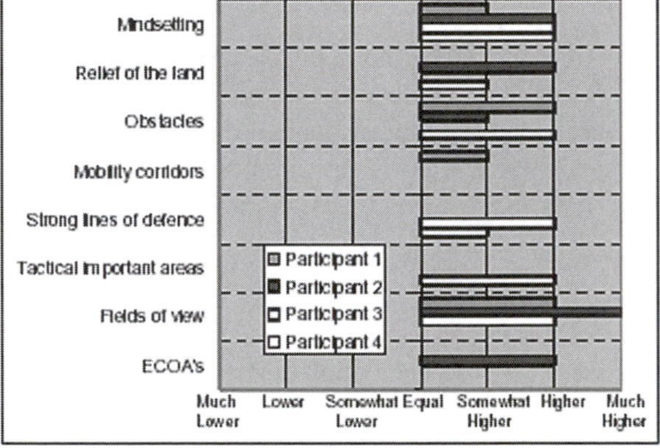

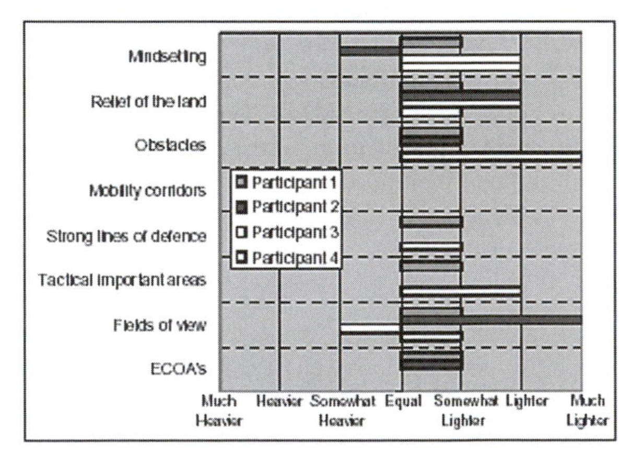

Figure 11.5 Evaluation results for interaction (speed, accuracy and mental workload respectively)

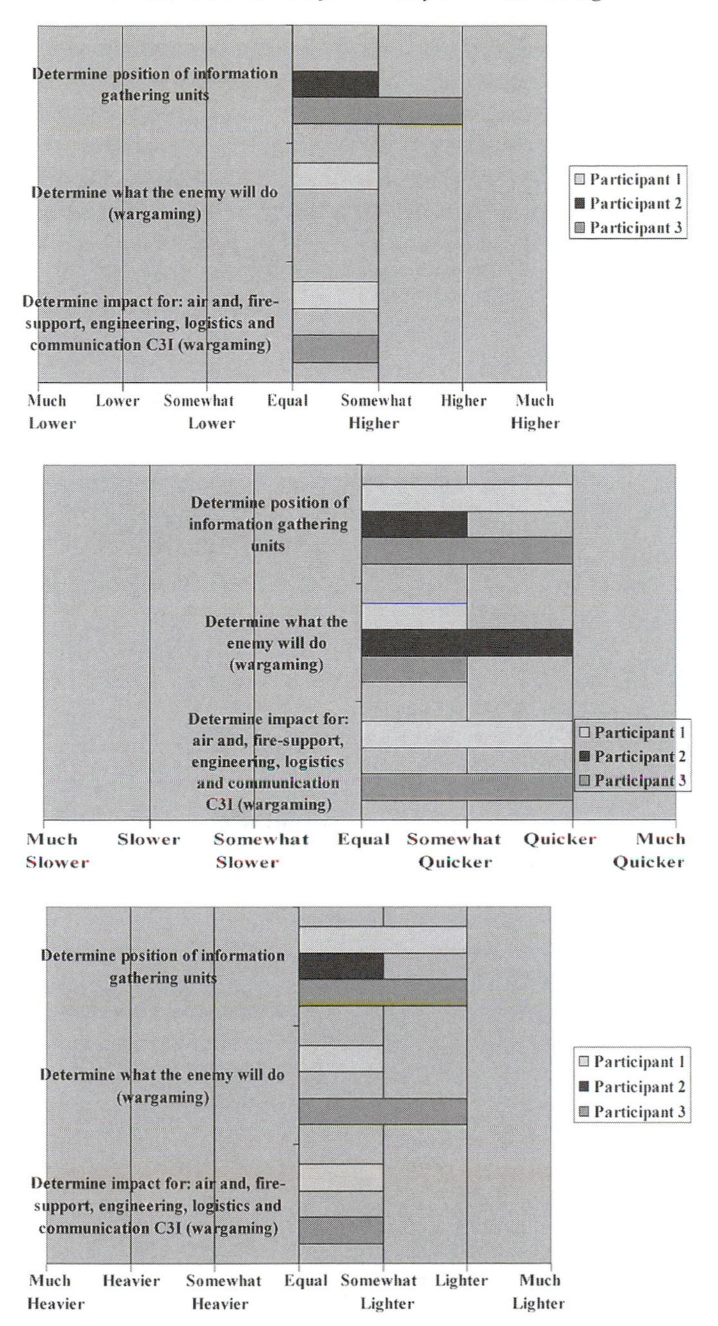

Figure 11.6 Evaluation results for group-working (speed, accuracy and mental workload respectively)

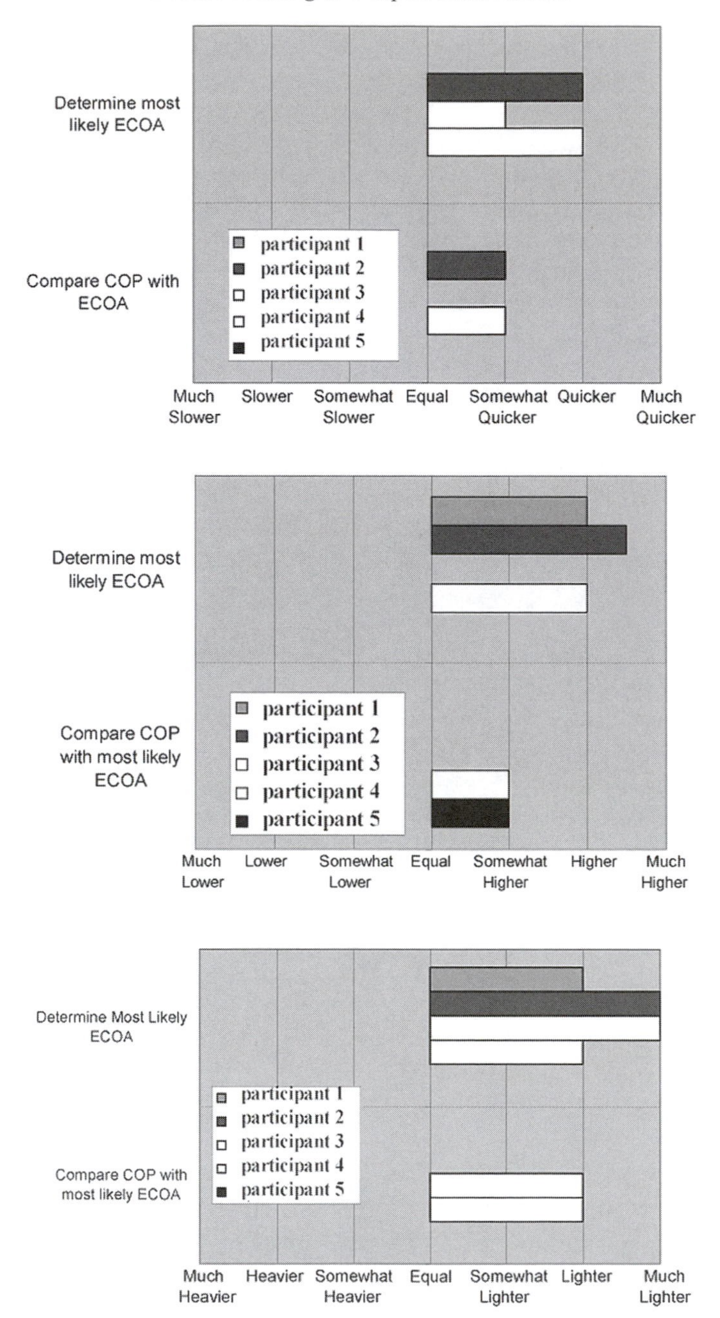

Figure 11.7 Evaluation results for decision support (speed, accuracy and mental workload respectively)

Conclusions

The MARVEL project has demonstrated the application of 3D/VR techniques to military intelligence decision making. These techniques have been evaluated positively by military intelligence experts and are to be incorporated into next generation battle management and C^4I systems.

Acknowledgements

The Consortium MARVEL (Military Applications foR VE Techniques in Logistics and Intelligence) is the name of a European research programme that is carried out as part of the EUCLID (European Co-operation for the Long-Term in Defence) RTP 6.14 programme. The MARVEL consortium consists of four different companies from four countries:

- ATOS ORIGIN, The Netherlands
- DATAMAT, Italy
- IFAD, Denmark
- INTRACOM, Greece

who have all contributed in the successful accomplishment of the project.

INTRACOM S.A. – Defence Programs
Many thanks to INTRACOM's Defence Programs Department for kindly providing us with the valuable time to produce this chapter.

The authors would also like to thank the Geoloket, Meetkundige Dienst, Rijkswaterstaat for providing the terrain elevation data of Limburg, the Topografische Dienst (TDN) of the Netherlands for providing the topographic map, DKLN Eurosense B.V. for providing the aerial photographs and the Instituto Geografico Militare (IGM) of Italy for providing terrain elevation data of the Friuli area. And last but not least, the authors would like to thank the members of the Hellenic User Focus Group for their participation in the evaluation workshops and their valuable feedback.

References

Detsis, G., Dritsas, L. and Kostaras, J. (2001). "Information filtering and control for managing the information overload problem". *IEE Conference & Exhibition – People in Control*, Vol. (481), 14-18, UMIST/Manchester, UK.

Disz, T. L., Olson, R., Papka, M. E., Stevens, R., Szymanski, M. and Firby R. J. (1997). "Two implementations of shared virtual space environments". *Argonne National Laboratory*, Argonne ANL/MCS-P652-0297.

Endsley M. R. (1995). "Towards a theory of situation awareness in dynamic systems". *Human Factors*, 37, 32-64.

Kirwan, B. and Ainsworth, L. K. (1992). *A guide to task analysis*. London: Taylor & Francis Ltd.

Mouzouris, A., Grammenos, D., Filou, M., Papadakos, P. and Stefanidis, C. (2003). "Virtual prints: An empowering tool for Virtual Environments".. In Harris, D., Duffy, V., Smith, M. and Stephanidis, C. (eds), *Human Centred Computing*, 3, 1426-1430, *Proceedings of HCII 2003*, Heraklion, Crete.

Perrie, A. A., Crisp, H. E., McKeely, J. A. and Wallace, D. F. (2000). "The solution for future command and control: human-centered design". *Naval Surface Warfare Center Dahlgren Division*. From http://www.manningaffordability.com/S&tweb/PUBS/SolutionFutureCommand-HCD/SolutionFutureCommand-HCD.pdf.

Shallman, H. S., John, M. St., Oonk, H.M. and Cowen, M.B. (2000). "Track recognition using two-dimensional symbols or three-dimensional realistic icons". *Technical Report 1818*, SSC San Diego, CA.

Shallman, H. S., John, M. St., Oonk, H. M. and Cowen, M.B. (2001). *Searching for tracks imaged as symbols or realistic icons: A comparison between two-dimensional and three-dimensional displays*. Technical Report 1854, SSC San Diego, CA.

Thomas, L. and Wickens, C. D. (2000). "Effects of display frame of reference on spatial judgements and change detection". *Technical Report* ARL-00-14/FEDLAB-00-4. http://www.aviation.uiuc.edu/new/html/ARL/report_fulltext.html.

Thomas, L. and Wickens, C. D. (2001). "Visual displays and cognitive tunneling: Frames of reference effects on spatial judgments and change detection". In *Proceedings of the 45th Annual Meeting of the Human Factors and Ergonomics Society*. Santa Monica, CA: Human Factors & Ergonomics Society.

Thomas, L., Wickens, C. D. and Merlo, J. (1999). "Immersion and battlefield visualization: Frame of reference effects on navigation tasks and cognitive tunneling". *Technical Report* ARL-99-3/FED-LAB-99-2. From http://www.aviation.uiuc.edu/ new/html/ARL/report_fulltext.html.

Wickens, C. D., Thomas, L. and Young, R. (2000). "Frames of reference for the display of battlefield information: Judgment-display dependencies". *Human Factors*, 42, 660-675.

Woods, D. D. (1984). "Visual momentum: A concept to improve the cognitive coupling of person and computer". *International Journal of Man-Machine Studies*, 21, 229-241.

Chapter 12

ROLF 2010: A Swedish Command Post of the Future

Berndt Brehmer

Introduction

ROLF 2010 is a Swedish acronym standing for "Mobile, Joint Operational Command and Control Function for the year 2010". It refers to a C^2 concept (Command and Control) that has been developed by a group of engineers, officers and scientists at the Swedish National Defence College at the request of the Swedish Armed Forces. The project started in 1995, and the concept was embodied in a first prototype, ROLF Mark I, in our C^2 laboratory in 1997, using very simple technology in the form of a set of projectors. The subsequent version, Mark II, illustrated in Figure 12.1, constitutes the first version of a "Command Post of the Future" for the Swedish Armed Forces. Mark II has been used at the college since 1998 in exercises and experiments with the students at the college, who are the future users of the command post. Mark III is now being developed.

The term "Command Post of the Future" (CPOF) refers to the projects, in a number of countries, aimed at designing new arrangements for command and control using the possibilities offered by the developments in information technology. The term originally comes from a DARPA project called Command Post of the Future, but parallel developments are found in the US Navy at SPAWAR, in Great Britain, where the concept is called Command Post 21, and in Australia in the FOCAL project, to give but a few examples.

It is not the purpose of this chapter to compare the various CPOFs currently being developed, but one observation is warranted. When working towards an information technology intensive command post, there are at least two possible approaches. The first is to start with existing information technology and then design the command post using this technology. This approach will result in worse than planned obsolescence: it will result in *guaranteed* obsolescence, for when the command post is ready, information technology will have moved on, and the command post will be obsolete. We have therefore taken an alternative approach and started with developing the general concept of what is required for command control. We then wait for information technology to catch up, as it surely will. This means that not all of the systems that we envisage for ROLF 2010 exist at the time of writing, but we are sure they will be available in 2010, when we have to deliver the command post to the Armed Forces. Indeed, development has been even faster than we imagined when we started the project, and the technology we need (and technology we could not imagine that we would need) is now becoming available.

In developing the ROLF concept, our point of departure was the interaction between the commander and his staff and the interactions among the staff members. This interaction aims at *sensemaking* (Weick, 1995). Sensemaking is not just a new term for situation awareness (Endsley, 1995). It is a *process*, rather than a *state*, and it is a very special process in that it is guided not only by the need to understand and to create an adequate reflection of the situation. It is aimed at understanding for a purpose. In command and control, the process is guided by the commander's need to find a way of fulfilling the mission given by his superiors, and it aims at an understanding of the situation that can serve as a guide to action. The reader will recognise this as a pragmatic view of the cognitive processes involved, with its roots in the writings of William James (1950), and others. Thus, the commander and his staff are trying to make sense of the situation in order to produce a command concept (Builder, Banks and Nordin, 1999) that can then be translated into a plan and finally in useful mission orders for the subordinate commanders. In this process, all staff members must contribute on the basis of their expertise. The command post should be designed to support this interaction and make it as efficient as possible, *and* ensure that the commander and his staff have the requisite variety (Ashby, 1950) for the task.

General characteristics of the concept

Before showing the existing prototype, the general characteristics of the ROLF 2010 concept will be discussed. This is but a general overview; a detailed account is given by Brehmer and Sundin (2004). Moreover, the present chapter should be seen as a progress report, rather than a definitive account. The concept, as well as its various embodiments in prototypes, is still under development.

The characteristics of the concept are illustrated in Figure 12.1, which gives an artist's impression of ROLF 2010. Of course, with the pace at which information technology develops, there is only one thing that we know about what we see in the figure: the finished product will be different from what is in it, but we could do no better than this. It is, perhaps, important to mention here that the ROLF 2010 project is only concerned with the command post. How, and in what form, information enters the command post are not problems for our project.

We think of the command post illustrated in Figure 12.1 as an M3-system, or a Man-Man-Machine-System where the interactions among the persons (the commander and his staff and among the staff members) are as important as design considerations as the interactions between the persons and their equipment.

Three characteristics of the command post are obvious from the picture. The first is that the staff is very small, the second is the seating, and the third the various forms of information technology. Each of these will be discussed in turn.

Figure 12.1 An artist's conception of ROLF 2010

The staff is small

There are two reasons for the small size. The first is that the command post should be truly mobile. We see mobility as the only possible form of protection today. When a cruise missile can be targeted at a given window in a building, and actually hit it, no other form of protection seems possible. Our design goal is that it should be possible to house the command post in a flexible container that can then be transported by air, rail, truck or ship, and be just one of many other containers on the road or at sea. The second reason for wanting mobility is the desire on the part of the Swedish Armed Forces, our customer, to lead close to the front, and to have a command post that could follow the front.

The small size offers new opportunities to design a creative environment where the staff members are able to interact in ways that are difficult, or impossible, in traditional large staffs. Specifically, it makes it possible to create a staff that can work as a team and where there can be true interaction and complementarity among the members.

The command post can be augmented by deploying a number of identical containers with different functions. In one development, we considered a command post with four containers, one for planning, one for battle management, one for follow-up and evaluation, and one for receiving visitors, such as journalists or generals from headquarters, who should not be allowed to disturb the actual work.

In our current concept, we are considering having only two containers. We have kept the container for receiving visitors. In the second container, planning, battle management and evaluation are performed by the same staff. This is because we think that these three functions cannot be separated, nor should they be. Planning not only serves to provide a course of action. It is also a way of exploring the future. By developing a great many plans, the staff will have explored many possible futures, and the commander and his staff are more likely to be prepared for the future that actually materialises when the original plan fails (which, according to military wisdom, it will). But this will work only if those doing battle management are also the planners and those who evaluate the results of the plan being executed. Hence, we now want all three functions to be performed by the same unit.

The price to be paid is a high workload. One of the problems that has to be considered, therefore, is the need for shift work for the staff, because a shift may well be worn out very quickly. To have the commander work shifts seems impossible, but the rest of the staff should be able to do so, provided that we can find a way of supporting shift change, so that the new shift is able to get a grasp of what is going on. We are considering replays of the development during the last shift as a way of supporting shift change, but so far we have not developed the technology required for this.

The seating

The seating is, of course, chosen to facilitate the interaction among the staff members, and thus the sensemaking process. Here, we have taken our inspiration from history and the idea that people used to gather in a close circle around the camp fire when having to make important and difficult decisions (today's boardroom table is a descendant of the camp fire, we think). This allows for eye contact, thus minimising the "psychological distance" (Wellens, 1986). The need to minimise psychological distance has to do with the nature of the activities of the staff. As noted above, we see this as a process is sensemaking, and we think that this process will, in part, have the character of *negotiations*, negotiations about what is (achieving an agreed upon common understanding, or *sense*) and what should be done (the command concept). According to Wellens (1986; 1990) negotiations are more effective when psychological distance is minimised, as it is when there is face-to-face contact. We also think that when circumstances are stressful, as they surely will be in command and control in war, communication is more effective if it is close and physical.

Information technology

The small size means that it will be necessary to substitute for many members found in traditional staffs by means of information technology. Information technology is needed to ensure that the staff can muster the *requisite variety* (Ashby, 1950) needed to carry out the task even though it has few staff members. Four forms of information technology are illustrated in Figure 12.1.

The first is the various forms of information displays. The first of these is the display around which the commander and his staff are seated. It is called the *Visioscope*™. It is a three dimensional (3D) display, usually a display of the situation in the form of a map, we think, but the staff members can, of course, put whatever they want in the *Visioscope*™. Information can be moved directly from screen to screen in the command post by means of a mouse. This is the principal form of information display in the command post. It is chosen, in preference to the ordinary maps on the wall, so as to make a common focus possible, as well as eye contact among the staff members. A common focus is important to further shared sensemaking because everyone will then make sense from the same elements, something that cannot be insured when everyone has his/her own display.

The reason that the *Visioscope*™ is a 3D display is not to make it easier for the staff members to read the map. If there is one thing that officers know how to do, it is to read maps (although the 3D representation may be important in operations where some of the people in the staff are civilians with little map experience). Instead, there are two other reasons for wanting a 3D display.

The first is that we want to display the battle space as a *volume*, rather than as a battlefield. This is important in the joint operations for which ROLF is being designed. In such operations it is important to display everything from aircraft to submarines, that is, the commander and his staff have to think in terms of a battle volume rather than a battlefield. If the display is only a 2D display, each staff member would have to construct the battle volume for himself. It would then be a private mental concept, perhaps unique to every staff member. At the very least, each staff member's concept would be unknown to every other staff member. This is against an important guiding principle for the ROLF concept: that as much as possible should be out in the open and be public so as to avoid misunderstanding as much as that is possible.

The second reason for requiring a 3D display is that we want the staff members to be able to actually grasp and move the objects in the display, thus enabling them to use what we call a *visual interactive language* to show what they mean, rather than just talking about it. This would be important, we think, not only when the staff members communicate among themselves, but also when the commander and staff communicate with subordinate commanders. For example, the visual language would allow the commander to communicate his/her intent by means of a picture, illustrating the desired end state, rather than in words. We know that when the commander's intent is communicated in words, the intent as understood by the subordinate commanders may differ from the commander's actual intent (see Klein, 1994, for results relating to this). A one-person command post allowing for communication with subordinate commanders by means of the visual language is therefore also part of the ROLF concept (see Figure 12.2). The concept assumes a helmet mounted display to provide the information.

Figure 12.2 A mobile, one-person command post

The concept of a one-person command post has been developed as part of the ROLF concept for two reasons. The first is for communicating with the subordinate commanders as already mentioned. But we have also taken note of the experience from developing the new forms of control rooms in process industry where it has been observed that the operators have become unwilling to leave their control rooms and do their rounds in the plant for the simple reason that all the information is in the control room, and that is where they need to be if something happens. We have made similar observations when demonstrating our command post to high-ranking visitors, some of whom have expressed the view that our command post is where they would want to be when things got difficult. Although this is flattering to us, we think that a commander's place is not always in his command post. S/he also needs to be with the troops, and we therefore need to be able to furnish him/her with the information s/he needs, wherever s/he wants to be on the move. The commander must be able to command from wherever s/he is. S/he cannot go to some designed place to exercise command. The one-person command post is an attempt to make it possible for the commander to leave his/her command post, without losing an understanding of what is happening, thus enabling him/her, not only to dare to leave the command post, but also to exercise command from wherever s/he happens to be. The development of this aspect of the concept is still in a very early stage, however.

Creating a 3D display where the virtual objects can be manipulated is not an easy task. We simply do not know enough about how information about virtual objects must be displayed to make it possible to "grasp" these objects. One of the problems is that the visual system does not seem to be designed for extracting 3D information from a horizontal display; it works much better with vertical displays. We have therefore started a separate research programme in collaboration with Uppsala University (Sweden) to handle this problem and to provide a basis for developing the 3D displays we need. Developing such displays is thus not only a technical problem, as some might think. It also requires basic information about how the visual system works that we simply do not have today. This is possibly the reason why some earlier attempts to develop 3D displays for command control, such as the SEADRAGON concept, have not been successful.

As noted, the *Visioscope*™ also provides for a natural seating around a table where the commander and his staff can communicate directly while maintaining eye contact. For this to be possible, we cannot use 3D visualisation techniques that rely on polarised light and require special glasses. Such glasses make it impossible to look each other in the eye. We are therefore trying out autostereoscopic displays based on holography. We have a small working prototype and we are now working towards developing a larger display that allows multiple users to see the display via "view ports". Although we have solved the technical problems in principle, we do not yet have a large working prototype. For such a display to be useful, it is not enough to solve the technical problem of producing a 3D display; we also need to solve the perceptual problems mentioned above so that we know what to put into the display. Meanwhile, a four-person version of the *Visioscope*™, based on polarised light, has been developed. It is called *Mimer's Well*, and we are now starting experiments with this display to explore problems relating to how 3D should be used to provide information for the staff. No results are available from this experimentation yet.

The walls in the command post are large displays, constituting the *Visionarium*™. These displays have two functions. First, they make possible a change in perspective, from looking *into* the *Visioscope*™ to looking *out* over a representation of the battle field, based on representations downloaded from UAVs and the like. This makes it possible for the staff to assume the traditional position of the commander on his *Feldherrenhügel*.

The second function of the *Visionarium*™ is to make possible the display of additional information, such as television broadcasts (BBC, CNN and the like), and whatever is required.

The third form of information technology is represented by the individual work stations behind each staff member. These are for e-mail communication with the subordinate commanders and support staff, as well as for individual access to the decision support that the staff member may require (this can also be displayed in the *Visioscope*™, of course). The individual work stations are placed behind each staff member so that, when using his or her personal decision support, he or she will signal to the other staff members that he or she is currently not taking part in the common discussion.

The Roman soldier in the *Visioscope*™ is an avatar. It represents the fourth kind of information technology. The avatar is the mouthpiece of a critiquing system that listens to the plans developed by staff and critiques them, pointing out aspects that they may have overlooked (see Silverman, 1992, for a discussion of critiquing systems). The critiquing system is being developed in collaboration with Linköping University.

This leads us to the topic of decision support. Our analysis suggests that staff members need help with three aspects of the command and control problem: complexity, dynamics and uncertainty. The trick is to find a unitary approach to decision support that can handle all three problems. If there are many different systems, we do not think that they will be used. We need one system that the staff members can use and feel comfortable with.

Our general point of departure is that the only acceptable form of decision support is support for *testing* sense, command concepts and plans. Decision support for developing ideas, and systems that suggest what decisions should be made, will not be acceptable. The creative part of the command and control process is the human's prerogative, while the more tedious detail and complexity of testing ideas can be left to the decision support.

Our general approach to an integrated form of decision support is called *STRATMAS*™ (STRATegicMAnagement System; Woodcock, Hitchins and Cobb, 1998). *STRATMAS*™ is a simulation based on agents moving in a substrate of cellular automata, and using genetic algorithms for optimisation. This makes it possible to represent both the spatial and the temporal aspects of a problem, that is, it makes it possible to handle both problems of complexity and problems of dynamics. A working prototype of the system, designed to help the staff members solve force composition problems, has been installed in ROLF Mark II. This is an important problem in the Swedish Armed Forces which will rely on forces tailored to the situation at hand, rather than standing units that are always used in a given form; that is, forces will be composed and recomposed as needed. The *STRATMAS*™ system is now being further developed to allow for the testing command concepts and rules of engagement.

Decision support for coping with uncertainty has proved difficult. We have given up our initial approach based on, first, probability theory, then fuzzy sets, in favour of an approach that relies on findings from studies of so called naturalistic decision making. Such studies (see Lipshitz and Strauss, 1997) have revealed the strategies that officers use to handle uncertainty naturally, such as making assumptions. We are running a series of similar studies ourselves, using both the questionnaire approach of Lipshitz and Strauss and observations and analyses of how staffs handle uncertainty in exercises. So far our results have been consistent with those of Lipshitz and Strauss, although they have given us a more detailed picture.

This suggests that a useful approach to uncertainty might be one enabling sensitivity analysis. Specifically, our current approach to the problem is that the staff members should be allowed to use their natural ways of handling uncertainty, but that they should be made aware of the consequences of assumptions made to make the uncertainty "go away". This awareness will come from *STRATMAS*™ simulations where the consequences of different assumptions are examined, and the extent to which the sense and plans developed by the staff are sensitive to the assumptions the staff members make. Our work here is at a very early stage, however.

The current embodiment of the concept: ROLF Mark II

ROLF 2010 left the Powerpoint world (where many new concepts in information technology seem doomed to live) in 1998 when we constructed ROLF Mark II in our command and control laboratory. Before that we had, as mentioned, experimented with a very simple realisation of the concept based on four projectors mounted in the ceiling and projecting down on a table which constituted an early form of the *Visioscope*™. Mark II is illustrated in Figure 12.3.

Figure 12.3 ROLF Mark II in an exercise

As can be seen from Figure 12.3, ROLF Mark II has much of the functionality that has been envisioned for the final version of ROLF 2010. It has a first version of the *Visioscope*™, a first version of the *Visionarium*™ in the form of four large touch sensitive screens, and it has the individual work stations. The *Visioscope*™ in ROLF Mark II is, however, only a 2D representation, and only two of the *Visionarium*™ screens currently provide for 3D representations, albeit using polarised light displays so that special glasses are required. As noted above, we are now developing a 3D version of the *Visioscope*™, and those displays are planned as part of ROLF Mark III, which will start building in 2005. (The reason for the delay is that the college is moving to a new location, and it is simply too expensive to build Mark III at our current location and then move it; the mobility aspect of ROLF is still in the future.)

Experience so far

ROLF Mark II has now been used in experiments and exercises at the National Defence College since 1998. We are fortunate to be able to employ the future users of the system, who are our students in the Staff programme (for captains who are to become majors) and the Command programme (for majors who are to become lieutenant colonels), as our participants in these experiments and exercises.

Although it is now close to six years since ROLF Mark II was first constructed in the laboratory, it has been modified continuously, and our empirical work so far has had the character of general feasibility studies, with the aim of finding out whether specific ideas work at all. It is only now that we feel ready to embark on more systematic experimentation to evaluate the concept.

Our experience with ROLF 2010 so far has of course given us a wealth of specific ideas to be considered in Mark III. Three very general conclusions from our work include the following:

The first is that it is possible to exercise effective command and control at the operational level from the ROLF command post.

The second is that those who have worked in the ROLF command post like doing so and they think that is represents a better way of exercising command and control than traditional command posts do. This is important for our participants as they are the future users of the system, and if they did not like the concept, it would never be used.

Thirdly, we have found that old habits do not die easily. We do not automatically achieve the creative atmosphere we have envisaged. Instead, the officers working the command often slip back into their old, hierarchical habits. This is dangerous, for the ROLF command post is an almost perfect environment for group think: high cohesion, stress, and a strong commander (Janis, 1992). It is therefore necessary to develop staff procedures for the new command post, and to train the future staff members to work as a team. It is also necessary to develop the commander's role so that he or she can function as a team leader and coach, while retaining the traditional role of a commander. After all, the responsibility rests with the commander, even if he or she functions as a coach, and it may sometimes be necessary to exercise command in the traditional way, also for the ROLF staff, for example, when time pressure is high and the staff cannot come to a consensus in the time available to them. This is one of the greatest challenges now facing the project. Yet, it is an important aspect of the ROLF concept that the commander should be one of the team. One of the many *desiderata* of modern command and control systems is that they should allow for faster command and control than traditional systems. Having the commander take part in the staff work is a step towards this goal, since it does away with the usual cycle of briefings that take up so much time in ordinary staff work.

Conclusions

In this chapter, the ROLF 2010 concept and the current prototype, Mark II, have been described. As should be obvious to the reader, we still have a long way to go. Three aspects require further work: the 3D displays, the decision support and the requisite staff procedures, including the role of the commander. These problems will probably be solved, one way or the other. What we cannot foresee is how the command post will actually be used by the commanders and staffs who will work in it. Information technology differs from other forms of technology in that it is impossible to foresee its consequences in any detail. This is because information technology provides opportunities, but it does not (or should not) shape behaviour in any predetermined direction. In developing our prototypes, we have been very much aware of this. We are confident that the future users will discover possibilities and opportunities that we have not foreseen. Because we have not been able to foresee exactly how the command post will be used, our work will necessarily fall short of providing for all of the needs of the future users. Designing information systems is a process with no clearly defined end, and the ROLF concept will certainly go through many development cycles even after we deliver a working prototype to the Swedish Defence Forces in the year 2010.

References

Ashby, W. R. (1950). *An introduction to cybernetics.* London: Hutchinson.

Brehmer, B. and Sundin, C. (2004). *ROLF 2010.* Stockholm: Swedish National Defence College.

Builder, C. H., Banks, S. C. and Nordin, R. (1999). *Command concepts – A theory derived from the practice of command and control.* Washington, DC: Rand Corporation.

Endsley, M. (1995). "Towards a theory of situational awareness in dynamic systems". *Human Factors*, 37, 32-64.

James, W. (1950). *Principles of psychology.* (Vols 1 and 2). New York: Dover (originally published 1890).

Janis, I. (1992). *Victims of groupthink.* Boston: Houghton-Mifflin.

Klein, G. A. (1994). "A script for the commander's intent statement". In I. H. Levis and I. S. Levis (eds), *Science of command and control: Part III. Coping with change* (sid.75).

Lipshitz, R. and Strauss, O. (1997). "Coping with uncertainty: A naturalistic decision-making analysis". *Organizational Behavior and Human Decision Processes*, 66, 149-163.

Silverman, B. G. (1992). *Critiquing human error. A knowledge based human-computer collaboration approach*. New York: Academic Press.

Weick, K. (1995). *Sensemaking in organizations*. Thousand Oaks, CA: Sage.

Wellens, A. R. (1986). "Use of a psychological distancing model to assess differences in telecommunications media". In L. Parker and C. Olgren (eds), *Teleconferencing and electronic media* (pp. 347-361). Madison, WI: Center for Interactive Programs.

Wellens, A. R. (1990). *Assessing multi-person and person-machine distributed decision making using an extended psychological distancing model*. (Report AAMRL-TR-90-006). Dayton, OH: Armstrong Aerospace Medical Research Laboratory, Wright-Patterson Air Force Base.

Woodcock, A. E. R., Hitchins, D. K. and Cobb, L. (1998). "The strategic management system (STRATMAS) and the deployment of adaptable battle staffs. I: Command and control decision making in emerging conflicts". In *Proceedings of the 4th International Symposium on Command & Control Research & Technology, Stockholm*. Washington: CCRP.

PART 3
Complex Decision Making in Civil Applications

Chapter 13

Human Information Processing Aspects of Effective Emergency Incident Management Decision Making

Jim McLennan, Mary Omodei, Alina Holgate and Alexander Wearing

Introduction

Significant advances in communications and fire suppression technologies over the last 10 years have changed dramatically the nature of firefighting operations, in both urban and wildland settings. However, emergency management operational activity, especially local incident command activity, remains a fundamentally *human* endeavour. It involves hierarchical teams of trained individuals, using specialised equipment, whose efforts must be coordinated via command, control, and communication processes to achieve specified objectives under conditions of threat, uncertainty, and limited resources, both human and material. The command and control function exercised by the on-scene, or local, fireground incident commander is crucial to success.

In the programme of research reported here, we endeavoured to focus on identifying those decision processes associated with more versus less effective incident command at fires. Effective incident command obviously requires considerable technical knowledge of fire chemistry and physics, suppression equipment capabilities, and standard operating procedures. However, previous work by Omodei and Wearing (1995) suggested the importance of another set of psychological capabilities which could perhaps best be described as information processing competencies.

We explore these by presenting some general conclusions reached about the decision processes of experienced fire officers in local command at fires and related emergencies. The conclusions were arrived at by comparing processes of more versus less effective incident commanders in a range of studies employing diverse methodologies.

Methodologies

Six methodologies were employed during the research programme. For each, an illustrative reference has been provided and the method of determining incident commander effectiveness has been noted:

1. Analyses of 20 fire-related death and injury investigation reports (United States Department of Agriculture, Forest Service, 2001) – consensus judgements by domain experts.
2. Naturalistic field observations and 46 structured post-incident interviews with urban fire officers (McLennan and Omodei, 1996) – judgements by peers.
3. Twenty structured retrospective interviews with wildland fire officers (Holgate, 2003) – judgements by peers and superior officers.
4. Twelve head-mounted, video-cued, post-incident recall interviews in urban operational settings (McLennan, Omodei and Wearing, 2001) – judgements by peers.
5. Twenty-nine head-mounted, video-cued, post-incident recall interviews in two urban field experimental settings (McLennan, Pavlou and Omodei, 2005) – performance ratings made by an expert panel of observers.
6. Six laboratory experimental studies of incident command team processes for managing wildland fires, utilising the *Networked Fire Chief* computer-generated microworld simulator (Omodei, Wearing, McLennan and Clancy, 2005) – objective measure of team performance in terms of proportion of total assets saved.

Findings

Effective emergency management decision making

We begin with a negative finding: we have so far failed to find evidence of an obvious "personality type" associated with good incident command. We found our good incident commanders to range from calm and phlegmatic to excitable and talkative, from aloof to gregarious, and from reserved to extroverted. It seems that good incident command is less a matter of what kind of person a commander *is* than what he or she *does* while in command.

At a surface level of description, first, effective incident management involved rapid extraction of the most relevant (and not necessarily the most salient) features from the information array. Sometimes this required an active search for information deemed "need to know". Often the information array comprised conflicting items and items of doubtful reliability in terms of timeliness or accuracy. Secondly, there was a rapid "good enough" conceptualisation of the deep structure of the problematic situation. Thirdly, a response was speedily chosen which had a high probability of success, having regard to the threat posed and the resources available. Fourthly, the situation was monitored closely in anticipation of possible deterioration in the threats/resources balance. Table 13.1 summarises behavioural markers of effective incident command identified by McLennan et al. (2005).

Table 13.1 Behavioural markers of effective incident command

Anticipation and Planning	Used 'dead time' to study site plans and diagrams Prepared for 'worst case' scenario early, took precautions, called for additional resources Warned crews (radio, face-to-face) of likely developments and tasks
Communication	Used site maps and diagrams to explain intentions to subordinates Clear, controlled speech to subordinates Maintained eye contact when speaking/listening to subordinates face-to-face Radio: paused after subordinate acknowledged call before giving orders/asking questions
Leadership and Assertiveness	Spoke clearly, firmly, decisively (radio, face-to-face) Greeted key (role) 'players' (e.g. building supervisor) warmly but decisively
Management of Workload	Used white board to record incoming information, to write 'reminder notes', and make sketches Incoming radio traffic: asked sender to "wait" until current task completed Requested new arrivals at the Control Centre to wait outside until ready to speak with them Gave 'complete' rather than 'open-ended' orders so not required to remember short-term crew assignments
Re-evaluation of Situation	On first indication of deterioration of the situation raised the alarm 'level' so as to call-out more resources
Use of Available Information	Used multiple sources: subordinates, local 'experts', site plans, diagrams

We identified four categories of actions which effective incident commanders took which contributed to their effectiveness as incident managers. First, they took active steps to control both the type of incoming information about the situation and the rate at which it was presented. They did so (face-to-face and radio) by (a) asking for specific information they believed to be relevant, (b) delegating particular individuals to find out and communicate need-to-know information, and (c) delaying receipt of less immediately relevant information. On occasions, they cut off eager subordinates wishing to provide information by stating that the issue being raised could be dealt with later. Secondly, they reduced the demands on their working memory capacity, for example by writing things down and drawing sketches. Thirdly, they tried to anticipate developments rather than being forced to react to changes in the situation. As one wildland fire officer interviewed put it: "You don't fight the fire in front of you, you fight the fire you're going to have in an hour from now." Fourthly, they monitored their level of cognitive arousal and their level of emotional stress, and they used active processes to prevent arousal level and stress from disrupting their decision processes, such as physical activity, deliberate physical methods of relaxation, calm breathing and positive self-talk.

Effective incident commanders functioned as if they had a good practical understanding of the limitations of their information processing system. The foundation of their ability to manage their information processing load appeared to be prior learning from past experience and training. They had developed a rich network of decision rules organised in schemas which enabled them to use, mostly, fast, rule-based, robust *recognitional* decision processes rather than slow, vulnerable, knowledge-based *analytical* problem solving processes, which involve heavy demand of working memory capacity. Such use of rule-based decision making forms the basis of what Adams and Ericsson (2000) characterise as *procedural* expertise.

Some emergency incidents were too complex for simple recognitional rule-based decision making. Such situations exhibited one or more of three features: novelty, opacity, or resource inadequacy. In such circumstances, effective incident commanders were often able to transcend their (necessarily limited) range of specific past experiences and use fast, robust, analogical decision processes to apply prior learning to novel situations. For example, an incident commander was confronted by a serious leak from a large container of liquid oxygen in the grounds of a hospital. He had never previously encountered an incident involving cryogenic material. He reasoned by analogy that the best thing to do was to handle the emergency by treating it as he would a volatile, flammable toxic, chemical leak. The emergency was speedily resolved, though some of the precautions he took were, in fact, unnecessary (for example, ordering his crews to wear breathing apparatus).

In other situations characterised by high levels of uncertainty, incident commanders were sometimes forced to use analytical knowledge-based problem solving processes in order to choose a course of action from among alternatives. Under such circumstances good commanders used a small number of simple and robust heuristics to guide rapid decision making. Two heuristics in particular were used. The first was that of *minimaxing* (Newell and Simon, 1972), that is, selecting the action least likely to lead to the worst outcome: "Anyone in the warehouse was probably dead by now. I'll start a crew in breathing apparatus organising the evacuation (from an adjacent childcare centre), I don't want a kindergarten of dead kids." The second was that of *means-ends analysis* (Newell and Simon, 1972), that is, when unable to deal with the total situation immediately, use available resources in such a way as to contribute to an overall solution later. "Even though there are people unaccounted for (in a large motel complex), I won't start a search yet. There are 140 rooms. I'll put my crew to containing the fire and when the next appliances come on-scene I'll start those crews on search." Processes such as the above probably underlie adaptive expertise (Adams and Ericsson, 2000).

Less effective incident command

Less effective incident commanders seemed to have fewer decision rules to draw upon. This forced them to make greater use of slow, vulnerable problem solving processes. The lack of available rules to guide situation assessment meant that they were likely to be overwhelmed by items of information, all of which had to be attended to to some degree, thus slowing situation understanding. Very salient (but less relevant) information was likely to be given undue weight. Where relevant information was not immediately at hand, its absence was often not noted and no steps were taken to search for the missing information. Situation conceptualisations were thus both slow to be developed and likely to be inadequate in important respects.

Table 13.2 illustrates the cognitive problems faced by less effective incident commanders as identified by McLennan et al. (2005).

Table 13.2 Relative frequency of cognitive activity category

	Less Effective Commanders	More Effective Commanders
Situation assessment and understanding	32%	38%
Intention formation and action decisions	37%	51%
Self-monitoring (mostly noting level of overload)	31%	11%

These data show that the less effective commanders devoted a much greater proportion of their cognitive activity to noting their level of overload, while more effective commanders devoted proportionally greater cognitive activity to intention formation and action decisions.

Less effective commanders appeared often to be overwhelmed by the situation. They reacted to events in an *ad hoc* manner and found it difficult to formulate a coherent plan to coordinate activity. Many described how difficult it was to concentrate on the tasks at hand because task-irrelevant, self-critical thoughts kept intruding upon, and interfering with, their attention to the tasks at hand. It appeared that their cognitive resources were fully occupied with the immediate situation: they had no spare cognitive capacity to devote to planning or anticipating. Table 13.3 illustrates this using extracts from head-mounted, video-cued recall interviews obtained by McLennan et al. (2005).

Table 13.3 Illustrative extracts from head-mounted, video-cued recall protocols

Effective Incident Commander: Fire (simulation) in a Hospital	Less Effective Incident Commander: Fire (simulation) in an Underground Railway Station
"So, at this stage I thought 'Right, that's the next thing I have to do is I have to give him (designated Forward Operations Officer) some manpower for a start so he can start operations'. I wanted to establish early on that he was going to be in charge over there so that's why I said to him 'You're the Operations Officer'. So I could just send him resources and he would delegate the tasks because he had the big picture and he could see what was going on."	"...at this stage I've sort of lost it too because I think I should have gone back and spoken to the Station Master and got everyone evacuated through the emergency evacuation system and started smoke ventilation straight away. I wasn't thinking clearly. I'm focussing on things in general and I'm not clearly identifying tasks and carrying them out. Then confusion reigns because in the short time I've let things build up and I haven't been able to prioritise things. I've just let it get away a bit."

Disastrous incident command

The causes of disasters are often complex, with organisational systemic failures frequently being a major contributor (Reason, 1990). However, from careful analysis of post-incident investigation reports and interviews with those involved in "near misses", some characteristics of disastrous incident command were identified. The most common appeared to be seriously flawed situation conceptualisation and inappropriate choice of tactics. Often, this resulted from key information being overlooked or misinterpreted through lack of relevant experience. For example, a commander of a fire team with experience in fighting structure fires but no experience in fighting forest fires is likely to underestimate the danger of being trapped in a high fire intensity burnover in a forest. Another source of flawed situation conceptualisation was faulty preconceptions and associated failure to seek confirmatory evidence. For example, a (false) belief and a failure to seek confirmatory evidence that a predicted wind change has already passed through the area is likely to lead to increased risk that a wildfire crew may be trapped by a sudden future change in wind direction and/or intensity. Finally, we found evidence suggestive of persistent biases in the way information is processed by incident commanders to arrive at a decision. We found evidence of a sunk costs effect, that is, persisting with a tactic which is obviously (to a detached observer) ineffectual, simply because time and resources have been invested in the tactic. Another such could be described as an optimism bias, that is, choosing a course of action which necessitates nothing going wrong if it is to succeed; for example, ordering a crew to make entry to a smoke-logged building without sound evidence that there is no risk of structural collapse. A final bias associated with disastrous incident command at wildland fires could best be described as a linear rate of change bias. Humans seem to be incapable of accurately predicting non-linear rates of change (Doerner, 1980). Fire spread rates change dramatically with only modest changes in wind speed or ground slope. The history of wildland firefighter fatalities is replete with incidents in which a team of firefighters knew that they were in danger, but failed to appreciate how immediate that danger was. They delayed escape, in some cases apparently reluctant to abandon tools and equipment (MacLean, 2003), and the fire overtook them, sometimes quite close to safety.

Discussion

We do not claim to have made new discoveries about incident command decision processes on the fireground, rather we draw attention once more to important psychological processes underlying effective incident command decision making which are easy to neglect in the face of emerging new technologies. We conclude that the most important psychological "drivers" of an incident commander's performance involve three aspects of the human information processing system:

- Rule-based decision making is appreciably faster than knowledge-based decision making (Rasmussen, 1983).
- The limited capacity of working memory (Baddeley, 2001).
- Effective decision making is dependent on regulation of arousal level and negative emotions (Omodei and Wearing, 1995).

Accordingly, we propose the following information processing competencies as necessary for effective incident command decision making:

- Acquiring through experience an extensive set of simple and robust rules to guide situation assessment and decision making across a wide range of operational circumstances.
- Developing effective means of preventing working memory capacity being exceeded in spite of the high mental workloads likely to be associated with emergency incident command operations.
- Developing self-awareness in order to monitor both arousal level and negative emotions.
- Learning effective ways of controlling arousal level and negative emotions.
- Developing a habit of watchfulness against processes likely to interfere with accurate situation assessment, such as preconceptions and decision biases.

Several other tentative conclusions follow from the findings, including:

- Proposed new Command and Control (C^2) information/communication systems should be viewed sceptically if they seem likely to simply present a commander with more information and allow him or her to be more readily interrogated and micro-managed by superiors.

- Rather than relying on stereotypes of what constitutes "commander material", Promotion Boards should seek evidence that candidates can (a) manage complexity, (b) learn quickly, and (c) retain a degree of self-control under stressful circumstances. Carefully constructed and evaluated field challenges are probably superior to pencil-and-paper tests of knowledge, aptitude, or personality.
- Incident Command training unit personnel must keep in mind a fundamental distinction between recognitional knowledge ("knowing about") and procedural knowledge ("knowing how to"). They must resist cost pressures to substitute classroom recognitional knowledge acquisition in place of learning experiences in the field. Novice commanders need to be trained the way they will be required to function during fireground operations. Training exercises which capture the deep-structure of emergency incidents (rather than merely reproducing surface appearances) are essential. However, beyond this, considerable planning and effort must be devoted to enhancing learning from such exercises by providing effective feedback and facilitating reflective self-appraisal of performance (McLennan, Pavlou, Klein and Omodei, 2005).
- If decision support/aiding tools are to be introduced, these tools should be aimed at supporting commanders' front-end situation assessments rather than back-end decision choices. Commanders are more likely to benefit from improved understanding of the situation rather than being constrained in their choices.

Conclusion

While communication and information technology systems used in emergency incident management will continue to evolve, the basic architecture of the human information processing system will almost certainly not. Care needs to be taken to ensure that emerging new communication and decision support systems match the operating characteristics, both strengths and weaknesses, of the human information processing system.

Acknowledgements

We acknowledge support from the Australian Defense Science and Technology Organisation, the Melbourne Metropolitan Fire and Emergency Services Board, and the Victorian Country Fire Authority. However, the views expressed here are those of the authors.

References

Adams, R. J. and Ericsson, K. A. (2000). "Introduction to cognitive processes of expert pilots". *The Journal of Human Performance in Extreme Environments*, 5, 44-62.

Baddeley, A. D. (2001). "Is working memory still working?" *American Psychologist*, 56, 851-864.

Doerner, D. (1980). "On the difficulties people have in dealing with complexity". *Simulations and Games*, 11, 87-106.

Holgate, A. (2003). *Decision making processes in effective fireground command and control*. Melbourne: CFA Victoria.

MacLean, J. N. (2003). *Fire and ashes: On the front lines of American wildfire*. New York: Henry Holt.

McLennan, J. and Omodei, M. M. (1996). "The role of prepriming in recognition-primed decision making". *Perceptual and Motor Skills*, 82, 1059-1069.

McLennan, J., Omodei, M. M. and Wearing, A. J. (2001). "Cognitive processes of first-on-scene fire officers in command at emergency incidents as an analogue of small-unit command in peace support operations". In P. Essens, E. Tanercan and D. Winslow (eds), *The human in command: Peace support operations* (pp. 312-332). Amsterdam: Mets & Schilt: Breda: TNI.

McLennan, J., Pavlou, O., Klein, P. and Omodei, M. M. (2005). "Using video during training to enhance learning of emergency incident command and control skills". *Australian Journal of Emergency Incident Management*, 20(3), 10-14.

McLennan, J., Pavlou, O. and Omodei, M. M. (2005). "Cognitive control processes distinguish between better versus poorer decision making by fireground commanders". In H. Montgomery, R. Lipshitz and B. Brehmer (eds), *How professionals make decisions* (pp. 209-222). Mahwah, NJ: Lawrence Erlbaum.

Newell, A. and Simon, H. A. (1972). *Human problem solving*. Englewood Cliffs, NJ: Prentice-Hall.

Omodei, M. M. and Wearing A. J. (1995). "Decision making in complex settings: A theoretical model incorporating motivation, intention, affect, and cognitive performance". *Sprache und Kognition*, 14, 75-90.

Omodei, M. M., Wearing, A. J., McLennan, J. and Clancy, J. (2005). "'More is better?' Problems of self-regulation in naturalistic decision making situations". In H. Montgomery, R. Lipshitz and B. Brehmer (eds), *How professionals make decisions* (pp. 29-42). Mahwah, NJ: Lawrence Erlbaum.

Rasmussen, J. (1983). "Skills, rules, and knowledge: Signal, signs, and symbols". *IEEE Transactions on Systems, Man, & Cybernetics*, 15, 234-243.

Reason, J. (1990). *Human error*. Cambridge: Cambridge University Press.

United States Department of Agriculture, Forest Service (2001). *Thirtymile fire investigation report*. Retrieved 22 March 2003 from: http://www.fs.fed.us/rb/wenatchee/fire/Thirtymile-Final-Report.pdf.

Chapter 14

Decision Making by Operational Incident Commanders in a Nuclear Emergency Response Organisation: Decision Strategy Selection

Margaret Crichton, Peter McGeorge and Rhona Flin

Introduction

Operational Incident Commanders (OIC) are members of an industrial Emergency Response Organisation (ERO) established in response to an emergency. Although the potential consequences of an emergency on a high-hazard site are severe, major emergencies in this industry in the UK are rare. Therefore, these on-scene OICs often have little or no direct experience of dealing with emergency incidents, and are seldom tested in a real-life high risk, hazardous incident. The aim of this study was to examine the decision making strategies used by OICs in an emergency situation. The non-technical skills relevant to OICs in nuclear emergency response teams had earlier been identified as being decision making, situation awareness, communication, leadership, teamwork and stress management/personal limitations (Crichton and Flin, 2004).

Naturalistic decision making

The Naturalistic Decision Making (NDM) approach has been used as the framework for this study (Zsambok and Klein, 1997). The role of experience in the decision making process is emphasised in many NDM theories (Drillings and Serfaty, 1997). Although OICs seldom have any actual experience of real incidents on which to build a repertoire of response patterns, this may not prevent the use of a rule-based form of decision making. They have domain knowledge and problem solving strategies stored in memory based on their normal operational role that can be implemented in an emergency. That is, OICs may have developed patterns of response based on their experiences in dealing with challenging, upset conditions, or ERO training opportunities, on which they can draw, particularly in an environment (such as managing an emergency), where dual or multiple tasks act as a stressor.

Different decision strategies (for example, rule-based and knowledge-based) require different levels of cognitive work, and make varying demands on cognitive components such as cue or situation interpretation, problem solving, and option generation (Orasanu and Fischer, 1997):

- Rule-based decision making relies on a prescriptive rule defining a situationally appropriate response (sub-divided into condition-action and go/no go strategies).
- Knowledge-based decision making relies on knowledge and experience (subdivided into choice problems, selection problems, situational management, or creative strategies).

Previous studies examining decision making processes have often used the Critical Decision Method (Hoffman, Crandall and Shadbolt, 1998) to determine decision typology (Calderwood, Crandall and Klein, 1987; Klein, Calderwood and Clinton-Cirocco, 1986). This study focuses on OIC decision making in response to presented problem events that could emerge during an emergency situation to determine the strategy used.

Working memory approach

A connection is emerging between the working memory theory in cognitive psychology and the construct of mental models, as fundamental to NDM (Fiore, Cuevas and Salas, 2003). The memory component of the human information processing system is proposed as comprising long-term memory (LTM) and working memory (WM). Long-term memory is an unlimited depository of both declarative and procedural knowledge (Best, 1992). Baddeley and colleagues (Baddeley and Hitch, 1974; Baddeley, 1990) proposed that WM is a multiple component system, with a limited capacity, involved in the temporary maintenance and manipulation of information. This system (Baddeley, 2001) consists of the central executive, which is a limited capacity attentional controller, the phonological loop, which is concerned with acoustic and verbal information, and the visuo-spatial sketch pad (VSSP), which relates to visual and spatial information.

Performance in a secondary task is assumed to be inversely proportional to the primary task resource demands (Wickens, 1992) if the resource demands of the secondary task are matched to the primary task. That is, a visual/spatial task will cause more interference with the VSSP than a verbal/auditory task. Hence, by creating a higher cognitive load in WM using a matched secondary task, attentional resources to the primary decision making task may be reduced (Wickens and Kessel, 1980).

Under stress, factors such as anxiety and noise can disrupt spatial working memory systems (Stokes and Raby, 1989). However, the retrieval of information from LTM is not disrupted, if that information is well rehearsed and memorised (Wickens et al., 1993). The working memory approach appears to provide a useful paradigm through which to examine OIC decision making under stress.

Based on the working memory approach, rule-based decision making strategies should be readily recalled from LTM, requiring little or no cognitive effort in WM (Stokes and Kite, 1994; Randel, Pugh and Reed, 1996). On the other hand, knowledge-based strategies would create a higher cognitive workload in WM. The introduction of a concurrent second (distractor) task that will compete for cognitive resources should interfere more when the decision maker is using a knowledge-based strategy than when they are using a rule-based strategy. In addition, more experienced personnel should have faster retrieval from LTM, and the distractor task would have less effect on their decision making responses. A proficient decision maker has been defined as being someone with relevant experience or knowledge in the decision's domain who relies on experience directly in making decisions. Experience enhances recognition freeing up cognitive processing space (Lipshitz, 2001).

The aim of this study was to determine the style of decision making used by OICs in emergencies by investigating the responses given, probe question data, and assessed decision quality. The specific hypotheses being examined are:

Hypothesis 1: OIC decision making is predominantly rule-based, which is less affected by the distractor task (dual condition) than knowledge-based strategy decisions.
Hypothesis 2: Rule-based decision making occurs more for typical decisions and those covered by a procedure. High experienced OICs show more use of rule-based strategies than less experienced OICs, especially in the dual task condition.
Hypothesis 3: The quality of decisions, as assessed by Subject Matter Experts (SMEs), is higher in the single task than in dual task, with the quality of typical and procedure decisions being higher than that of atypical and no procedure-based decisions.

Method

Participants

A total of 16 OICs (15 males, one female) from five UK power generation installations participated in the study. Years of experience as OICs ranged from 2 to 15 years (mean = 6.8 yrs; SD = 4.1). Participants were divided into two groups: low experience (mean = 3.5; SD = 1.41) and high experience (mean = 10.13; SD = 3.23). No significant differences in factual knowledge of the OIC role (multiple choice test) emerged between experience groups.

Materials

Decision making task: A total of 20 problem situations were aurally presented by computer. These problem situations were developed from a series of interviews with OICs. A card sorting task identified the two major influences on OIC decision making as being (i) typicality of the decision and (ii) whether the decision was covered by a procedure (Crichton, Flin and McGeorge, 2005). The problem situations (5 x typical/procedure; 5 x typical/no procedure; 5 x atypical/procedure; 5 x atypical/no procedure) were designed as open questions, such that participants had to state their decision rather than reply with a "yes" or "no".

Distractor task: The distractor task consisted of a spatial memory task (Corsi block task) where the participant has to reproduce correctly a sequence tapped on a 9x9 matrix of ceramic buttons in the correct sequential order. A VSSP task was selected as the OIC task involves a high spatial component, that is, location of specific pieces of plant or equipment, routes to be taken by emergency crews (Crichton, 2003). The Corsi block task is widely used to measure the capacity of VSSP memory (Fischer, 2001) and is often used in a dual task paradigm as a means by which to interfere with information in visuo-spatial memory, as remembering and reproducing the block-tapping sequences creates an immediate memory load. VSSP memory span can be measured by increasing the length of the sequences (Fischer, 2001). At most, three trials at any one level were administered, of which participants have to correctly repeat two. The span discontinuance criteria applied was when a participant missed two trials at a particular level, and then no further trials at that level were conducted. Span limit was designated as the level below that which the participant is unsuccessful in reproducing block-tapping sequences (Kessels et al., 2000). Only a complete correctly repeated sequence was scored as correct; self-corrections were permitted.

Probe questions: Probe questions, based on the Critical Decision Method (Hoffman, Crandall and Shadbolt, 1998), were asked following the presentation of each problem situation, that is, cues, goals, available options. Further probes included whether the decision was based on a known procedure or rule, analogous to a previous decision event (actual or training), or novel.

Procedure

At the beginning of each session, each participant's working memory span was ascertained using the spatial memory task. After presentation of each problem situation, participants pressed a key when they were ready to state their solution to the problem situation. Both the time to respond and the actual response were recorded.

Participants practised six trials of the computer-based decision making task, which could be repeated if required. Participants were then randomly assigned to either Group A, who received the single task followed by the dual task, or Group B, who received the dual task followed by the single task. The procedure for the single and dual task is described as follows:

Single task: Participants complete ten decision events on the computer-based task, including responding to the probe questions for each decision event. The participant also rates the difficulty of the decision.

Dual task: Prior to each decision event, the participant observes a spatial sequence displayed on the Corsi blocks equivalent in length to the participant's individual span. They are then required to reproduce this sequence after responding to a decision event on the computer-based task. Probe questions are also completed.

Analysis

The 20 problem situations were examined for differences between the low and high experience groups. Rule-based decisions were defined as those where the participant did not consider any alternative, and specifically referred to a rule/procedure as guidance. Knowledge-based decisions involved option comparison or no rule/procedure applied. Based on Orasanu and Fisher's taxonomy (1997), rule-based decisions were further categorised as Go/No go or Condition-Action; knowledge-based decisions were categorised as Choice, Selection problems, or Situational management. Decisions made were transcribed, and the quality of each decision was assessed by three SMEs. They were requested to read over the responses and to rate the quality of the decision relating to its suitability as an action for dealing with each decision event. The quality of the decision was rated on a 6-point scale: of 1 = very low to 6 = very high.

Results

A total of 320 decisions events (16 participants x 20 decisions) resulted in a total of 62 courses of action being generated (range: 1 to 5 options per decision event; mean = 3.1; SD = 1.07).

Working memory span

Working memory span did not significantly differ between the low and high experience groups (mean = 5.75, SD = 0.68, range 4 to 7). No differences in memory span emerged between the high and low experience groups on single or dual task. A comparison of memory span and span correct in the dual task condition indicated a significant deterioration in spatial memory performance between the single and dual tasks, $t_{(14)} = 7.077$; $p<.001$.

Decision strategy selection

Figure 14.1 indicates the results of decision strategy selection across levels of experience in both the single and dual task conditions. A significantly higher level of rule-based strategies (80 per cent) than knowledge-based (20 per cent) emerged in all categories of decisions (that is, typicality or procedures), $\chi^2_{(3)} = .243$; $p<.001$.

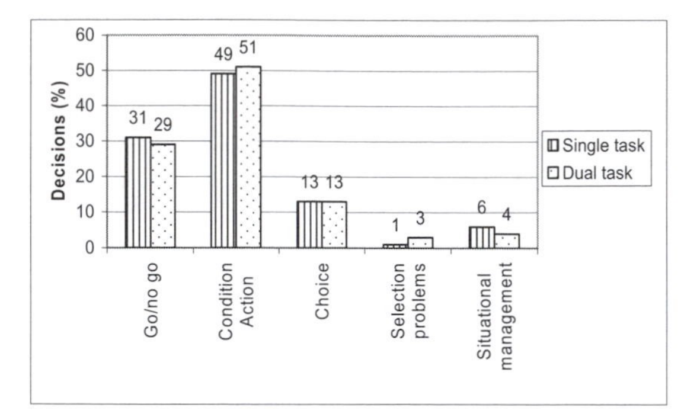

Figure 14.1 Overall decision strategy (single/dual task)

No significant differences emerged in terms of rule-based strategy use by experience group, as, for all categories of decisions (that is, typicality and procedures), both groups of OICs were more likely to use a condition-action strategy than a go/no go strategy. No differences emerged between experience groups for knowledge-based strategy use, although high experience OICs showed a greater tendency to use a situational management strategy for knowledge-based decisions than low experience OICs. A significant association emerged between having had a previous specific decision making episode and rule-based strategies, $\chi^2_{(1)} = 25.643$; $p<.001$.

Decision quality

A median value per response, from all SME ratings, was used in further analyses. A significant difference emerged in overall quality of decisions between the single and dual task conditions ($t_{(14)} = 2.072$; $p<.05$), with the quality of decisions in the single task condition (mean = 4.41; SD = .48) being greater than in the dual task condition (mean = 4.07; SD = .54). Overall decision quality did not differ between the low experience or high experience OICs, nor between types of decisions (typicality and procedures).

Discussion

How do OICs make decisions in emergencies? These results suggest that OIC decision making is predominantly rule-based, especially condition-action rules (that is, *if...then...*). This frequency of rule-based strategy use (80 per cent) was comparable to that found by Carvalho, dos Santos and Vidal (2005) in a study into decision making by shift supervisors in nuclear control rooms. The first hypothesis was only partly supported as no significant differences emerged between the single and dual task conditions. The second hypothesis was not supported as, although the high level of rule-based decision making may have been anticipated for typical or procedure decisions, unexpectedly, a similar strategy was applied for atypical decisions and those not based on procedures. In addition, high and less experienced OICs showed similar levels of rule-based decision making. The suggestion is that the OICs access a generalised knowledge base, developed during normal operational role, from which they can abstract prototypes or response patterns to guide the course of action (Ericsson and Polson, 1988).

The higher cognitive workload condition did interfere with the quality of decisions, which were lower in the dual task condition, indicating that the distractor task affected the decisions that OICs were making, thus supporting the third hypothesis. The implication is that OICs made a satisficing decision (Simon, 1978), that is, selecting the first solution that appears to be reasonably close to satisfactory rather than continuing to search for others, in the higher workload condition, but the quality of these decisions was reduced.

This study also suggests that length of experience in the OIC role does not necessarily equate to increased rule-based decision making. Decision making by OICs in emergencies appeared to be based on prior episodes of a specific decision (identification and recognition of situational cues and access to patterns stored in LTM) combined with domain-specific knowledge and problem solving strategies often developed through normal operational role. Breadth of experience, rather than length, reduced the load on working memory, and decision making was more resilient to distractor effects.

In practical terms, practice in decision making skills, especially specific decision making events (that is, the kind of experience), is critical for effective performance in emergencies. Experience of emergency decision making, that is, schema development, and guided feedback in training, leads to the development and sustainment of available response patterns. The availability of a procedure did not enhance decision making performance. Two possible explanations for this finding are: first, that merely having a written procedure did not assist the decision making process in an emergency; secondly, that the appropriate procedure is not readily recalled and effort is spent searching through memory. Both of these factors can be improved through extensive training of emergency procedures so that they become easily retrieved. Training should also focus on supporting existing decision making strategies rather than trying to train new, more formal strategies (Klein, 1997).

The working memory approach has provided a controlled paradigm through which to study decision making, especially under stress (from increased cognitive workload). The use of a computer-based task has circumvented one of the criticisms of NDM, that is, rigour vs relevance (Lipshitz et al., 2001). Although some contextual features of the phenomena in the real world may not have been fully simulated, OIC decision making was studied in context. Nevertheless, further research is required to examine OIC decision making under stressors such as time pressure, and alternative cognitive workload conditions, such as a verbal distractor task, or with decisions based on uncertainty or ambiguity. These are two further factors that have an impact on OIC decision making but were not examined in this study.

References

Baddeley, A. D. (1990). *Human memory. Theory and practice*. Hove, UK: Lawrence Erlbaum Associates.

Baddeley, A. D. (2001). "Is working memory still working?" *American Psychologist*, 56(11), 851-864.

Baddeley, A. D. and Hitch, G. (1974). "Working memory". In G. A. Bower (ed.), *Recent advances in learning and motivation.* , New York: Academic Press.

Best, J. B. (1992). *Cognitive psychology*. St Paul, MN: West Publishing.

Calderwood, R., Crandall, B. and Klein, G. A. (1987). *Expert and novice fireground command decisions*. Yellow Springs, OH: Klein Associates Inc.

Carvalho, P. V. R., dos Santos, I.L. and Vidal, M.C.R. (2005) "Nuclear power plant shift supervisor's decision making during microincidents", *International Journal of Industrial Ergonomics*, 35, 619-644.

Crichton, M. (2003). *Decision making in a nuclear emergency response organisation: The Access Controller*. Unpublished PhD thesis. Department of Psychology, University of Aberdeen.

Crichton, M. T. and Flin, R., (2004). "Identifying and training non-technical skills of nuclear emergency response teams". *Annals of Nuclear Energy*, 31(12), 1317-1330.

Crichton, M., Flin, R. and McGeorge, P. (2005). "Decision making by on-scene incident commanders in nuclear emergencies". *Cognition, Technology & Work*, 7(3), 156-166.

Drillings, M. and Serfaty, D. (1997). "Naturalistic decision making in command and control". In Zsambok, C. E. and Klein, G. (eds), *Naturalistic decision making*. Mahwah, NJ: Lawrence Erlbaum Associates.

Ericsson, K. A. and Polson, P. G. (1988). "A cognitive analysis of exceptional memory for restaurant orders". In M. Chi, R. Glaser and M. Farr (eds), *The nature of expertise*. Hillsdale, NJ: Lawrence Erlbaum Associates.

Fiore, S. M., Cuevas, H. M. and Salas, E. (2003). "Putting working memory to work: Integrating cognitive science theories with cognitive engineering research". In *Proceedings of the 47th Annual Meeting of the Human Factors Society.* Denver, Co.

Fischer, M. (2001). "Probing spatial working memory with the corsi blocks task". *Brain and Cognition*, 45, 143-154.

Hoffman, R. R., Crandall, B. and Shadbolt, N. (1998). "Use of the critical decision method to elicit expert knowledge: A case study in the methodology of cognitive task analysis". *Human Factors*, 40(2), 254-276.

Kessels, R. P. C., van Zandvoort, M. J. E., Postma, A., Kappelle, L. J. and de Haan, E. H. F., (2000). "The corsi block-tapping task: Standardisation and normative data". *Applied Neuropsychology*, 7(4), 252-258.

Klein, G. (1997) "Developing expertise in decision making". *Thinking and Reasoning*, 3(4), 337-352.

Klein, G., Calderwood, R. and Clinton-Cirocco, A. (1986). "Rapid decision making on the fire ground". In *Proceedings of the 30th Annual Meeting of the Human Factors Society, 1*. San Diego, CA: HFES.

Lipshitz, R. (2001). "Acquisition of proficiency in complex decision making: A knowledge-driven decision making approach". In *Research conference on subjective probability, utility, and decision making*. Amsterdam, NL.

Lipshitz, R., Klein, G., Orasanu, J. and Salas, E. (2001). "Taking stock of NDM". *Journal of Behavioural Decision Making*, 14, 331-352.

Orasanu, J. and Fischer, U. (1997). "Finding decisions in natural environments: The view from the cockpit". In Zsambok, C. E. and Klein, G. (eds), *Naturalistic decision making* (pp. 343-358). Mahwah, NJ: Lawrence Erlbaum Associates.

Randel, J. M., Pugh, H. L. and Reed, S. K. (1996). "Differences in expert and novice situation awareness in naturalistic decision making". *International Journal of Human-Computer Studies*, 45, 579-597.

Simon, H. A. (1978). "Rationality as process and product of thought". *American Economic Association*, 68, 1-16.

Stokes, A. and Kite, K. (eds) (1994). "Flight stress: Stress, fatigue, and performance in aviation". Aldershot, Hants: Avebury.

Stokes, A. and Raby, M. (1989). "Stress and cognitive performance in trainee pilots". In *Proceedings of Human Factors and Ergonomics Society*. Santa Monica, CA.

Wickens, C. D. (1992). *Engineering psychology and human performance* (2nd ed.). New York: Harper Collins.

Wickens, C. D. and Kessel, C. (1980). "The processing resource demands of failure detection in dynamic systems". *Journal of Experimental Psychology: Human Perception and Performance*, 6, 564-577.

Wickens, C., Stokes, A., Barnett, B. and Hyman, F. (1993). "The effects of stress on pilot judgement in a MIDIS simulator". In O. Svenson and J. Maule (eds), *Time pressure and stress in human judgment and decision making*. Cambridge, UK: Cambridge University Press.

Zsambok, C. E. and Klein, G. (eds) (1997). *Naturalistic decision making*. Mahwah, NJ: Lawrence Erlbaum Associates.

Chapter 15

Presentation of Verbal Material: The Impact of Modality on Situation Awareness and Performance on the Flightdeck

Alastair Nicholls, Tracey Milne, Eric Farmer and Anne Melia

Introduction

The type of technological revolution that is now facing military and civilian industry capability is likely to result in a qualitative shift in the manner in which information can be communicated. Coupled with the likely increase in volume of information, new technology may impose strains on the human and affect the ability to exploit information for construction of situation awareness (SA). Investment in sophisticated information systems will be wasted if their use does not maintain or augment the level of SA and performance attained from existing techniques or procedures.

The present chapter is concerned with the consequences of this widening technological opportunity for the manner in which verbal information can be conveyed. Critically, and from the systems' design point of view, the choice of presentation modality should not be determined by the available technology but rather by the impact on human performance. One such issue is faced currently by the aviation industry.

Presently, aircrew are privy to an open radio channel on which instructions from air traffic control (ATC) to all aircraft in the locality are broadcast. Pilots claim that this channel, or *party line*, assists in maintaining their SA of, amongst other things, the location and movements of other aircraft (for example, Midkiff and Hansman, 1993; Pritchett and Hansman, 1993). However, sharing the party line with other aircraft does have disadvantages such as information bottlenecking and errors in comprehension (due to the transient nature of speech). A solution to this problem has been proposed in the form of abandonment of open radio channels in favour of *datalink*: a system in which information, relating only to the specific aircrew to which it is sent, is conveyed in a textual format.

The aviation sector is just one of many domains that face the choice of whether to change the medium in which information is communicated. To date, however, relatively little empirical work has been undertaken into the effect of such a shift from spoken to textual presentation on an operator's SA of the surrounding environment. It is important, therefore, to try to ensure that SA (specifically, spatial awareness) is maintained and supported by any change in the manner that information is presented. The aim of the present research is to provide empirical evidence for the effect (gain or loss) on SA and performance from the use of speech and/or text interfaces.

The present study

The present experiment aims to investigate how a mental *spatial* representation (of aircraft manoeuvres) that requires constant updating can be affected by the modality in which information pertaining to movement is presented. The task faced by participants is designed to resemble the inference pilots make about risk of conflict amongst aircraft when listening actively to their movements transmitted on a party line channel. The design, adapted from that used by Pope, Houghton, Jones, Parmentier and Farmer (2002), involved participants tracking the location of three (non-visible) aircraft with the aim of detecting imminent conflicts (that is, one aircraft moves into a section of airspace already occupied by another aircraft).

Information concerning the movements of aircraft was presented in one of five ways: as speech, in text that disappeared from view after a short delay (on/off); in text that remained on the screen and scrolled up when new reports of movement arrived (scroll), or in two conditions that consisted of a combination of the speech and text conditions. During the trial, participants were not provided with visual feedback as to the location of the aircraft and were asked to maintain an internal representation of the location of each aircraft, as would be the case upon contemporary flightdecks.

It should be noted that, although the present task is set within the context of aircrew/ATC communications, the paradigm has been designed to enable generalisation of findings to any scenario in which maintenance of a (visuo-) spatial representation from verbal information is required. Nonetheless, within the context of the flightdeck the present experiment attempts to evaluate whether the current presentation of the party line over a radio channel can be supplemented by the presence of text in a manner that facilitates SA.

In addition, the inclusion of two text-only conditions (on/off and scroll) will permit investigation of how best to present text information, which may hint at future problems that the implementation of datalink could create. Although datalink has never been suggested as a means by which to present the position of other aircraft in the airspace, the present experiment includes this manipulation in an attempt to generalise the findings.

Method

Design

The factors of presentation mode [speech, text (on/off), text (scroll), combined (on/off), and combined (scroll)] and moves made by aircraft before conflict (two, three, four, six, and seven) were contrasted in a repeated-measures design. The factor of presentation mode was blocked and the order counterbalanced across participants. The factor of number of moves (before conflict) was counterbalanced amongst presentation blocks and participants.

Participants

Thirty participants, all reporting normal hearing and normal or corrected-to-normal vision, undertook the experiment. Participants were employees of QinetiQ and received no direct payment for taking part in the study.

Apparatus and materials

The program, run on a Windows-compatible PC, consisted of three virtual aircraft manoeuvring within a virtual three-dimensional airspace: Aircraft occupied a single volume of airspace at any one time within a four (flight level) by eight (latitude) by eight (longitude) matrix. Aircraft moved one unit of airspace at a time, in either a longitudinal, latitudinal, ascending or descending direction. Each aircraft followed a pre-determined flight path set by the experimenter.

Participants were presented with a display upon which was presented an eight by eight grid and a text window that displayed any textual messages during trials. The grid was intended to act as a reference to the longitudinal and latitudinal positions of aircraft (although no aircraft were represented visually within the grid during trials). Altitude was not represented visually. Figure 15.1 illustrates the visual display presented to participants.

For all conditions, the movement of each aircraft was updated one at a time in the same order and at a rate of one message every eight seconds. Thus, consecutive messages relating to the same aircraft were presented approximately every 24 seconds. All messages consisted of the aircraft's call sign and the direction in which the aircraft was about to move (for example, "Five Zero Whisky, heading North"; "Eleven Sierra, changing to Flight Level 2").

Spoken messages were recorded by three different speakers; a different speaker associated with each aircraft. The spoken duration of each message was approximately four seconds. New text messages were presented in the bottom of the text window, adjacent and to the right of the grid, in a bold, size 8, Tahoma font. In the text (on/off) condition, each new message was presented for four seconds at the bottom of the text window before being removed. For the text (scroll) condition, messages were presented as in the text (on/off) condition but were displaced up the text window by any subsequent message. The text window could accommodate up to six (old) messages. Combined conditions used the same text stimuli but coupled presentation with the same message in speech.

Within each condition, participants were required to identify when a conflict scenario occurred. A conflict scenario was defined as the moment that one aircraft moved into the same volume of space occupied by another aircraft. A conflict could occur after two, three, four, six or seven moves (per aircraft) from start locations.

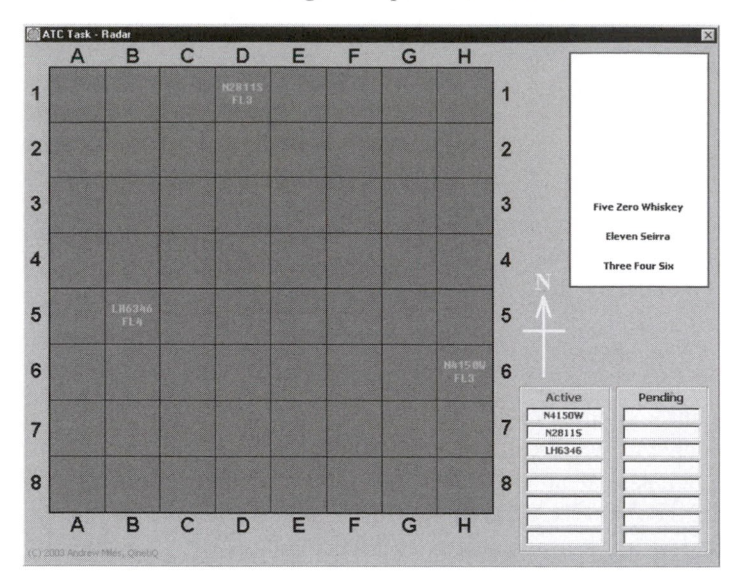

Figure 15.1 Visual display presented to participants at the beginning of trial
Note: Starting positions of each aircraft are displayed within the grid. When present during a trial, text information was displayed in the text box (positioned to the top right of the figure). The visual display was adapted from Pope et al. (2002).

Procedure

Participants were tested individually in a quiet room and sat approximately 50cm from the display screen. Participants wore headphones throughout the experiment that were set to a comfortable level, corresponding approximately to 60 dB.

At the start of each trial, the starting locations of the three aircraft were depicted visually and simultaneously for five seconds within three separate cells of a visible eight by eight grid (as shown in Figure 15.1). Aircraft were represented by their corresponding call sign (for example, "Five Zero Whisky"), which was suffixed with the aircraft's starting altitude (for example, "FL2": flight level two). Once the trial commenced all aircraft disappeared from the grid and remained out of sight for the remainder of the trial. The direction of each aircraft was updated one at a time and in the same order. Participants were instructed that they were not allowed to use their hands or the mouse to aid them in tracking the aircraft.

Participants were asked to track aircraft movements until confident that a conflict was present. A conflict was defined as the occupancy of one volume of space by two aircraft. When a conflict occurred, participants were asked to indicate as rapidly as possible with the mouse the grid square in which the conflict had occurred. On clicking on a grid square, participants received a prompt to indicate the flight level at which the conflict had occurred. After each presentation block, participants completed a 14-question Situation Awareness Rating Technique (SART) questionnaire (for example, Taylor, 1990).

Results

Performance was assessed with regard to conflict detection accuracy. A response was considered correct only if the participant identified correctly the precise three-dimensional co-ordinate of the conflict. Ratings of SA (Total or *T* Scores) under each presentation condition were also assessed.

Conflict detection accuracy

Figure 15.2 illustrates average conflict detection accuracy as a function of presentation condition. Accuracy was markedly lower in the text (scroll) condition than in all other conditions. Interestingly, scrolling presentation did not affect accuracy deleteriously in the combined (scroll) condition and, numerically, even appeared to improve accuracy in relation to the combined (on/off) condition. Performance in the speech condition was not substantially different from that in either of the combined conditions.

For the purpose of formal analysis, the proportion of conflicts detected was analysed within a five [presentation condition: speech; text (on/off); text (scroll); combined (on/off); and combined (scroll)] by five (moves prior to conflict: 2; 3; 4; 6; and 7) repeated-measures analysis of variance (ANOVA). Here and elsewhere adjusted F-values were adopted when sphericity was not assumed. There were significant main effects of presentation condition, $F(4, 116) = 3.8, p<0.01$, and move, $F(4, 116) = 19.6, p<0.001$. Planned comparisons amongst presentation conditions demonstrated that accuracy under the text (scroll) was significantly worse than all other conditions (all $p<0.01$). No other comparisons reached significance (all $F<1$). The main effect of move is attributable to the general decline in detection accuracy as the number of moves to conflict increased. No significant interaction was present between the two factors, $F(16, 464) = 1.2, p = 0.26$.

Figure 15.2 suggests strongly that the impact of text mode (on/off vs. scroll) varies depending on whether the text information is accompanied by speech. To investigate this effect, a further analysis, from which the speech only condition was excluded, was undertaken. A two (text mode: on/off; scroll) by two (modality: text only; combined) repeated-measures ANOVA on accuracy scores (averaged across move) demonstrated a significant main effect of modality, $F(1, 29) = 15.8, p<0.001$ but no main effect of text mode $F(1, 29) = 1.5, p = 0.23$. The significance of modality highlights that, on average, the same information presented in two modalities confers an advantage relative to text alone.

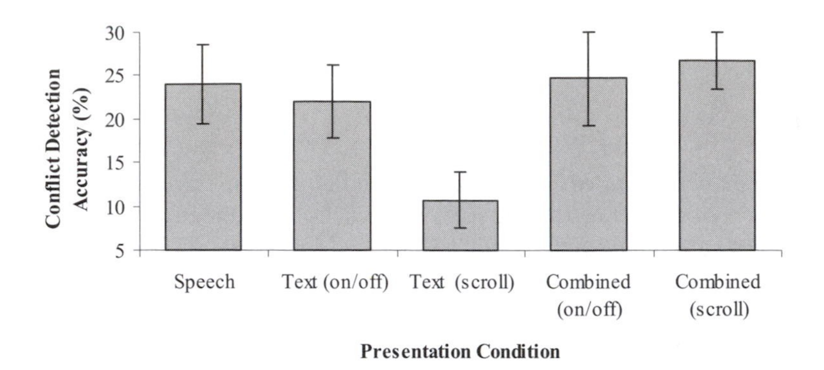

Figure 15.2 Detection accuracy as a function of presentation condition. Standard error bars shown

Importantly, there was a significant interaction between the two factors, $F(1, 29) = 4.8, p<0.05$. Clearly, the main locus of the interaction lies with the detrimental effect of scrolling text on accuracy in the absence of concurrent speech compared to text (on/off).

There is also an indication (albeit non-significant) that this trend is reversed in combined conditions such that conflicts are detected more faithfully when scrolling text, rather than text (on/off) is combined with speech.

Situation awareness

A one-way repeated-measures ANOVA on *T* scores revealed a non-significant main effect of presentation condition, $F(4, 116) = 1.48, p = 0.23$. Although not significant, differences between *T* scores (Table 15.1) demonstrated a similar trend amongst conditions to that shown amongst detection accuracy scores.

It should be noted that, in all conditions, participants rated their SA well below the maximum *T* score of 200. Although this suggests that the task, independent of presentation modality, was relatively difficult, it further implies that verbal information does not support the construction of complex spatial representations.

General discussion

The important finding from this study is that, on average, a combination of speech and text is preferable to text alone. Combining speech and text, however, does not seem to confer much more advantage than speech alone.

There is a trend for scrolling text in combination with speech to show the most accurate performance. Theoretically, this would suggest that during times of high input demand (message presentation), participants relied heavily on the speech signal to encode new information but, in scrolling text, were provided with a permanence not permitted with the speech signal, and a chance to cross-reference moves over time.

Table 15.1 Mean *T* score by presentation condition. Standard error (SE) shown

Condition	T Score	SE
Speech	48.6	7.7
Text (on/off)	45.6	6.6
Text (scroll)	41.8	8.1
Combined (on/off)	52.4	7.2
Combined (scroll)	47.0	7.3

However, the results suggest that the present combination of scrolling text and speech, despite showing superior performance to other conditions, is more vulnerable to disruption than the speech and text (on/off) condition. Indeed, failure of the speech signal in a dual-modality system would drastically impair performance if participants had to rely on scrolling text alone. Results suggest that a dual-modality system implementing a text (on/off) function would not demonstrate the same degree of susceptibility.

As a final point, it should be remembered that, although the present task is set within the context of aircrew/ATC communications, the findings should generalise equally to any scenario in which maintenance of a (visuo-) spatial representation is required from verbal updates.

Acknowledgements

The present work was funded by the Human Sciences Domain of the United Kingdom Ministry of Defence Scientific Research Programme. Thanks are due to Mark Channer and Natalie Rodriguez for their assistance in data collection. Special thanks are due to Dr Fabrice Parmentier for his role in software development.

References

Midkiff, A. H. and Hansman, R. J. (1993). *Identification of important "party line" information for situation awareness in the datalink environment.* SAE Technical Paper 922023. *Aerotech '92*, Anaheim, California, 5-8 October.

Pope, D., Houghton, R. J., Jones, D. M., Parmentier, F. and Farmer, E. W. (2002). *Report on Work Package 1: Possible Usefulness of Irrelevant Sound.* Care Innovative Action: Cognitive Streaming Project. CARE-IA-CS-CFU-WP1-D1-01-1.2.

Pritchett, A. and Hansman, R. J. (1993). "Preliminary analysis of pilot rankings of 'party line' information importance". In R. S. Jensen and D. Neumeister (eds), *Proceedings of the Seventh International Symposium on Aviation Psychology.* Columbus, OH: Department of Aviation, The Ohio State University.

Taylor, R. M. (1990). "Situational awareness rating technique (SART): The development of a tool for aircrew systems design". In *Situational Awareness in Aerospace Operations (AGARD-CP-478)*, pp. 3/1-3/17). Neuilly Sur Seine, France: NATO – AGARD.

Chapter 16

Air Traffic Controller Strategies in Holding Scenarios

Lynn Springall

Introduction

Air traffic in the UK is projected to grow at a rate of about 5 per cent per year. There is an increasing need to hold aircraft under Air Traffic Control (ATC) instruction in holding stacks, when the air traffic flow into airspace exceeds runway capacity. Currently air traffic controllers can experience difficulties with Secondary Surveillance Radar (SSR) when identifying aircraft in and around holding patterns on radar display. Controllers are not allowed to use the radar data around stacks when there is label overlap. In these situations it is necessary for controllers to use a procedural method of control obtaining verbal level reports from aircraft, which is more time-consuming.

Advances in datalink technology will provide controllers sequencing aircraft in and around the London Heathrow holding stacks with new information directly downlinked from aircraft avionics. A new tool has been designed to overcome the legibility problems of SSR displays. The altitude or flight level selected by the pilot will be displayed in the new tool.

Seamster, Redding and Kaempf (1997) highlighted the importance of identifying the controllers' decision making strategies and tactics when designing decision supports or Human Machine Interface (HMI) in order to understand how the controller as the decision maker performs the task to be supported. The tactics that the controllers use must be understood within the context of their goals and the decisions and judgements they make in order to accomplish these goals. The controller considers a range of factors whilst sequencing a traffic stream and their decision making strategies are expected to differ according to the level of complexity of traffic flows. The traffic streams around London Heathrow vary due to the different geographical layout of airways feeding traffic into the different holds. These traffic streams have conflicting and non-conflicting traffic flows, which the controller will sequence in different ways depending on the complexity of the situation.

Zsambok (1997) discussed "naturalistic" decision making processes and strategies, which differ from the traditional "rational" theories of decision making, such as multi-attribute utility analysis. Klein (1997) described the classical decision making methods as generating a set of options, assigning a weight or value to each combination and then selecting the combination with the highest value, in order to help people make better decisions. Klein reasoned that experienced decision makers in real world settings in complex, dynamic, high-risk environments adopt a "naturalistic" approach to decision making when under time pressure with shifting goals rather than the "traditional" methods. Naturalistic decision making theory proposes that experts use their previous knowledge to make decisions in a "naturalistic" way, by sizing up a situation quickly and refreshing their awareness of the situation through feedback in order to generate reasonable courses of action, rather than developing multiple options which are then compared. Each naturalistic decision made is specific to the context that it is made in.

Miller and Woods (1997) discussed the application of naturalistic decision making theory to system design. In their paper they acknowledged that adding new automation in the form of tools would affect the cognitive activities, strategies and tactics operators employ in order to perform the task. Understanding these changes allows the system designer to model error and expertise for the task affected.

Seamster et al. (1997) found that air traffic controllers are not "rational" decision makers. To sequence aircraft, controllers apply previous experience to identify patterns of relevant cues and sources of information. Controllers are also searching for any indications of problems or difficulties which are commonly encountered, to establish a course of action, and employ tactics to accomplish each goal in order to implement shortcuts or heuristics for frequently occurring tasks. The aim of this study was to understand the decision making strategies controllers use when sequencing traffic and make a comparison between simple or complex traffic flows in order to identify any differences in strategies for sequencing non-conflicting and conflicting traffic flows.

Method

Critical decision method (CDM)

In this study the Critical Decision Method (CDM) was applied. CDM is a widely used method of Cognitive Task Analysis (CTA). The CDM uses a semi-structured interview technique, which was derived from the critical incident technique, which used focused probes to elicit information from domain experts in emergencies. Seamster et al. (1997) explained that the CDM has been applied in complex and dynamic environments to provide information on higher cognitive tasks of experienced operators such as decision making. The CDM has been applied previously in ATC applications by Seamster et al. to develop a cognitive model of ATC decision making for a sequencing problem. Controllers viewed simulated scenarios to focus on previous experience and were interviewed to elicit their decision making strategies, cues, perceptual discriminations, pattern recognition and heuristics. Scenarios were devised with pre-recorded real traffic to simulate a dynamic operational setting to generate the cognitive processes of the controllers whilst the evaluator conducted a semi-structured cognitive interview. Two different traffic scenarios were constructed to represent different complexities of traffic flows. In this study sequences of aircraft were presented to controllers in two scenarios, one simple scenario and one complex scenario. Aircraft were presented to the controller in random vertical level order in both scenarios.

Study design

Twenty air traffic controllers with a mean of 13.9 years' operational experience in Terminal or Approach control, aged between 28 and 58, with a mean of 40.7 years, participated in this study. The controllers were randomly split into two equal groups, A and B. Group A viewed a scenario with a simple traffic flow and group B viewed a complex traffic scenario. Each controller viewed one scenario on a standard PC with a 17-inch monitor, which represented a radar display. Medium to high traffic samples were presented in the REVIEW software package with Track Data Block (TDB) information on the display which showed the position of the aircraft, the identity of the aircraft and the aircraft altitude. Accurate airspace maps were shown on the display and flight progress strips provided information on the aircraft in the scenario. The scenarios were scripted and constructed, to allow the traffic sample to be fast-forwarded and frozen at certain points to stimulate the controllers' memory and enable probe questions to be posed.

The probes were designed to investigate the decision options and goal alternatives the controller set when sequencing aircraft into the hold. In addition, probes were generated to find the critical cues controllers used, any missing data, patterns recognised or heuristics applied when they were sequencing aircraft. The probe questions used are shown in Table 16.1. The controllers were briefed before watching the scenario and the semi-structured interviews were recorded and transcribed.

Table 16.1 Critical Decision Making probe questions used in study

Type of probe	Question description
Decision Option	What will be your order of traffic into the hold?
Goal Alternatives	What different sequences could you have set?
Critical Cues	What cues/information are you looking at to make your decision? Could you prioritise which piece of information is most important to you when making these decisions?
Hypothetical	What would be your order if the flight level of aircraft A were significantly higher?
Pattern Recognition	Are you looking for particular patterns of traffic?
Heuristics	Are you using any shortcuts or rules of thumb?

Results

The controllers' responses to the probe questions in Table 16.1 were compared between group A (simple traffic) and group B (complex traffic).

Decision options – orders of aircraft in scenarios

Controllers within group A and group B set three different orders of traffic. Group A and group B controllers did not use the same information to sequence traffic. A majority 60 per cent of group B controllers, viewing the complex scenario, sequenced aircraft in a vertical manner according to actual flight level, whereas a minority 10 per cent of group A controllers sequenced aircraft in a vertical order.

Goal alternatives – different traffic sequences

The majority of controllers, 70 per cent in group A and 60 per cent in group B, displayed a degree of flexibility when ordering traffic and changed the order of two aircraft in the sequence.

Critical cues – prioritisation of information

Table 16.2 shows the cues identified by controllers in group A and group B, displayed in percentages. The most important cues are named priority cues; all other cues are non-priority. For example, 70 per cent of controllers in group A identified track miles as the most important cue (priority) and 100 per cent of controllers in group A identified track miles generally as a cue (non-priority). The standing agreement level refers to the flight level at which, according to procedures, the aircraft is agreed to enter the sector. Individual controllers in group A identified other cues for decision making: airspace restrictions; aircraft callsign to indicate aircraft type; and colour of flight progress strip holder to indicate wake vortex category of aircraft.

Hypothetical – change flight level of aircraft

The majority of controllers, 70 per cent of group A and 80 per cent of group B, redefined their order of traffic when told that one aircraft was higher and out of flight level order. The controllers delayed the higher aircraft until later in the traffic sequence.

Pattern recognition

A majority of controllers, 80 per cent of group A and 60 per cent of group B, identified patterns of traffic. In group A, 50 per cent of controllers identified visual patterns of traffic related to the geographical plan position of aircraft. Controllers in group A looked for groups of traffic, pairs of parallel aircraft or traffic in line.

Parallel aircraft were rated more difficult to control than traffic in line. In group B, 30 per cent of controllers discussed visual patterns of traffic they were looking for on the radar display, such as different streams of aircraft following airways, which were then classified as easy to control or more difficult according to the amount of vectoring required. When these group B controllers had identified the aircraft in one of the aircraft streams, the aircraft was given an order in the sequence according to the level at which the next sector would be expecting the aircraft, called the standing agreement level, to avoid the need to vector the aircraft.

Table 16.2 Priority and non-priority cues noted by controllers in group A and group B

Cues	Priority cue		Non-priority cue	
Group	**A**	**B**	**A**	**B**
Track miles	70	50	100	80
Actual flight level (Mode C)	0	0	30	60
Track miles and actual flight level	0	30	30	30
Aircraft speed	20	0	60	0
Aircraft type	0	10	40	10
Other traffic	0	0	20	0
Standing agreement level	0	10	0	20
Aircraft going to hold	10	0	10	0

In group A, 40 per cent of controllers, and in group B, 10 per cent of controllers identified non-visual patterns of traffic by aircraft callsign, which indicated the aircraft type and relative performance. The controllers used this information to ensure that no overtake situations would evolve. A minority 20 per cent of group B put aircraft into groups according to aircraft type and wake vortex category in order to sequence.

Heuristics – rules of thumb

A majority 70 per cent of controllers in both groups discussed heuristics they applied when sequencing traffic. Controllers used different rules in each scenario; 40 per cent of group A differentiated traffic according to the plan position of the aircraft on the display. The traffic on the east side of the airway was prioritised with less track miles to run. Other controllers in group A prioritised aircraft by flight level and track miles. In group B, 40 per cent of controllers assessed if an aircraft was going to hold, and sequenced according to the standing agreement level for the aircraft. A minority of group B controllers used wake vortex category rules to decide an order.

Discussion

The results showed that controllers used different decision options, goal alternatives, cues, priority of information, heuristics and traffic patterns to sequence aircraft in simple or complex traffic flows. Within group A and group B, controllers had three different decision options for aircraft order; this showed that the goals set by these controllers were different. A majority 60 per cent of controllers used a vertical sequence in the complex scenario whereas 90 per cent of controllers in the simple scenario did not use a vertical sequence. Approximately two-thirds of controllers in both groups were able to set alternative goals and were flexible in their ordering of traffic for normal operations. About three-quarters of the controllers in both groups defined a different traffic order according to the hypothetically raised flight level of one aircraft.

A majority of controllers in both groups used track miles as a cue; in the simple traffic flow this was a priority cue for a majority 70 per cent of controllers. In the complex scenario the actual flight level (Mode C), standing agreement level and aircraft type were also priority cues. Controllers in group A considered aircraft speed and whether the aircraft was going to hold as non-priority cues.

Approximately three-quarters of controllers in both groups identified patterns of aircraft when sequencing traffic. The patterns or pairs of aircraft were grouped according to the geographical plan position, aircraft level, aircraft type and performance. The controllers then rated the traffic groupings according to the difficulty in controlling, such as amount of vectoring required. In order to simplify decision making, the controllers applied heuristics to the different groupings of aircraft.

The difference in pattern matching highlighted one of the main differences between the sequencing of simple or complex traffic flows. The controllers sequencing the simple traffic flow mainly ordered the traffic according to patterns of traffic in a plan view. The controllers viewing the complex traffic flow considered the flight level of the aircraft more important and sequenced traffic according to the vertical view of the aircraft. Overall, controllers sequenced a simple traffic flow according to a plan view of traffic and considered the vertical position of aircraft more in a complex traffic flow.

Conclusions

The results from this study provide insight into the strategies which air traffic controllers use when decision making. The goals, tactics and cues that the controllers consider most important are outlined. In addition, this study showed that controllers are searching for patterns of aircraft and then apply heuristics, in order to simplify decision making when sequencing aircraft. These findings can be used in future system design to assess how the controllers' sequencing task in the complex and dynamic terminal control environment may be affected by the development of new tools displaying datalink information for different traffic flows. Further work based on these findings could predict where the tool would support the controllers' decision making strategies when sequencing different traffic flows and any potential error modes.

Acknowledgements

I would like to thank Dr Don Harris at Cranfield University for his supervision, and Jenny Weston and Laura Voller at NATS for their help in writing this chapter.

References

Klein, G. (1997). "An overview of naturalistic decision making applications". In C. Zsambok and G. Klein (eds), *Naturalistic decision making* (pp. 49-60). Mahwah, NJ: Erlbaum.

Miller, T. and Woods, D. (1997). "Key issues for naturalistic decision making researchers in system design". In C. Zsambok and G. Klein (eds), *Naturalistic decision making* (pp. 141-150). Mahwah, NJ: Erlbaum.

Seamster, T., Redding, R. and Kaempf, G. (1997). *Applied cognitive task analysis in aviation.* Aldershot: Ashgate.

Zsambok, C. (1997). "Naturalistic decision making: Where are we now?" In C. Zsambok and G. Klein (eds), *Naturalistic decision making* (pp. 3-16). Mahwah, NJ: Erlbaum.

Chapter 17

How Roles Change when Disaster Strikes: Lessons Learnt from the Manufacturing Domain

Carys Siemieniuch and Murray Sinclair

Introduction

The "High Reliability Organisation" (HRO) is a term that has become ever more popular over the last two decades, and refers to organisations which have shown very high levels of effectiveness in the avoidance of disasters, the recovery of errors, and in spreading learning from these among the organisation's personnel. Oft-quoted examples are usually in the defence domain, such as the US nuclear aircraft carrier fleets, submarines, and the nuclear industry as a whole. Certainly, there are incidents, but the response to them is immediate, effective, and often results in organisational amendments to address the root causes of these incidents. As Reason (2001) has argued, there are two perspectives by which to view humans in organisations; the "human as hazard", in which people are seen as sources of variability, errors, and unexpected events upsetting the smooth running of a given process, and the "human as hero", in which people are seen as protectors of the system, interceding between a changing and unpredictable environment in order to maintain system integrity and effectiveness. The systemic differences between these two perspectives are largely ones of boundaries, and in a given organisation both can co-exist; the behavioural and cultural differences, however, are enormous. Needless to say, the HRO perspective tends towards the latter view of humanity; the former view is incorporated by recognising that it is human nature to act in ways that can be detrimental ("errors"), but that human nature has also equipped us with extremely efficient and effective error-retrieving capabilities too. This applies, so long as people have time to think, time and the means to retrieve errors, and time to learn from these occurrences. Characteristics of HROs include the following:

- An organisational culture of reliability
- Commitment to, and effectiveness in, continuous learning and dissemination of good practice
- Provision of resources and full support for communication, over extended periods
- Transparent human resource management practices that support the goals of individuals, as well as organisational goals

- Strong support for the maintenance of the organisation's technology
- Adaptable decision making and problem ownership (Weick, Sutcliffe and Obstfeld, 1999)
- Flexible organisational structures that allow *ad hoc* groupings to solve problems
- Recognition of the importance of "slack" (Lawson, 2001)
- Leadership that continuously emphasises safety and reliability

In the sequel, we discuss a "disaster" that happened in an organisation, and compare the events to these requirements.

The scenario

During a team-working study in a steel rolling mill, a "disaster" happened; a hot, flat plate became trapped under a set of rollers. It took three days to restore production, at a reputed cost of £1 million per day.

The purpose of the team-working study was to predict likely outcomes of a new staffing plan for the steel mill, and this real-life incident provided useful information and insight for the exploration and evaluation of this staffing plan. We outline some of the issues that emerged from this analysis.

Steel mills provide an excellent example of a complex, heavily automated socio-technical system, with emergent properties. Because of the high operational cost of operating a steel mill, allied to severe global competition in the market, there is a strong need for reliability of production to make any profit at all. The mill in which the study took place extended over some 400 metres, in two sections. The rear section was the cool end, where coils of steel plate were sorted and dispatched to customers. This was not part of the study. The front end, the "Hot Mill", was the focus of the study, where steel slabs were selected from storage according to customer demand, reheated, rolled to near-specification, partly-cooled and rolled to specification, and then cooled again, coiled and banded. This is a continuous process where the properties of the product are strongly influenced by the cooling and rolling processes.

This system was designed with cost-efficiency and agility in mind, and from a strongly engineering perspective. The result was a line controlled by computers having two shop floor managers and six shop floor operators, with supervisory control distributed over 300 metres of the hot mill. Additionally within each 12-hour shift there were some 30 support people (electricians, crane drivers, and so on). Production was intended to be 24 hours, seven days a week, utilising a total of 21 shifts. In addition, there was a "Day Support Team" and two senior managers working normal hours each day, essentially acting as the interface between the shift teams, the company and the outside world.

The cultural assumption and managing ethos of the company was that "normal" performance would be the standard state, with a few perturbations from time to time. Normal performance was visualised as full production with the automation being in control moment by moment, reflecting the organisation's deep understanding of steel making, with the operators adjusting parameters from time to time as required by contingent circumstances, and with planned maintenance of the line. This was not the case in reality as the staffing indicates: frequent human interactions and interventions were necessary for "normal" performance. It should be noted that this culture departs somewhat from that of the HRO; in this, shop floor personnel are seen as acolytes to the automation, there to maintain conditions necessary for the automation to control production, and to deal with untoward events outside the scope of the automation. In effect, they are handy bolt-ons to the process.

The teams in the study

The current arrangement was that the hot mill shop floor crew for each shift comprised one single team, divided loosely into operators and support crew. The plan was to create two semi-autonomous teams in the hot mill in each shift, which would attend to their coherent area and be self-organising. The hot mill would be divided into a "hot" team, who select the slab, heat it in the reheat furnace, and roll it to near-specification. The "finishing" team would take the rolled plate, finish-roll it to specification, coil and band it. Support roles, for example, electricians and so on, would be allocated to one or other of these two teams. This represented a significant change to the current situation of one big team, hierarchically controlled, with paternalistic management and a very strong unionised workforce. However, because business conditions in this industry have been poor for so long, unions and the management had to learn to work together for their mutual salvation.

As this plan was being put into effect, it became possible to explore the implications of the planned structures. During the period in which the exploration took place, the "disaster" happened. When a steel plate is undergoing finishing, it passes once through a set of six rollers, with a guillotine machine at the beginning of the set to trim the plate. By the time the plate reaches the finishing rollers, it is about 70 metres long, weighs 20 tonnes, is at a temperature of about 900°C and is travelling at 50-70 km per hour. For some reason, the leading edge of one particular plate stopped in the middle of the rollers, causing the rest of the plate to buckle in the rollers, and the trailing edge to wrap around the guillotine. Fortunately, no personnel were near the rollers at the time.

Fixing the disaster

Immediately, management was informed and the line was cleared of all other product. Over the next day, the rollers and guillotine machine were dismantled, the solidified plate cut away, replacement rolls were put into the rollers while the old set were re-machined off-line, and the line was reassembled. At the same time, the software was re-analysed to discover if a bug was the cause. By the third day it was possible to restart production.

Observed organisational changes

While senior management may well be dealing with the external interfaces, their authority (and presence) on the shop floor is characterised by remoteness. This is exemplified in Figure 17.1, which plots the lines of communication in the mill. This shows the personnel in any shift, the proposed structure of two teams, the flow of product (solid black line, left to right across middle of page), and the operational channels of communication (grey lines) under "normal" conditions. Viewing this with half-closed eyes shows a top triangle of communication involving senior management and the Shift Manager, and a bottom triangle along the flow of product and including the Shift Manager and Floor Roller, the two supervisors of actual production. What is also noticeable is the exclusion of the shop floor support staff from operational information.

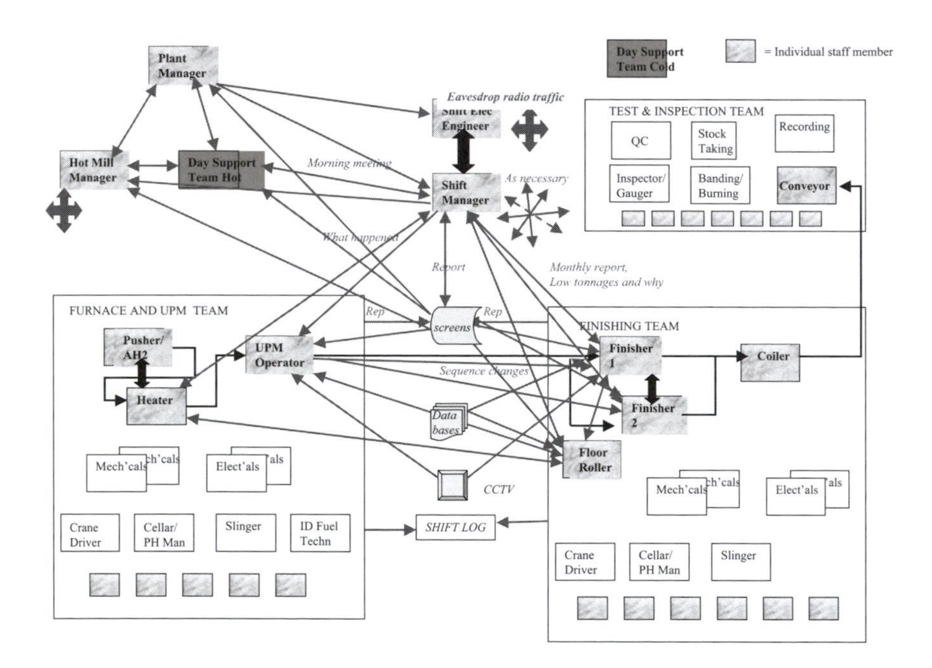

Figure 17.1 Diagram showing lines of communication for "normal operating conditions"

Note: The flow of product is left to right across the middle of the diagram. All the other arrows show well-established lines of communication to control and co-ordinate the flow of product. The Shift Manager (top centre) links the senior management communications triangle to the shop floor communications triangle; computer screens are the only other online communications medium between management and operators. Support staff are not included in these communication patterns.

Figure 17.2 shows the original organisational set-up, under "normal" production conditions, plotted on a "Role Matrix", showing relationships between roles. The axes of the matrix are "Degree of discretion in achieving target and executing operations", and "Degree of responsibility in planning resourcing and scheduling operations to achieve targets"; these form a bi-dimensional map of decision authority and, by implication, official status. Hence, top right is the acme of power, and bottom left is where nobody should be. The blobs are roles, and arrows are known, well-used communication paths connecting the roles, by which authority is translated into activity. PM is the Plant Manager, top right; FR on the right is the Floor Roller, supervising the operation of the mill on the shop floor. Between these two are management and engineering roles; other shop floor operators are all grouped towards the bottom left.

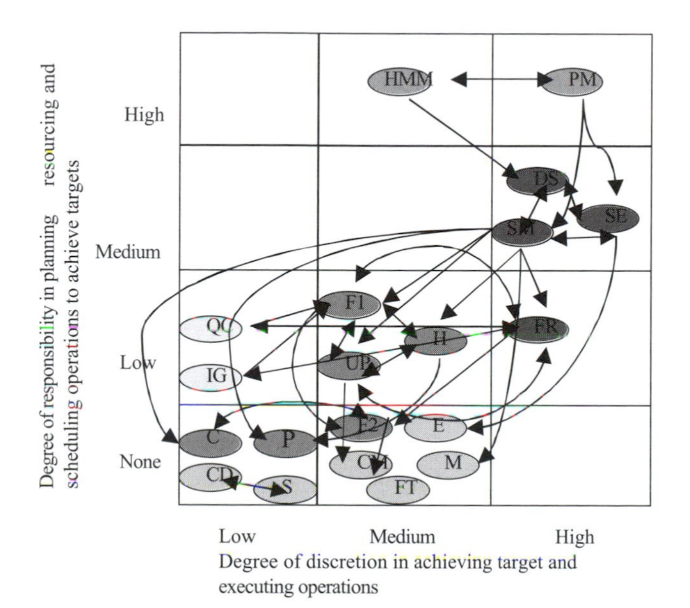

Figure 17.2 A Role Matrix, showing more detailed communication patterns between shift personnel, and capturing authority relationships between the roles in "normal" operations

A re-plot of the Role Matrix for the "disaster retrieval" scenario is in Figure 17.3. Support staff, particularly fitters, have their roles enhanced (that is, their blobs have moved upwards and to the right), markedly changing their relationships with other roles compared to "normal" operations; the senior management are now in less authoritative roles, as far as the mill is concerned, though it is likely they were very busy in negotiations outside the mill operational boundaries.

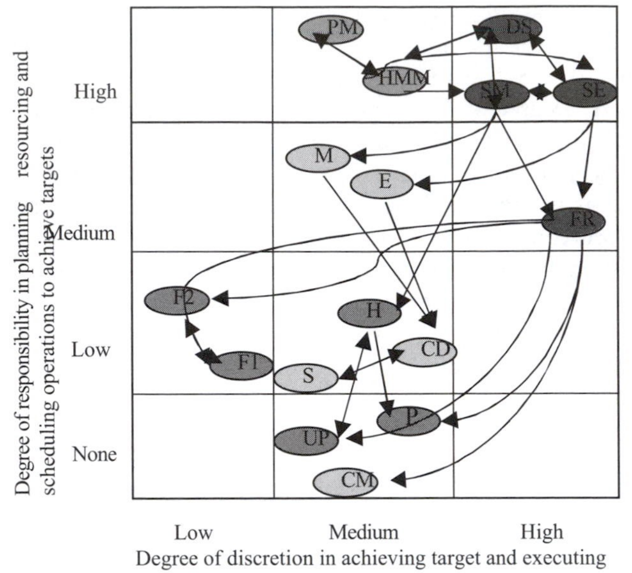

Figure 17.3 A Role Matrix for "disaster retrieval", showing how the authority relationships have changed

Note: The lines of communication remain the same; it is noticeable how senior management have been shifted sideways, and other roles have moved into far more authoritative positions.

Organisational implications

These are discussed from two aspects; first, in relation to the needs of the company, and secondly, in relation to the tenets of HROs.

- Under "normal" operations, the line is still dependent on its older, pre-automation staff for smooth running. But their "process-tweaking" skills are being lost due to retirements and absence of use with increasing automation. On the other hand, their understanding of the actual behaviour of the automation (as opposed to its designed behaviour) is increasing. However, there is a continuous introduction of new software into the mill, with two outcomes: first, the changes mean that new behaviour is continually appearing, vitiating some of the understanding that the shop floor operators have gained; and secondly, the unwritten rule is that every time a significant introduction of software is made, out of every four operators whose jobs are affected by the introduction, three are redeployed and one remains *in situ*. Net, this represents a departure of intelligence from the shop floor, and some loss of knowledge. In HRO terms, it is the latter which is the more serious; the knowledge that vanishes is tacit knowledge, and it is tacit knowledge which addresses the interface between the changing environment and its unexpected events, and the needs of the automation.

- These older workers are also very important in disaster scenarios; their tacit knowledge of the behaviour of the line, gained from manual control days, is very helpful in minimising down time and for foreseeing problems. Repeating the point above, the loss of these operators could mean an overall loss of recovery time and reliability.
- Furthermore, the group cohesion built over a long period of time, with the experience gained of each other under different operating scenarios, indicates that the *ad hoc* role changes that occur for different scenarios of operation are not likely to cause undue friction and difficulties. This is one of the basic characteristics of HROs, and needs to be recognised in the design of roles and teams.
- Their understanding is related to the behaviour of the whole line; but new teamworking plans envisaged placing a boundary across the line, before the finishing rollers. Either side of this, the teams would have their own performance metrics, goals and (eventually) loyalties. In effect, an artificial, organisational fence could be created, splitting in two a continuous process whose main operational communication flows run across the fence (see Figure 17.1). It is not obvious how this will assist normal operations, and may interfere with the current distribution of organisational knowledge and decision making structures that evidently swing into place when disaster strikes.
- The existing divisions between management and the shop floor, as indicated by the pattern of communication links, are unlikely to change, given the plans as enunciated. Decision paths and structures are not reflected in communication paths and new team boundaries and roles are likely to increase this dichotomy at the expense of response times and disaster avoidance. High power distances (Hofstede, 1991) between the two groups will remain. This is likely to hinder performance from the HRO perspective.
- The new structure will impair situation awareness (Artman and Garbis, 1998; Endsley, 1998) since it fails to address the need for coherence of communication flows and tacit knowledge sharing (for example, reports may still be "massaged" and many key shop floor roles remain the passive recipients of information with little opportunity to provide real input to the decision making process). This describes a culture alien to the HRO tenets, with the implication that responding to disaster efficiently will depend largely on the older workers and the unofficial shop floor culture that has been in existence. Any damage to this from the proposed team structure may have unfortunate effects in the future.
- The intended blurring of roles so that team members can take on more than one role will allow more flexible operation within each team. However, this benefit will depend heavily on management evangelism and trust, allied to

quality training over an extended period. Furthermore, such benefits will only be garnered within teams rather than across teams with possible negative effects on tacit knowledge transfer for the process as a whole.

- Both shop floor groups, the older workers and the younger workers, expressed concern about the overall loss of skills as the older workers left the business. They both commented that while the steadily increasing level of automation meant that better quality, more consistent steel could be produced, they felt ever less in control of it. The younger workers were definite in saying that, in the event of a breakdown of the automation, they could not as a team operate the line manually; the older workers were worried that those still in work were losing their skills due to lack of use. Many of them said that they no longer trusted themselves to operate the line manually.
- Given the role changes that are indicated above, requiring wider knowledge and authority than under normal operations, it seems evident that simulation of the line in a variety of decision making scenarios would be a very necessary skills-building tool. Given the level of knowledge as evinced by the automation, it should not be hard to develop a simulation model of the line on which skills could be developed. In turn this would enhance both disaster avoidance and disaster management skills/roles for all personnel involved.

In summary, the current behaviour of shop floor personnel conforms in many ways to the tenets of HROs, but the organisational structure and the proposed team boundaries do not. The aim of multi-skilling the workforce would be beneficial, and would probably improve the reliability of performance within the teams, but unless care is taken, overall efficiency and reliability might become impaired. It would appear that the time is ripe for reflection on the adequacy of the management structure, given the characteristics of the business environment, the vulnerability of the whole process to financial loss, and the ever-greater formalisation of the process of automation.

Acknowledgements

The authors would like to thank the company concerned for providing this field research opportunity, and to recognise the support provided by EPSRC (Engineering and Physical Sciences Research Council), without whom this research would never have happened.

References

Artman, H. and Garbis, C. (1998). "Situation awareness as distributed cognition". In *Proceedings from the 9th European Conference of Cognitive ergonomics*. Limerick, Ireland.

Endsley, M. C. (1998). "Situation awareness, automation and decision support: designing for the future". *CSERIAC Gateway*, 9(1), 11-13.

Hofstede, G. (1991). *Cultures and organisations*. London: McGraw-Hill.

Lawson, M. B. (2001). "In praise of slack: Time is of the essence". *Academy of Management Executive*, 15(3), 125-135.

Reason, J. (2001). "The dimensions of organisational resilience to operational hazards". *British Airways Human Factors Conference: Enhancing Operational Effectiveness*. No publisher.

Roberts, K. H. and Bea, R. (2001). "When systems fail". *Organizational Dynamics*, 29(3), 179-191.

Weick, K. E., Sutcliffe, K. M. and Obstfeld, D. (1999). "Organizing for high reliability: processes of collective mindfulness". *Research in Organisational Behaviour*, 21, 23-81.

PART 4
Complex Decision Making in Military Applications

Chapter 18

A History Lesson on the Use of Technology to Support Military Decision Making and Command and Control

Robert Bolia, Michael Vidulich, Todd Nelson and Malcolm Cook

Introduction

Network-centric warfare is a concept of operations which seeks to improve the efficiency of military operations by promoting speed of command and self-synchronisation of forces, enabled by a dense network of geographically distributed sensors and shooters (Alberts, Garstka and Stein, 1999; Alberts and Hayes, 2003; Cebrowski and Garstka, 1998). While this concept is as intriguing as it is intuitive, it does have its discontents. Barnett (1999), for example, raised seven "deadly" issues that proponents of network-centric warfare have failed to address, while Vego (2003) resurrected the Clausewitzian argument that technology does not change the character of war, and suggests that over-reliance on technology will lead to a degradation in operational art. In spite of these voices of caution, network-centric operations will undoubtedly play a major role in future combat.

The idea that a technological advantage can influence the outcome of battle is not a new one. In 1298, it was the English use of the longbow that broke the Scottish line at Falkirk (Featherstone, 2003). Nearly 700 years later, advances in weapons technology played a role in Operation Desert Storm, as American and British armoured units engaged Iraqi tanks from well outside the range of the Iraqi guns (Scales, 1998).

Yet it is important to remember that superior technology alone cannot guarantee results. The Scots, for example, learned a lesson at Falkirk, and when they met the English at Bannockburn in 1314, maintained a reserve of cavalry to attack the unprotected archers on their flank (Featherstone, 2003). The result was an English defeat.

More recently, the Prussian Army soundly defeated the French in 1870-71, despite the fact that the French *Chassepot* rifle far outranged its Prussian counterpart. This was accomplished by a shift in infantry tactics from massed assaults to more fluid manoeuvres that allowed the Prussian forces to feel their way along the front to the French flanks, which could then be enveloped. While this approach resulted in higher levels of Prussian casualties, it led to a series of rapid French defeats (Wawro, 2003).

Countermeasures are one way to deal with a technological advantage, but not the only way. Indeed, history is replete with examples suggesting that superior force, whether due to manpower or to firepower, is neither necessary nor sufficient to win battles. At least equally important are "the moral dimensions" (*die moralische Größen*), the significance of which was recognised and stressed by Clausewitz (2002). It was these "moral dimensions", coupled with superior training, which allowed Israel to defeat technologically and numerically superior Arab armies in 1967 and 1973 (Bolia, 2004; Eshel, 1989).

There is an increasing emphasis in military circles on the application of advanced technology not merely to provide offensive or defensive systems with enhancements in range, accuracy, or lethality, but also to improve decision making in combat environments (Bolia, 2005a). The concept of "network-centric warfare" is but one manifestation of this phenomenon, albeit the most conspicuous one. Will the same classes of problems associated with over-reliance on technology in weapons systems emerge as a result of the use of technology to support decision making?

A concept often designated as critical to understanding the role of technology in military decision making is that of mediation: each successive development in military technology has resulted in a more elaborate relationship between those applying force and the force they are opposing. This view requires several caveats, however. First, mediation in war is unavoidable, except in the very rare instances of hand-to-hand combat without weapons. Even Alexander the Great, more than 2000 years ago, required runners to send orders to or receive information from parts of the battlefield not under his direct observation (Keegan, 1987). Secondly, increased mediation due to technology is not necessarily a negative consequence. It would be difficult to make the argument that mediation by radio between a commander and his troops would produce more deleterious effects than mediation by runners. The problem is not mediation as such but rather the failure to understand its effects. There are some who are optimistic about the mediating role of technology (Britten, 2001) but the arguments presented are distinguished by the absence of contrary and antagonistic considerations.

In spite of the conceptual momentum generated by public discourse on network-centric operations, or perhaps because of it, it is important to consider the consequences of relying too heavily on technology in the process of decision making, and in other processes central to command and control (C^2). The purpose of the present chapter is to adopt a historical perspective on the problem, presenting examples from military history to illustrate a series of general principles for consideration prior to the application of technology to complex military decision making and C^2 problems. The idealised potential for sensors, computers and communications to support a Revolution in Military Affairs should not be accepted unquestioningly. It needs to be recognised that a strategically useful leverage of technology will require talented, highly trained personnel as well as the development of appropriate tactics and doctrine to deliver the expected benefits (O'Hanlon, 2000). These considerations are often easily dismissed in rhetorical arguments by the advocates of technological solutions.

The principles

1. Understand the limits of the technology

Helmuth von Moltke, Chief of the Prussian General Staff from 1857 to 1887, is frequently praised for his innovative application of technology, especially the telegraph and the railroad. However, it was not the technology *per se* but rather Moltke's thorough understanding of its limitations that allowed it to play a decisive role in the campaign of Königgrätz in the Austro-Prussian War of 1866, and in subsequent campaigns against the French in 1870-71 (Howard, 1961; van Creveld, 1985; Wawro, 1996; 2003).

This assertion is especially true of the telegraph. Moltke recognised the fact that, due to its limited bandwidth and the inherent fragility of a system of lines and poles stretched over open country, the telegraph could be relied upon neither for the delivery of detailed tactical orders nor for the construction at headquarters of an accurate real time situational picture. Moltke possessed a keen awareness of the geographical limitations of telegraphic and railway technology, as well as their vulnerability to interception and sabotage. To compensate for this, he personally educated staff officers to promote a shared mental model within the general staff. When the time came for war, he used the telegraph to issue *Auftrag* orders, that is, directives that conveyed the general objective of the mission rather than detailed execution orders, and relied on the staff to carry them out. This facilitated the interpretation of command intent in the field and allowed the Prussian Army to operate in a mode of "centralised command, decentralised execution" (Howard, 1961; Wawro, 2003). It is noteworthy that Admiral Togo had a similar appreciation of the tactical limitations of wireless communications, that is, technology that he had used effectively for strategic and operational communications, as he led his fleet against the Russians at Tsushima Bay in the Russo-Japanese War (Palmer, 2005).

The Battle of Britain provides a somewhat analogous example. While it is often suggested that radar was the critical technology that led to the United Kingdom's victory in the battle, it was actually the elaborately networked system of fighter control, of which radar was only a component, which was significant. At Pearl Harbour, for example, the incoming Japanese planes were detected and tracked by radar, but due to the lack of an informed command centre the information was misconstrued and did not contribute to the defence of the anchored fleet (Prange, 1981).

That the fighter control system worked so well during the Battle of Britain was due precisely to the fact that it took into account the limits of the technology. The Germans, who also had radar, had no such system, relying on the radar station local to the squadron for fighter control. It was not merely their lack of a networked system, but their lack of understanding of the British system and thus failure to develop an appropriate strategy to counter it, that proved decisive (Deighton, 1977). If the Germans had developed a comparable network, they would have been better prepared to defend against the Allied strategic bombing campaign. Moreover, had they better understood Britain's fighter control system, they could have enhanced the effectiveness of their own bombing campaign by first striking at the heart of the British C^2 system.

2. Always consider interoperability

As Barnett (1999) has pointed out, the United States military spends more on information technology than most countries spend on their entire military, creating an environment that makes it as difficult for allies as it is for adversaries to stay abreast. Network-centric warfare begins to fall apart when coalition partners and even services within a single nation are not part of the network (Richter, 2002). Thus, it is important to keep in mind joint and coalition interoperability when proposing advanced technology solutions to C^2 problems.

Despite being regarded as one of the most successful operations in military history from the perspective of joint and coalition warfare, Operation Desert Storm yielded several examples of breakdowns in interoperability affecting C^2. The inability of the United States Navy to receive electronic versions of the air tasking order, forcing a printed version to be delivered to each of the carriers by helicopter, is one example (Marolda and Schneller, 2001). Another is the lack of secure voice communications and data links by most coalition air forces (Hunt, 1998). While neither of these failures proved to be a showstopper in an operation conducted against an enemy that was not prepared to sacrifice its air force, a different foe might have tested the limits of these circumventions.

A more disastrous consequence of the failure to require interoperable systems was the collision of a helicopter and a transport aircraft during the attempt to free American hostages held in Tehran in April 1980. Here unexpected weather conditions, combined with a lack of communications between forces of different services, led to reduced situation awareness, total mission failure, and a number of US casualties (Kyle, 1995).

3. More information is not always better

The fact that an increase in the number of sensors and the complexity of the network uniting them has engendered an increase in the amount of available data does not necessarily mean that officers using that data will make better decisions. There are several reasons for this.

First, an increase in the quantity of information available does not necessarily mean an increase in the amount of *relevant* information. In the Battle of Britain, much more information was available than the quantity that was passed to any single fighter pilot. Presenting the pilots with all of the raw radar returns would have only produced confusion. The purpose of the filter rooms was to filter the incoming returns and consolidate the results into a relevant picture, analogous to the contemporary notion of sensor fusion (Deighton, 1977).

Secondly, not all data are good data. The possibility of deliberately fabricated information being fed into the network in order to deceive the operators is very real, and is exemplified by the deception campaign undertaken by the Allies prior to the invasion of Normandy in 1944. Bogus radio traffic, false news reports, and compromised German intelligence agents were all part of the plan to convince the Germans that the landings would take place at the Pas de Calais instead of at Normandy. The plan worked (see Howard, 1995).

While deception is one means by which inaccurate data may make its way into a system, it is not the only one. The possibility exists, for example, that a data fusion algorithm might decide that two sensor inputs represent a single entity, when in fact they correspond to two distinct vehicles. Furthermore, there are temporal issues to consider, such as at what point data is no longer to be regarded as "good". Indeed, the premature removal of the track of the USS *Liberty* from the Israeli Air Force's (IAF) command and control centre led to its inadvertent bombing by IAF aircraft during the Six-Day War (Yonay, 1993).

Thirdly, no matter how much data is available, the picture will never be complete. While this may be taken as a purely epistemological point, it is well illustrated by the fact that, even with a composite picture provided by five Airborne Warnings And Control System (AWACS) aircraft, the operators in the air operations centre during the Gulf War could not track the F-117s (Clancy with Horner, 1999).

A less trivial example comes from the campaign of Gallipoli in the First World War, in which British, Australian and New Zealand troops mounted a massive joint operation to take a hill in Turkey, Achi Baba. Their maps told them they would then command the Dardanelles fortresses, the capture of which would hasten Turkish withdrawal from the war and the eventual defeat of the Central Powers. After months of some of the bloodiest fighting of the war, resulting in thousands of casualties, the British high command ordered a withdrawal. After the war, troops visiting the site discovered that Achi Baba did not command the Narrows at all, and was in fact of no strategic importance whatever. The entire operation had been for nought (Carlyon, 2001).

Finally, data is only as good as its interpretation. The absence of Enigma intercepts in the days leading up to the Battle of the Bulge was interpreted by the Allies as evidence of inaction due to a general acceptance of defeat by the Wehrmacht. In fact the Germans had been instructed not to use radio transmission as a means of command. On the one hand, it was not necessary; the distances over which messages had to be transmitted were very short, and the number of people who knew of the attack limited. On the other hand, the Germans did not want to do anything to compromise the element of surprise (Toland, 1999). The attack that resulted was a shock to the Americans and, although eventually repulsed, resulted in thousands of preventable casualties.

Similarly, in the autumn of 1973 Israel had most of the relevant information to conclude that an attack by Egypt and Syria was imminent, but interpreted it in the context of overconfidence in the Israeli intelligence services and underestimation of the Arab desire to regain the occupied territories, as well as an inappropriate model of Arab decision making. The result was the Yom Kippur War (Black and Morris, 1991; Bolia, 2004).

The issues associated with overabundance of information can be interpreted in terms of Reason's theory of human error (Reason, 1992), which suggests that mistakes made by humans, aside from failures in automatic processes such as typing, are the result of one of two causes: 1) bounded rationality; or 2) incomplete or inaccurate mental models. Bounded rationality refers to the human's limited capacity for information processing. Even if all of the information required for solving a problem exists, if the amount of information is too great or if the processing required to solve the problem is too complex, one is more likely to accept an error or side-step the information processing demands by adopting a heuristic strategy. An inaccurate or incomplete mental model contributes to the potential for creating a faulty plan due to a lack of understanding of the true causal relationships present in the environment.

Bearing Reason's theory in mind, it is not difficult to envision that a decision loop supported by advanced technology would be at least as prone to mistakes as a human unsupported by technology. On the one hand, the information available will almost certainly exceed the capacity and processing power of the human operator; on the other, there is no guarantee either that all of the required information will be available, or that the available information will be accurate.

4. Technology can disrupt traditional roles

One likely consequence of the use of technology to support C^2 and decision making is the alteration of how tasks are distributed within the chain-of-command. In particular, the availability of a common operational picture at all levels of the command structure has the potential to reduce the delegation of authority and promote micromanagement. In this case, senior players without operational knowledge or experience relevant to current warfighting technology may try to manage or interpret information on a level at which their expertise is of dubious advantage (Hamzideh, 2003). This raises issues of trust, competence, authority and responsibility.

For example, advancements in communications technology have enabled high-level commanders to give tactical- and operational-level orders to field commanders, regardless of their level of competence or the presence of adequate information. Thus President Lyndon Johnson, referring to the fact that targets bombed by US aircraft in Vietnam had to be cleared through Washington, could boast that "they can't even bomb an outhouse without my approval" (Karnow, 1983).

A less dramatic illustration is drawn from the Falklands War, where operational decisions were often made by Admiral Sir John Fieldhouse and his staff, commanding from Joint Force Headquarters at Northwood in the UK, instead of being delegated to a commander in the theatre of operations. Such decisions often led to operations which from a purely military viewpoint were unnecessary (Thompson, 2001). One such operation was the attack on the Argentine positions at the twin settlements of Darwin and Goose Green. Although occupied by a full regiment of Argentine infantry, these positions were of no strategic importance, and their capture would not lead to a timelier recapture of the islands. However, it was considered crucial from the perspective of the War Cabinet that the land forces be seen to be doing something. The result was an eventual British victory, but only at the expense of numerous lives on both sides, among them the commanding officer of the British parachute battalion and several junior officers (Adkin, 1992; Bolia, 2005b).

While the behaviours exemplified by Johnson and Fieldhouse demonstrate that technology may lead to micromanagement, they do not imply that it has to. That commanding officers need to be aware of this possibility in order to avoid its occurrence has been pointed out by, among others, General Chuck Horner, the Joint Forces Air Component Commander in Operation Desert Storm (Clancy with Horner, 1999). The fact that this is not an entirely novel problem, and that it is indeed one which commanders have had to come to grips with before, is demonstrated by its discussion in the writings of Moltke more than a century ago (Hughes, 1993).

5. Beware of complacency

Over-reliance on technology should be among the greatest of concerns for users of technology designed to support decision making. Despite its ubiquity, technology is neither faultless nor safe from interdiction, and the expectation that it will always work is a bad one. This was discovered by the paratroops of the British 1st Airborne Division at Arnhem in 1944, when they made the discovery that their radios did not work well in the Dutch countryside. Needless to say, this had negative impacts on C^2, as commanders were unable to maintain communications with either the troops under their command or with their superiors (Middlebrook, 1995).

This principle is also illustrated by considering failures caused by relying on Identification, Friend or Foe (IFF). In 1982, the HMS *Cardiff* shot down a British Army Gazelle helicopter, believing it to be an Argentine C-130. While this blue-on-blue incident was attributable to a number of factors, it could have been prevented had the Gazelle's IFF system been working properly (Middlebrook, 2001). Another instance in which IFF technology interacted with operator error to produce fratricide was the Black Hawk helicopter shootdown over Northern Iraq on 14 April 1994 (Snook, 2000). In 2001, terrorists planning to fly planes into the World Trade Center and the Pentagon used over-reliance on IFF to their advantage when they simply turned the transponders off, essentially blinding controllers as to their intentions.

Support for this principle also comes from research on human interaction with automated systems, which has demonstrated that the introduction of automation can change the cognitive demands placed on the operator and occasion reductions in performance and situation awareness, unbalanced workload, mistrust, skill degradation and complacency (Parasuraman and Riley, 1997). Recently, a number of researchers have proposed models of human interaction with automated systems which attempt to provide a framework for the investigation of these issues, through which a better understanding of how to deal with them may develop (Parasuraman, Sheridan and Wickens, 2000). Despite the subsequent proliferation of studies adopting these frameworks (Galster, Bolia and Parasuraman, 2002; McGarry, Rovira and Parasuraman, 2003; Rovira, Zinni and Parasuraman, 2002), and the continued evolution of the models (Parasuraman and Miller, 2003), many of the fundamental problems associated with technology integration remain unsolved.

Conclusions

The five principles discussed herein are neither orthogonal nor exclusive. It is clear, for example, that inappropriate levels of information may engender the disruption of traditional command roles, or that complacency can lead to the employment of technology without an appreciation of its limits. On the other hand, it should also be clear that these principles address qualitatively different issues in the use of technology to support military decision making and C^2, and as such it makes sense to consider them separately.

If the innovative technologies proposed for network-centric warfare are to have the impact that many expect from them, it is apparent that doctrinal issues relating technology to the users and the mission roles they are expected to accomplish must be explored comprehensively. It will be equally important that system designers understand the issues raised by technology insertion that cannot be managed by doctrinal change. The aspiration is that these five principles may impart guidance to the development and implementation of technological support systems for decision making and C^2, as well as the doctrine for their use.

Acknowledgements

The authors would like to express their appreciation to Gloria Calhoun and Mark Draper of the Air Force Research Laboratory for their careful readings of earlier drafts of this chapter.

References

Adkin, M. (1992). *Goose Green: A battle is fought to be won*. London: Cassel and Co.

Alberts, D. S., Garstka, J. J. and Stein, F. P. (1999). *Network centric warfare: Developing and leveraging information superiority*. Washington: Command and Control Research Program.

Alberts, D. S. and Hayes, R. E. (2003). *Power to the edge: Command... control... in the information age*. Washington: Command and Control Research Program.

Barnett, T. P. M. (1999). "The seven deadly sins of network-centric warfare". *US Naval Institute Proceedings*.

Black, I. and Morris, B. (1991). *Israel's secret wars: A history of Israel's intelligence services*. New York: Grove Press.

Bolia, R. S. (2004). "Over-reliance on technology in warfare: The Yom-Kippur War as a case study". *Parameters*, 34(2), 46-56.

Bolia, R. S. (2005a). "Intelligent decision support systems in network-centric military operations". In *Intelligent decisions? Intelligent support? Pre-proceedings for the International Workshop on Intelligent Decision Support Systems: Retrospect and prospects* (pp. 3-7).

Bolia, R. S. (2005b). "The battle of Darwin – Goose Green". *Military Review*, 85(4), 45-50.

Britten, S. M. (2001). "Directing war from home". In W. C. Martel, *The technological arsenal: Emerging defense capabilities*. Washington: Smithsonian.

Carlyon, L. (2001). *Gallipoli*. Sydney: Pan Macmillan Australia.

Cebrowski, A. K. and Gartska, J. J. (1998). "Network-centric warfare: Its origin and future". *US Naval Institute Proceedings*.

Clancy, T. with Horner, C. (1999). *Every man a tiger*. New York: Berkley Books.

Clausewitz, K. von (2002). *Vom kriege*. Berlin: Ullstein.

Deighton, L. (1977). *Fighter: The true story of the Battle of Britain*. New York: HarperCollins.

Eshel, D. (1989). *Chariots of the desert: The story of the Israeli Armoured Corps*. London: Brassey's Defence Publishing.

Featherstone, D. F. (2003). *The bowmen of England: The story of the English longbow*. Barnsley, UK: Pen & Sword Military Classics.

Galster, S. M., Bolia, R. S. and Parasuraman, R. (2002). "Effects of information automation and decision-aiding cueing on action implementation in a visual search task". In *Proceedings of the Human Factors and Ergonomics Society 46th Annual Meeting* (pp. 438-442). Santa Monica, CA: Human Factors and Ergonomics Society.

Hamzideh, K. (2003). "The issue of decision up-creep in network centric warfare". Unpublished research paper, US Naval War College.

Howard, M. (1961). *The Franco-Prussian war*. London: Routledge.

Howard, M. (1995). *Strategic deception in the Second World War*. New York: W. W. Norton and Company.

Hughes, D. J. (ed.) (1993). *Moltke on the art of war: Selected writings*. Novato, CA: Presidio Press.

Hunt, P. C. (1998). *Coalition warfare: Considerations for the air component commander*. Maxwell AFB, AL: Air University Press.

Karnow, S. (1983). *Vietnam: A history*. New York: Viking Press.

Keegan, J. (1987). *The mask of command*. New York: Viking Press.

Kyle, J. H. (1995). *The guts to try: The untold story of the Iran hostage rescue by the on-scene desert commander*. New York: Ballantine.

Marolda, E. J. and Schneller, R. J. (2001). *Shield and sword: The United States Navy in the Persian Gulf war*. Annapolis: Naval Institute Press.

McGarry, K., Rovira, E. and Parasuraman, R. (2003). "Effects of task duration and type of automation support on human performance and stress in a simulated battlefield engagement task". In *Proceedings of the Human Factors and Ergonomics Society*

47th Annual Meeting (pp. 548-552). Santa Monica, CA: Human Factors and Ergonomics Society.

Middlebrook, M. (1995). *Arnhem 1944: The airborne battle*. London: Penguin Books.

Middlebrook, M. (2001). *The Falklands War, 1982*. London: Penguin Books.

O'Hanlon, M. (2000). *Technological change and the future of warfare*. Washington D.C.: Brookings Institution Press.

Palmer, M. A. (2005). *Command at sea: Naval command and control since the sixteenth century.* Cambridge, MA: Harvard University Press.

Parasuraman, R. and Miller, C. A. (2003). "Beyond levels of automation: An architecture for more flexible human-automation collaboration". In *Proceedings of the Human Factors and Ergonomics Society 47th Annual Meeting* (pp. 182-186). Santa Monica, CA: Human Factors and Ergonomics Society.

Parasuraman, R. and Riley, V. A. (1997). "Humans and automation: Use, misuse, disuse, abuse". *Human Factors*, 39, 230-253.

Parasuraman, R., Sheridan, T. B. and Wickens, C. D. (2000). "A model for types and levels of human interaction with automation". *IEEE Transactions on Systems, Man, & Cybernetics*, 30, 286-297.

Prange, W. (1981). *At dawn we slept: The untold story of Pearl Harbor*. New York: McGraw-Hill.

Reason, J. (1992). *Human error*. Cambridge, UK: Cambridge University Press.

Richter, A. C. (2002). "Alongside the best? The future of the Canadian Forces". *Naval War College Review*, 56, 67-107.

Rovira, E., Zinni, M. and Parasuraman, R. (2002). "Effects of information and decision automation on multitask performance". In *Proceedings of the Human Factors and Ergonomics Society 46th Annual Meeting* (pp. 327-331). Santa Monica, CA: Human Factors and Ergonomics Society.

Scales, R. H., Jr. (1998). *Certain victory: The U.S. Army in the Gulf War*. Washington: Brassey's.

Snook, S. A. (2000). *Friendly fire*. Princeton: Princeton University Press.

Thompson, J. (2001). *No picnic.* London: Cassell & Co.

Toland, J. (1999). *Battle: The story of the Bulge*. Lincoln: University of Nebraska Press.

van Creveld, M. (1985). *Command in war*. Cambridge, MA: Harvard University Press.

Vego, M. (2003). "Net-centric is not decisive". *US Naval Institute Proceedings*.

Wawro, G. (1996). *The Austro-Prussian war: Austria's war with Prussia and Italy in 1866*. Cambridge, UK: Cambridge University Press.

Wawro, G. (2003). *The Franco-Prussian war: The German conquest of France in 1870-71*. Cambridge, UK: Cambridge University Press.

Yonay, E. (1993). *No margin for error: The making of the Israeli Air Force*. New York: Pantheon Books.

Chapter 19

The Intuitive vs. Analytic Approach to Real World Problem Solving: Misperception of Dynamics in Military Operations

Bjørn Tallak Bakken, Stig Johannessen, Dag Søberg and Morten Ruud

Introduction

According to Brewster (2002), many studies within military and scientific communities conclude that (military) commanders actually rely more heavily on an intuitive versus an analytic approach when in a field environment. The intuitive approach to decision making appears to be chosen when facing: ill-structured problems; uncertain or dynamic environments; time stress; and/or high stakes. The intuitive approach is based on pattern recognition and experience, and goes within the military profession under terms such as "fingerspitzengefuhl" and "coup d'oeil". The research field of Naturalistic Decision Making (NDM) is largely concerned with intuitive decision making, and defines it as "the way people use their experience to make decisions in field settings" (Klein, 1998, page 1).

Intuition and analysis represent the end points of the "cognitive continuum" (Dunwoody et al., 2000), but the cognitive mode is rarely purely intuitive or analytical. More often, it is a mixture of both. This is referred to as "quasi-rationality" (Brunswik, 1956). This "middle course" is characterised by a robust and adaptive decision making/problem solving process, and is closely associated with "common sense". Tasks that contain uncertainty, dynamics and many (redundant) perceptual cues (thereby making the task hard to analyse) will benefit from a largely intuitive approach (Dunwoody et al., 2000). Intuitive decision making is quick (almost instantaneous), and happens with low cognitive control and low conscious awareness. Perhaps paradoxically, good intuition is not commonplace, and in constant need of improvement!

That people have problems when applying common sense (or intuition) to static situations involving simple probability judgement is well known (see Kagel and Roth, 1995 for a comprehensive review). Several authors now point to decision makers' failure to consider feedback in complex, dynamic systems. Let two recent studies illustrate the magnitude of this problem: in his studies of management of renewable resources, Moxnes (1998) observed that experienced decision makers over-invest and over-utilise their resources. He attributes this behaviour to systematic misperceptions of stocks and flows, and of non-linearities. Sweeny and Sterman (2000) took a different approach when they gave system dynamics case problems to students at an elite business school. The students, who were highly educated in mathematics and science (but had received no prior schooling in system dynamics concepts), were found to have a poor level of understanding of the basic system dynamics concepts: stock and flow relationships, and time delays.

Inspired by the Sweeny and Sterman study, Ossimitz (2002) has conducted an investigation where 154 participants were given different tasks in dynamic thinking in general, and in the interpretation of stock-flow related graphs in particular. The results were alarming, in that the mean performance of the participants was approximately at the level of tossing a coin for each answer. He suspects that the lacking ability is to grasp that a positive net-flow results in an increase in the corresponding stock. In a related study, Kainz and Ossimitz (2002) find that a 90-minute crash course introducing basic stock-flow concepts between pre- and post-test was suitable to bring about an improvement in performance.

Jensen and Brehmer (2003) let a collection of laypeople attempt to establish equilibrium in a simple predator-and-prey simulation. They find that even though the task was structurally simple, it was still perceived as difficult. They conclude that participants may have a low ability to apply indirect reasoning, at the same time as they resort to thinking in terms of discrete time steps rather than in terms of continuous time.

It is thus a general observation that people perform quite poorly in systems with even modest levels of complexity. Sterman (2000) labelled this kind of cognitive dysfunction "misperceptions of feedback". The solution would be to develop "systems thinking" abilities.

Tversky and Kahneman (1987, page 90) recognised the shortcomings of a static, one-shot approach to learning. They described the prospects for learning in dynamic environments like this: "Effective learning takes place only under certain conditions: it requires accurate and immediate feedback about the relation between the situational conditions and the appropriate response. The necessary feedback is often lacking for the decisions faced by managers, entrepreneurs, and politicians because:

- outcomes are commonly delayed and not attributable to a particular action;
- variability in the environment degrades the reliability of the feedback, especially where outcomes of low probability are involved;
- there is often no information about what the outcome would have been if another decision had been taken;
- most important decisions are unique and therefore provide little opportunity for learning (see Einhorn and Hogarth, 1978)".

Although decision making errors may disappear with guided experience and reflection, little research has focused on the improvement in decision strategies; in particular, changes in decision making with experience have not been revealed to any extent. Notable exceptions exist, however. Bakken (1989) found that in a simulated economy, participant's performance improves over trials. Paich and Sterman (1993), in their investigation of human performance in a product lifecycle task, found that performance suffers as dynamic complexity increases. Brehmer (1988) found that action lags decrease performance in a simulated forest fire, and at the same time conditions for learning get worse.

The aim of the present experiments is to demonstrate how a structurally simple, deterministic task may induce complex dynamic behaviour; and that this behaviour is not fully perceived even by fairly competent decision makers. The importance and generality of the task (within a defence organisation) makes this a strong case for improvement, and we suggest that a training programme based on practice with simplified and focused simulations may be the way to go.

A real world decision problem

The experimental task is an extremely simplified model of the logistics chain in a "Peace Support Operation", and focuses on the dynamic interrelations between budget allocations and cost of operations (initial purchases and running sustenance), and the short- and long-term consequences of a limited time budget expansion. The paradox encountered is that a one-time budget increase may lead not only to a performance boost in the short term (primary effect), but also to a significant penalty in the longer term (unintended side-effect). The primary cause of this behaviour is the "chaining" and thus preservation of discrete age classes (cohorts) in the model, which through time lag and negative feedback conserves an imbalance of structure from the one-time increased purchases and throughout the lifetime of the resources. In fact, the one-time perturbation sets off a pattern of oscillations in performance, which, though dampened, could continue for the whole duration of the operation. Note that the task is fully deterministic, and complete information is contained on a single page. Based on the task description alone, participants will be asked to depict the performance curve as a function of the budget input.

The experimental task and associated system dynamics simulation model is described in detail in Bakken and Gilljam (2003). The task description (experimental stimulus) is provided in an appendix.

Study 1

The first study involved seven defence analysts as experimental participants. When prompted, the participants did not manage to reproduce the cyclic performance resulting from the one-time budget increase. Instead, graphs sketched (by hand) resembled more the "plain" budget profile, but stretched out in time. Only one participant correctly perceived and rendered the "unintended" performance penalty, but he also had a background in system dynamics (SD) (see Figure 19.1). Detailed inspections of the graphs reveal that participants stipulated an accumulated performance boost of more than three times the actual increase.

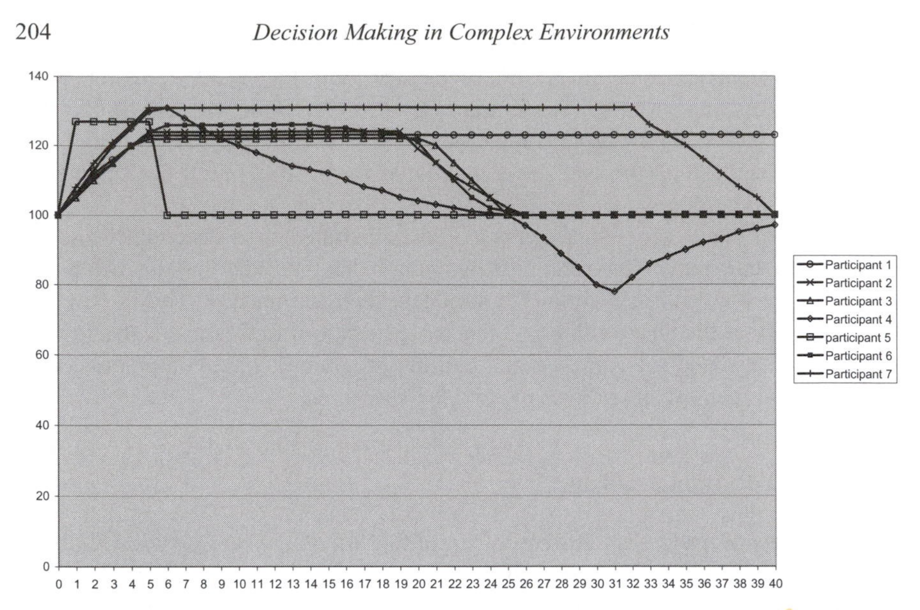

Figure 19.1 Sketches of performance ("Total effect") as drawn by the participants
Note: The sketches were transferred to MS Excel for readability.

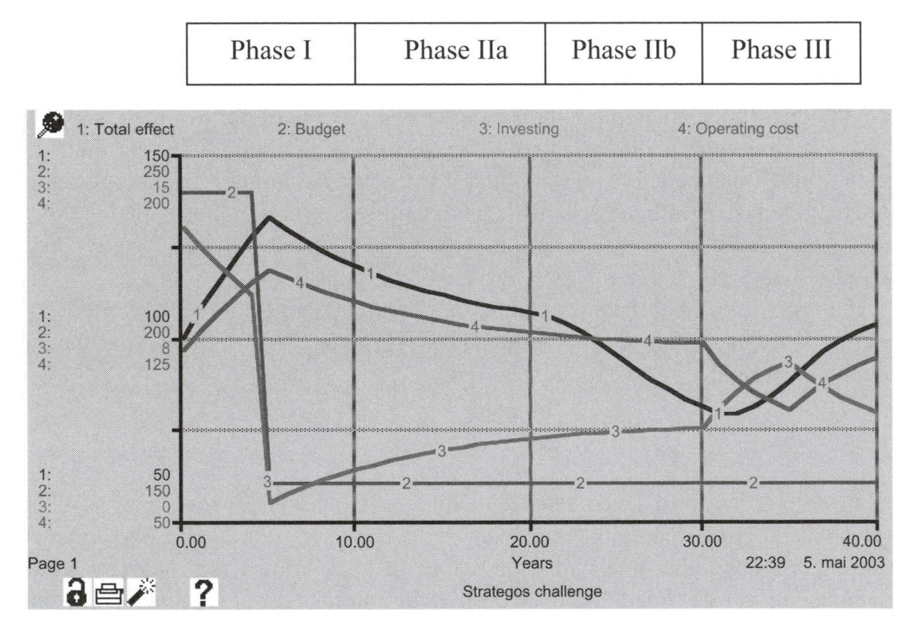

Figure 19.2 Simulated solution ("1: Total effect")

When the sketches are compared to the simulated solution (Figure 19.2), we may make detailed comparisons across the three phases of the operation.

In phase I, total effect experiences a steep concave increase for the duration of the extra funding (five days). Only Participant 5 suggests that the increase is completed by day 1, which is wrong. Thus, most participants get this phase right.

In phase IIa, total effect undergoes a gentle downward slope after extra funding stops. Five participants misperceive this, and instead assume that the new, increased level of efficiency is maintained for (at least) another 15 days.

In phase IIb, total effect trajectory approaches original baseline between days 23 and 24. There are four participants who estimate that the total effect will be back to baseline by that time, and thus get this roughly right.

In Phase III, total effect curve drops below baseline and reaches an all time low between days 31 and 32, before turning upward again. Only one participant perceives this pattern correctly. This is Participant 4, who happens to have an educational background in system dynamics.

Other anomalies observed: Participant 5 depicts a distinct "plateau" between day 1 and 5, and drops down to baseline from day 6. Participants 1 and 7 project a much too long period of increased total effect.

Overall, Participant 4, with a background in system dynamics, has the "most correct" perception of development in total effect. He (or she) predicts a steep concave increase in total effect from day 0. S/he also matches the gentle downward slope from day 6, the slight increase in drop and crossing of baseline at approximately days 23-24, and the subsequent upturn from days 31-32.

The most common (five participants) misperception is the "plateau" depiction of the period with increased effectiveness, where a pointed peak would be most correct. The sketched pattern thus resembles the budget increase pattern, but is more stretched out in time, probably made to correspond with the unit lifetime.

Study 2

The second study involved 10 military college cadets as experimental participants. The experimental task was repeated as in Study 1, but this time with a debrief session immediately following the task itself. The participants were also presented with four questions aimed at assessing participants' confidence in own judgement as well as their understanding of the problem at hand. To simplify, participants were this time presented with six template graphs (Figure 19.3), and instructed to choose individually the one that in their opinion best resembled the actual performance profile. None of the participants managed to pick the best one (Graph E).

Moreover, participants believed the accumulated performance boost to be more than twice (200 per cent) the actual value, while at the same time they believed their estimate could be no more than 50 per cent off the actual value.

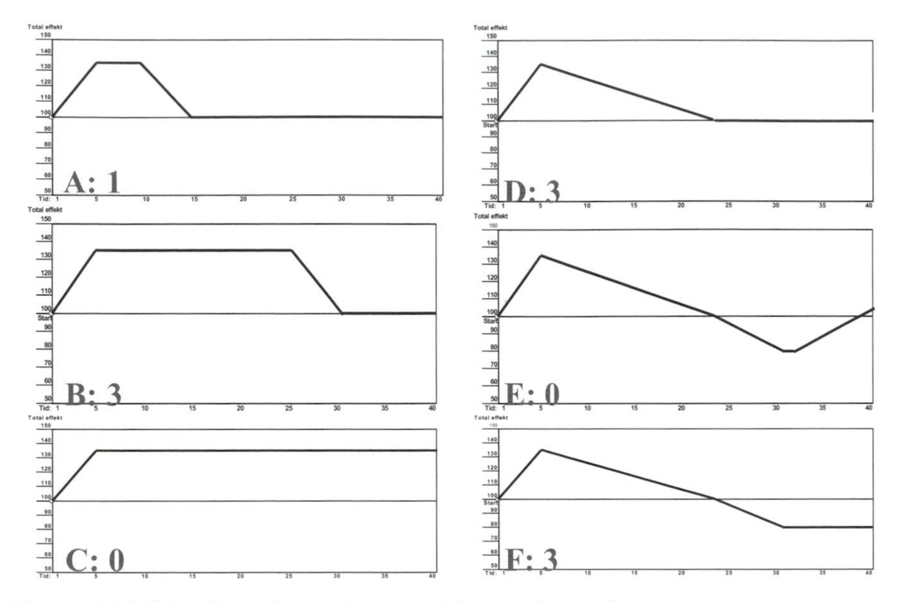

Figure 19.3 Selection of template graphics to choose from

Note: Number of participants choosing the actual graph is indicated.

Results

Despite the low number of experimental participants, the obvious conclusion is that intuition is a poor judge of dynamic (feedback) processes.[1]

The participants appear to use the first order intuitive judgement as an "anchor" to which subsequent adjustments may be made (processing in spreadsheet or systems dynamics modelling software might reveal the analytic solution, but this approach usually takes valuable time which otherwise might be spent on environment monitoring and other management/executive activities). When the necessary system dynamics insight is lacking, in this case the ability to recognise that a significant time lag coupled with negative feedback tends to produce oscillations around a baseline, such adjustments rarely happen. This kind of misperception results in over-optimism when it comes to estimate the performance of the system, because over-shoots are perceived as extended for a longer time than appropriate, and under-shoots are not detected at all. The probable roots for this will be discussed in the following. What is the nature of these problems, and how may they be remedied?

1 Data from follow-up studies (not yet analysed in detail) involving a total of more than 100 military officers indicate that, on average, less than 10 per cent choose the correct graph among the six alternatives presented.

Discussion

How commonplace is it to misperceive the complexity of planning tasks, and subsequently underestimate costs/overestimate effect? In a study undertaken at the University of Aalborg (Denmark), there was found to be a consistent tendency to underestimate the costs of larger, public projects. This tendency is observed in many countries, and with undiminished magnitude between the years 1910 and 1998 (Flyvbjerg, Holm and Buhl, 2002). The average cost overrun for 258 larger projects observed in the study was 28 per cent. The study indicates either that one is incapable of exploiting previous experience, or that costs deliberately are underestimated in order to get a project approved for funding. In the latter case, it is the decision makers who are incapable of learning from previous experience. In any case, there might be a decision making bias at work, so that when costs are initially estimated, the sheer complexity of the task will render many costs undiscovered. These initially undiscovered costs, which later turn up as unexpected costs, might well result from an incomplete perception of feedback, time lags and non-linearities.

Clark (1987) has identified the public budgeting process as one of the major causes of non-planned side effects in structural planning. Regarding the 1970s US Defense budget reductions, he commented (page 24): "Because unplanned budget changes are absorbed by the acquisition portion of the budget [...] the result was a significant unplanned annual reduction in the funding of system acquisitions [...]."

McKean (1965) was probably one of the first to suggest taking a "systemic" approach to public management, and promoted the use of quantitative analysis in comparison of alternative courses of action, since it was found that analytic methods are wanting. But the point of this study is not the lack of mathematical methods and the knowledge or time to use them, but the faulty intuition applied by decision makers when first confronted with a non-trivial decision problem. If first-order intuitive judgements had been adjusted in light of subsequent quantitative analysis, all might have been well. But all too often major decisions are made without troubling to get more detailed data, or running the appropriate computer-based planning tool. It seems like confidence in the accuracy of own judgements is superior to reason (for a review of decision making biases, see, for example, Lai, 1999).

According to Kleinmuntz (1993), it is a very real possibility that faulty mental models of the task environment cause the misperception phenomenon. First of all, it is believed that decision makers are more likely to detect feedback loops if: they perceive that there is connection between past action and future effects; the time lag is short; actions and effects are of similar kind; and exogenous variables are proved to be irrelevant. All these point to the benefits of building "dynamic intuition" in a controlled, interactive environment, such as a decision game.

Brehmer (2000) points to two main principles when it comes to interpreting people's problems in handling dynamic settings:

- Overemphasis on the present: Decision makers tend to attend to only the information currently at hand, and as a result experience difficulties in accommodating feedback delays. The world is perceived "here and now", and the information is not processed any further.

- Lack of systems thinking: The tendency to think linearly, that is, to believe that actions and results are directly related and ignore the side effects of actions. This tendency can also be viewed as an over-reliance on information that is readily available, along with a tendency to ignore what must be inferred such as side effects.

In a discussion of so-called "command concepts", Brehmer (2002) issued a warning against relying solely on (direct) feedback control of a command process. Without taking into account inherent delays, the command system would inevitably become reactive and lagging, resulting in loss of initiative and subsequently loss of personnel and equipment. A (mental) model needs to be able to produce reliable predictions of future events if it is to provide the basis for a functional command concept.

To challenge improper beliefs people have about causal relations has been a major focus in improvement research. But the same line of research shows in essence that learning from experience in complex, dynamic tasks is a troublesome and demanding undertaking. It also tells us that even though much effort may be put into training in realistic settings (and usually realism is positively related to complexity), little or no learning outcome may be expected. One key to effective learning seems to be found in environment simplification.

The most salient problem with the experimental task is that grants are given for discrete periods (with no inter-temporal transfer possible). It requires only a very simple model to illustrate the severe effects of, for example, an unplanned budget cut, or the bad economy of giving extra grants as lump sums rather than more evenly distributed. Another problem is the failure to anticipate escalating costs and declining efficiency with age. In the experimental task, such information was given explicitly, but participants apparently did not manage to integrate that information. It is probably not common knowledge that when a public grant is given for a shorter time than the lifetime of the structure it is meant to finance, then there is a risk that the rest of the organisation (eventually) will "pay" the residual. Yet another source of problem is related to "illusion of control", that is, the (wrong) belief that you have more control than is really the case. This could be a perception that the mode of control is more direct than it really is (for example, that fixed resources can be manipulated as if they were movable), or that time between issuing a directive and until implementation is shorter than it really is. Finally, it is the question of controlling with the right goals in mind. For example, it could be wrong to invest heavily in new resources when getting rid of the older and less efficient ones would produce the same effect but at a lower cost.

It is left for further research to investigate whether some kind of system dynamics training may contribute to more robust dynamic mental models in the minds of military commanders and analysts, thus enabling more reliable intuitive assessments to be made and higher quality command concepts to be constructed.

Another equally interesting approach would be to take into account decision makers' prior level of experience (or rather "expertise") within the field of long-term and/or operational planning. If it turns out that more experienced decision makers are more likely to infer the correct dynamic behaviour pattern, it would imply that dynamic intuition might develop from experience.

References

Bakken, B. E. (1989). "Learning in dynamic simulation games: Using performance as a measure". In P. M. Millin and E. O. K. Zahn (eds), *Computer-based management of complex systems: Proceedings of the 1989 International Conference of the System Dynamics Society*. Stuttgart: Springer.

Bakken, B. T. and Gilljam, J. M. (2003). "Misperception of dynamics in military planning: Exploring the counter-intuitive behaviour of the logistics chain". In *Proceedings of the 2003 International Conference of The System Dynamics Society*. New York.

Brehmer, B. (1988). *Strategies in real time dynamic decision making*. Chicago: University of Chicago. (Einhorn Memorial Conference).

Brehmer, B. (2000). "Dynamic decision making in command and control". In McCann and Pigeau (eds), *The human in command: Exploring the modern military experience*. New York: Kleuwer.

Brehmer, B. (2002). "Utveckling av ett 'Command Concept' som en problemlösningsprosess". In Henrik Friman (ed.), *Command concepts*. Stockholm: Swedish National Defence College.

Brewster, F. W. (2002). "Using tactical decision exercises to study tactics". *Military Review*, Nov-Dec, 3-9.

Brunswik, E. (1956). *Perception and the representative design of psychological experiments*. Berkeley: University of California Press.

Clark, R. (1987). "Defense budget instability and weapon system acquisition". *Public budgeting and finance*, Summer.

Dunwoody, P. T., Haarbauer, E., Mahan, R. P., Marino, C. and Chu-Chun Tang (2000). "Cognitive adaptation and its consequences: A test of cognitive continuum theory". *Journal of Behavioral Decision Making*, 13, 35-54.

Einhorn, H. J. and Hogarth, R. M. (1978). "Confidence in judgement: Persistence of the illusion of validity". *Psychological Review*, 85, 395-416.

Flyvbjerg, B., Holm, M. S. and Buhl, S. (2002). "Underestimating costs in public works. Error or lie?" *Journal of the American Planning Association*, 68, 279-295.

Jensen, E. and Brehmer, B. (2003). "Understanding and control of a simple dynamic system". *System Dynamics Review*, 19(2), 119-137.

Kagel, J. H. and Roth, A. E. (1995). *Handbook of experimental economics*. New Jersey: Princeton University Press.

Kainz, D. and Ossimitz, G. (2002). "Can students learn stock-flow thinking? An empirical investigation". In *Proceedings of the 2002 International Conference of the System Dynamics Society*. Palermo, Italy.

Klein, G. (1998). *Sources of power. How people make decisions.* Cambridge, MA: The MIT Press.

Kleinmuntz, D. N. (1993). "Information processing and misperceptions of the implications of feedback in dynamic decision making". *System Dynamics Review*, 9, 223-237.

Lai, L. (1999). *Dømmekraft*. Oslo: Tano-Aschehoug.

McKean, R. (1965). *Efficiency in government through systems analysis*. (RAND) New York: John Wiley.

Moxnes, E. (1998). "Not only the tragedy of the commons: Misperceptions of bioeconomics". *Management Science*, 44, 1234-1248.

Ossimitz, G. (2002). "Stock-flow thinking and reading stock-flow related graphs: An empirical investigation in dynamic thinking abilities". In *Proceedings of the 2002 International Conference of the System Dynamics Society*. Palermo, Italy.

Paich, M. and Sterman, J. D. (1993). "Boom, bust, and failures to learn in experimental markets". *Management Science*, 39, 1439-1458.

Sterman, J. D. (2000). *Business dynamics: Systems thinking and modelling for a complex world.* Boston: McGraw-Hill.

Sweeny, L. B. and Sterman, J. D. (2000). *Bathtub dynamics: Preliminary results of a systems thinking inventory*. Presented at the International System Dynamics Conference, Bergen, Norway.

Tversky, A. and Kahneman, D. (1987). "Rational choice and the framing of decisions". In R. M. Hogarth and M. W. Reder (eds), *Rational choice: The contrast between economics and psychology*. Chicago: University of Chicago Press.

Appendix: Task description

Here is presented a very simplified problem of relevance to military operations. The task is to be completed within 15 minutes, and without the aid of calculator or computer.

The state Utopia has on request from the Alliance agreed to contribute a number of military units to an international peace support operation. The contribution constitutes a minor fraction of Utopia's standing forces, for which there is no alternative domestic usage for the duration of the operation.

To equip one military unit with fuel, spare parts, provisions and ammunition for deployment costs 10 million NOK.[2] Consecutive sustenance (salaries, supplies, and so on) costs 1 million NOK per unit per day, as long as the unit resides within the area of operations. Each unit yields full effect during the first 20 days of operation, thereafter the effect gradually declines to zero during the subsequent 10 days (the

2 1 USD = approx. 7 NOK. 1 Euro = approx. 8 NOK.

daily sustenance costs remain the same, however). Every unit can therefore remain in the area of operations a maximum of 30 days, before it returns home to Utopia. The duration and costs of transportation between Utopia and the area of operations can be disregarded.

The Alliance will cover all costs to equip and sustain Utopia's military units, for the whole duration of the operation. At present time a daily sum of 160 million NOK is allocated. This amount covers consecutive sustenance of 120 operating units, as well as the initial equipment of four units daily. At the same time, four units return daily to Utopia (after 30 days of operating). This pattern of operation, which has been stable for more than two months, yields a total effect corresponding to 100 fully effective units.

The diagram below shows the situational picture:

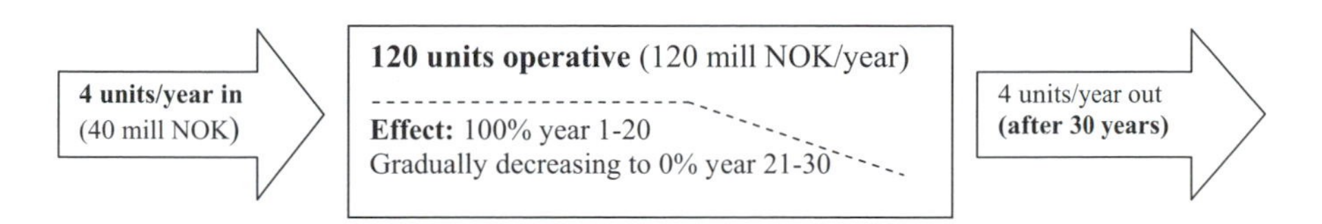

Task:[3]

The following directions for spending the daily allowances from the Alliance are given: *Of the allocated funds, the daily sustenance should first be covered. The remaining funds (if any) should be used to equip new units for deployment.* Until today the daily allowance has been 160 million NOK. Now the Alliance wants to increase the effectiveness in the area of operations for some time, and allocates 80 million NOK extra per day for five days (note: after five days the daily allowance is down to 160 million NOK again).

Assume that the above directions are followed; that there are available units for deployment; and that less than a single unit (for example, ½ unit) may be equipped. Use the supplied grid, and sketch the development in total effect from the time the extra allowance starts (day 1), and 40 days ahead in time. It is not important to draw an exact diagram, so long as the main features are present.

3 A second task was the question: "Is there a way to increase the total effect, without allocating extra funds? If that is the case, how?" Only the participant with SD background answered this question correctly; that is, to cut back on the number of days each unit is in operation.

Chapter 20

Critical Thinking in Tactical Decision Games Training

Karel van den Bosch and Anne Helsdingen

Introduction

As a result of changes in the international political situation, military missions are more and more focused upon peace-enforcing operations in regional conflicts. There is often uncertainty about the intentions, capabilities and strategies of the parties involved. Successful preparation, execution and management of military operations in complex and unstable conditions therefore requires competent commanders and staff personnel.

Recent studies have shown that experts in military tactical command treat decision making as a problem solving process (Cohen, Freeman and Thompson, 1998). Experts have large collections of schemas, enabling them to recognise a large number of situations as familiar. When faced with an unfamiliar tactical problem, experts collect and critically evaluate the available evidence, seek for consistency, and test assumptions underlying an assessment. They then integrate results in a comprehensive, plausible, and consistent story that can explain the actual problem situation.

Being able to interpret a tactical situation requires the adequate recognition and judgement of relevant factors (for example, weather, terrain, time of day,and so on). Insight into the nature of a particular problem is not so much the result of knowledge of individual factors, but more the appreciation of the specific combination of factors in the specific context. Experts capture such interrelated and contextualised knowledge in the form of mental tactical schemas (Schmitt, 1994). Novices do not (yet) have elaborated mental tactical schemas. They are therefore more inclined to focus on isolated cues and tend to take these at face value. Further, they are often not aware of assumptions they implicitly adopt to fill in missing parts; hence, they cannot be critical about them, and are more likely to "jump to conclusions".

If we want novices to become experts, training tactical command therefore needs to address two components: (a) expansion and refinement of tactical schemas, and (b) practice in solving complex and unfamiliar tactical problems.

Training tactical schema acquisition

Experienced decision makers can quickly and accurately achieve situation awareness in critical situations due to their large knowledge base of tactical patterns. Their experience enables them to make fine discriminations between cues and to detect anomalies in "prototypical" cases (Klein, 1998; Stokes, Kemper and Kite, 1997).

Acquiring expertise in a high-level complex skill like command and control is a matter of intensive, deliberate and reflective practice over time (Ericsson, Krampe and Tesch-Römer, 1993). It requires active engagement in situation assessment and decision making in representative and relevant cases. Studying such cases from different angles, and acknowledging the relevance of cues and their intercontingencies, help students in the build-up of mental tactical patterns.

Exposure to command and control situations can take place in operational and in training settings. Although the value of experiencing operational missions is undisputed, such missions are seldom ideal for learning tactical patterns. For one, commanders participate in only a small number of missions. Furthermore, the decision to assign a mission is usually based on operational considerations, not on a commander's training needs. Therefore, commanders are likely to be assigned to the type of missions with which they are already familiar. Secondly, the emphasis of current military missions is on peace-enforcing. Bringing these missions to a success depends heavily on a commander's competencies in for example, management, administration, logistics, negotiating, administration, but seldom requires handling combat situations. Thirdly, the course of such missions is normally too uncontrolled and unstructured for effective learning. Taken together, experiences in operational missions are valuable, but do not provide the conditions for effective and efficient learning.

In a training setting, the nature and difficulty of live and simulation exercises can be controlled, thus enabling the delivery of training tailored to identified training needs. However, again, this potential is seldom achieved, not in live training exercises nor in simulator exercises. Live exercises are expensive in terms of costs and organisational efforts, and are therefore organised scarcely. Furthermore, achieving training goals is only one of multiple goals in live exercises, and generally not the most important. More often than not, the focus is on determining operational readiness or team building.

Training tactics in simulators also requires a high overhead in terms of personnel. A popular way to solve this problem is to let students play the role of a team member. Sometimes, students receive more training as support player than as commander. Thus, even simulator exercises require substantial logistic and organisational efforts, making exploitation of simulator training costly and inefficient.

In sum, current operational and training practice provides commanders with insufficient opportunities to build up a framework of mental tactical patterns. What is required is a method by which to provide a commander, either along with the command team or not, with intensive, reflective practice over time. The skill of decision making can improve by learning to deal with specific cases and to approach problems from different angles. The method should enable the expansion and sophistication of an individual's database of mental tactical patterns.

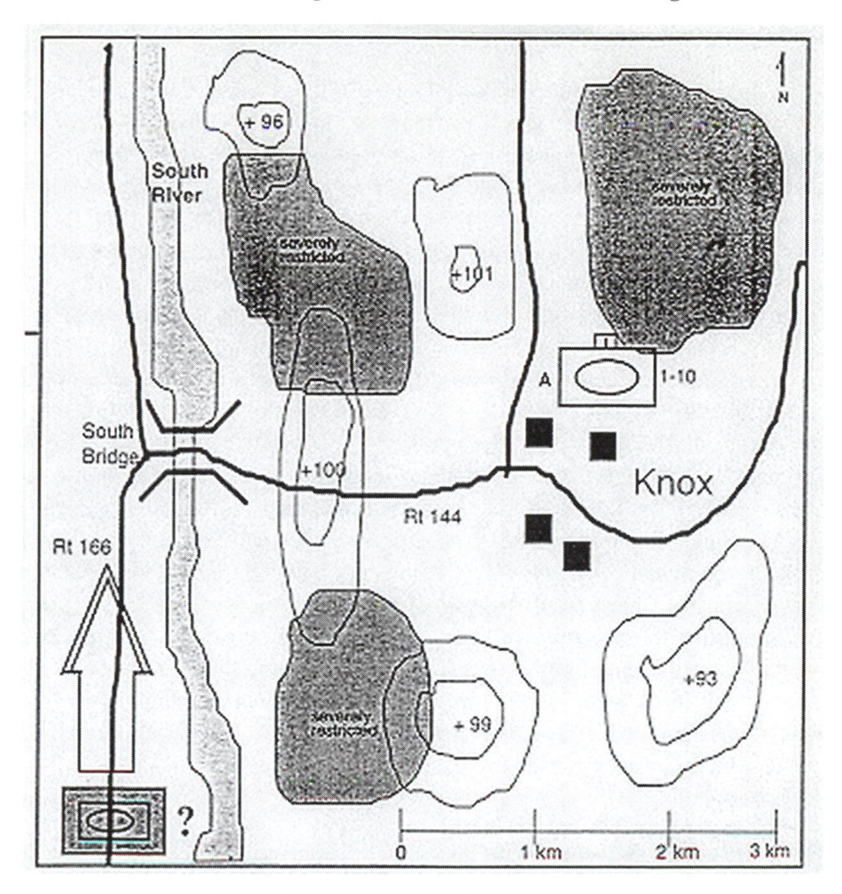

Situation: You are CO TM A, TF 1-10 AR. The TF is occupying hasty defensive position north in preparation for a morning attack to the south. They are approximately 5 km to the north. You and a TF scout section are the screening force for the TF. You have two tank platoons and one mech platoon. Currently you have halted your company north of Knox. TF scouts are ahead of you conducting a route recon south along Rt 166. Your mission is to provide early warning to the TF. You have permission to engage the enemy, but are not to become decisively engaged. The enemy, which has the ability to mass up company size units of T-62s and BMP-Is, is not expected to attack. You have priority of mortars and FA. As you survey the terrain to your front, you watch the scouts cross South Bridge and head south along Rt 166. Suddenly you hear MG and tank main gun firing west of the bridge. You try to contact the scouts but receive no answer. What is your plan?

Figure 20.1 Example of Marines-TDG

Tactical decision games

With this goal in mind, the US Marines adopted a low fidelity training technique to present tactical problems to trainees: Tactical Decision Games (TDGs) (Schmitt, 1994). A TDG is a tactical problem consisting of a short written scenario, a sketch or map, a requirement, and (optionally) a time limit. The scenario tells the players who they are, what they have for assets, defines their mission, and presents some type of enemy situation. The enemy situation is often vague and incomplete, forcing the players to make assumptions. Scenarios should be tailored to a commander's or a unit's training needs, and is preferably realistic and challenging. The written scenario is usually no more than a few paragraphs. See Figure 20.1 for an example (Gonsalves, 1997).

Participants make decisions, consider the consequences of a selected course of action by mental simulation (Klein, 1998), and compare this with other possible courses of action. The TDGs are not script-driven, but stimulate participants to review and discuss the reasons for why a particular decision was made rather than focusing upon the decision itself. The TDGs can be administered individually or to groups. They can be static, requiring trainees to develop a detailed and founded plan. However, they can also be dynamic by introducing events upon which trainees must respond.

The TDGs have been used successfully to present a wide variety of relevant tactical situations to trainees, and to enable them to practise situation assessment and tactical decision making (Gonsalves, 1997). The use of TDGs has been further developed and refined for civil emergency management training. Case studies show that TDG-training enhances planning, communication and decision making (Crichton, Flin and Rattray, 2000).

Training tactical problem solving

The approach of expert decision makers when handling difficult, unfamiliar and new situations has been used to develop a new training concept: critical thinking (CT) (Cohen and Freeman, 1997; Cohen, Freeman and Thompson, 1998). Critical thinking involves a problem solving approach to new and unfamiliar situations. It is a highly dynamic and iterative strategy, consisting of a moderately sized set of methods to build, test and critique situation assessments. These methods are to some extent general but they can only be taught if grounded in a specific domain and trainees already have a certain level of knowledge of that domain.

Effective training in critical thinking combines instruction with realistic practice (Cohen, Freeman and Thompson, 1998). The design of scenarios is very important since these have to provide opportunities to practise critical thinking processes, in particular:

- Producing different explanations for events
- Recognising critical assumptions of situation assessments
- Critiquing and adjusting assumptions and explanations
- Mentally simulating outcomes of possible decisions

One principal aim of critical thinking training is to keep trainees from assessing tactical situations solely on isolated events. Instead, trainees are taught when to collect additional information, and how they can integrate the available information into its context, which may include such elements as: the history of events leading to the current situation, the presumed goals and capacities of the enemy, the opportunities of the enemy, and so on. Trainees are instructed how to identify inconsistency and uncertainty, and how to adjust or refine their story by deliberate testing and evaluation. CT training also includes a procedure for handling time constraints.

Field studies showed positive effects of CT-training on the *process* of tactical command as well as on the *outcomes* (Cohen and Freeman, 1997; van den Bosch and Helsdingen, 2001, 2002). It stimulates trainees to produce a founded situation assessment, and helps them to anticipate alternative courses of events by developing contingency plans. The method supports not only individual commanders in situation assessment and decision making, but is also particularly suitable for team members to clarify their assumptions and perspectives on the situation to other team members (van den Bosch and van Berlo, 2002).

CT/TDG: a Navy application

Recently, we applied the concept of critical thinking training and the use of TDG in the training for CIC (Command Information Centre) officers. The Operational School of the Royal Netherlands Navy experienced a gap between theoretical classroom lessons and practical exercises in the tactical trainers. Theoretical lessons emphasised learning tactical procedures and properties of sensor and weapon systems. The relevance of the presented materials to tactical situation assessment and the implications for decision making often remained implicit. When, later in the training course, students were required to bring this knowledge into use during exercises in the tactical simulators, they often lacked the skills to do so. In order to solve this bottleneck, classroom lessons had to be redesigned in such a fashion that their content and form are congruent with the knowledge and skills associated with expert task performance. We aimed to achieve this by embedding critical thinking into Tactical Decision Games exercises. A series of four paper-based exercises were developed. See Figure 20.2 for an example of TDG.

Prior to the actual training sessions, scenario leaders and instructors were instructed extensively on the concept and principles of critical thinking. Observation protocols and performance measures were designed to support instructors in their tasks.

For the trainees, we developed an instruction book on critical thinking within the context of surface warfare, including self-study questions and exercises. In a two-hour classroom session on critical thinking, we familiarised students with TDGs and explained what was to be expected from them in the TDG-sessions. The TDGs were administered to groups of four students (see Figure 20.3).

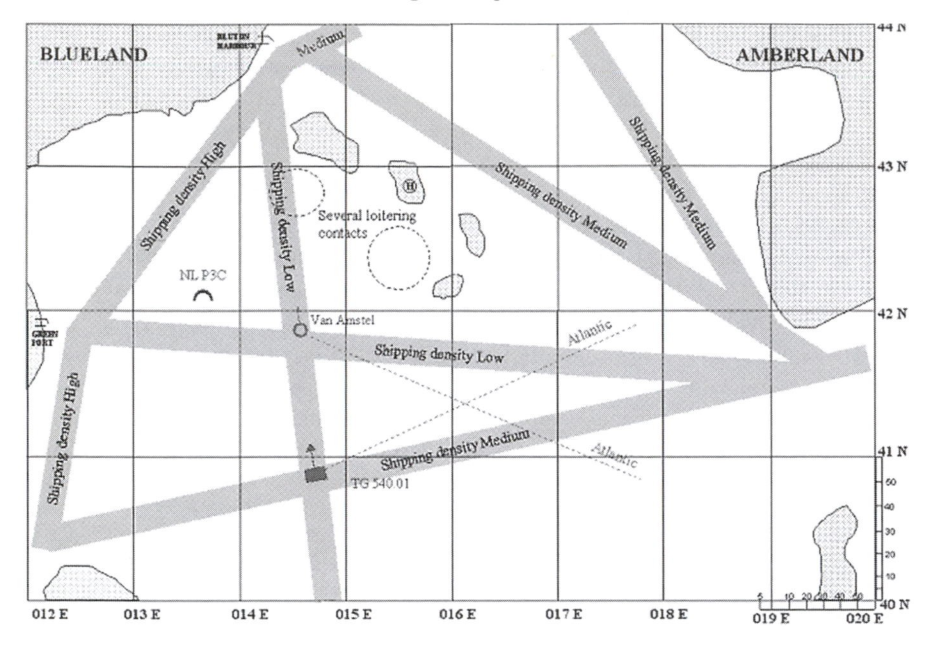

Mission: The mission of the Task Group (TG) 540.01 is the safe arrival of Her Majesty Rotterdam in the harbour of Bluton.
History and setting: NATO is at war with Amberland. Several land, sea and air battles have taken place. Amberland is determined to keep the islands off Blueland's coast occupied. Greenland's position is officially neutral, but intelligence information indicates they sympathise with Amberland. There is a busy merchant shipping lane along Blueland's coast. The TG picked up an enemy MPA's radar signal.
Task: You are the ASuW (Anti Surface Warfare) officer of the 'van Amstel', picket of the Task Group, sailing 60nm ahead of the TG. Develop your plan and at least one contingency plan.
Tactical issues: When students develop (contingency) plans, instructors observe whether the following tactical issues are taken into account: the loitering contacts in the area (who are they; what cues are used to assume their identity?); is our task group located and/or identified?; is the 'van Amstel' identified/classified?; if the helicopter is used for reconnaissance, what instructions regarding course, speed and emission are given?; do students take into account that towed array works only to the rear?

Figure 20.2 Abridged example of ASuW TDG

Figure 20.3 Group performing a TDG

In turn, one of them was assigned the role of observer using a scoring form to evaluate the group on the following dimensions:

- Information selection and acquisition
- Argumentation and reasoning
- Planning and contingency planning

In addition, an experimenter-observer also evaluated the group's performance. Students were asked to clarify their assessments, thus giving observers and the instructor access to the assumptions and reasoning underlying their decisions. In order to enhance critical thinking processes, the instructor guided the session by specific CT-exercises, like "now try to finalise your initial assessment into a story", or "now test your story upon conflicting, unreliable or incomplete information", or "identify a critical assumption in your story and apply the advocate-of-the-devil technique". After completion, each group presented their assessments, plans and contingency plans to the other groups. Tactical key decisions were discussed collectively.

Results and conclusions

After the series of exercises was completed, we interviewed students and staff about their experiences and asked them to fill in a questionnaire. The majority of students were enthusiastic about the training programme and tried to follow the instructions to their best. They responded to appreciate the exercises as a suitable method for consolidating and applying their tactical knowledge, and for practising their skills in tactical assessment and decision making. A few individuals, however, considered the concept of critical thinking not useful. This may have been due to insufficient domain knowledge required to conduct critical thinking as intended. For instance, these students had difficulty identifying critical assumptions in assessments, and were often unable to judge the tactical relevance of ambiguous information.

Instructors were of the opinion that the exercises will enable students to successfully prepare for the practical simulator training. They also argued that the required elaboration on the tactical issues helps students to develop tactical schemas, and that critical thinking helps in shaping the necessary strategic skills.

We observed that proper facilitation of exercises brought about a high workload for the instructors as they had to simultaneously teach, guide, monitor and assess critical thinking processes and task performance. Although all the navy instructors were enthusiastic and motivated, it is important to realise here that they are domain-experts in the first place. In a three-week course they learn the basic didactic methods of how to transfer knowledge and skills to students. The methods addressed in the course do not include the training principles described in the present chapter. For that reason it is understandable that instructors tended to fall back on traditional teaching techniques and provided students with the expert-solution too quickly. The importance of adequate preparation and training of instructors for delivering CT-training can therefore not be overstated.

The shortage of personnel and the high costs involved with developing domain experts into good instructors induces the defence organisation to investigate new forms of training that are less dependent (or even independent) of location, time and staff. Thanks to their entertaining and involving qualities, computer games are potentially suitable for this type of training. However, using games successfully for training purposes requires control over the scenario. For decision making tasks, the control problem lies in the fact that an instructor must be able to respond adequately to any situation emerging as a result of the trainee's decisions. It turns out that this is hard to automate in a computerised instructor agent. In contrast to software-agents, human instructors have the ability to take the context into account when evaluating (on-line) the appropriateness of trainee behaviour and to assess whether the training scenario develops in the intended direction. That is probably the reason why human instructors are now still in control in simulation-based or game-based decision making training. However, human instructors have their disadvantages as well. One problem is that these experts tend to evaluate trainee performance intuitively, without being able to precisely point out which cues (or absence of cues) they use for diagnosing trainee behaviour. Furthermore, experts often differ in opinion on what is to be considered appropriate and inappropriate behaviour. It is clear that this hampers transparency of performance measurement and feedback. Finally, the need for domain experts elevates costs of training and requires high organisational and logistic efforts. If we can develop software agents that (semi)autonomously and intelligently evaluate tactical assessments and decisions, training can become more traceable, more systematic, and more cost-efficient. Promising research in that direction has taken the approach of developing agents whose behaviour is a function of simulated cognitive processes (for example, beliefs, intentions, goals) (Norling, 2003; Zachary et al., 2000). The heart of such agents is a cognitive model. A cognitive model represents the knowledge and cognitive processes of an individual or entity (for example, the instructor) in a certain domain, task or scenario. This representation needs to be so specific that, when provided with input, the cognitive model produces realistic behaviour as output. There is growing evidence that cognitive modelling can be used successfully to improve tactical training (van Doesburg and van den Bosch, 2005).

This pilot study has explored the effects of intensive and reflective practice in situation assessment and decision making in tactical decision games. Results are promising and have led to ideas on formalising these principles for use in game-based environments. More standardised and formal methods of performance are needed to fully evaluate the strengths and limitations of this approach.

References

Bosch, K. van den and Berlo, M. P. W. van (2002). *Training en evaluatie van tactische commandovoering*. (In Dutch) (Report No. TM-02-A025). Soesterberg, the Netherlands: TNO-TM.

Bosch, K. van den and Helsdingen, A. S. (2001). "Critical thinking in tactical command: A training study". In *Proceedings of Conference on Simulation Technology for Training (SimTecT)*. Canberra, Australia. SimTecT; 2001.

Bosch, K. van den and Helsdingen, A. S. (2002). "Improving tactical decision making through critical thinking". In *Proceedings of Human Factors and Ergonomics Society*. Held at: Baltimore, MA. HFES.

Cohen, M. S. and Freeman, J. T. (1997). "Improving critical thinking". In R. Flin, E. Salas, M. Strub and L. Martin (eds), *Decision making under stress: emerging themes and applications* (pp. 161-169). Brookfield, Vermont: Ashgate.

Cohen, M. S., Freeman, J. T. and Thompson, B. B. (1998). "Critical thinking skills in tactical decision making: a model and a training strategy". In J. A. Cannon-Bowers and E. Salas (eds), *Making decisions under stress: implications for individual and team training* (pp. 155-190). Washington, DC: American Psychological Association.

Crichton, M., Flin, R. and Rattray, W. A. R. (2000). "Training decision makers: tactical decision games". *Journal of Contingencies and Crisis Management*, 8(4), 208-217.

Doesburg, W. A. van and Bosch, K. van den (2005). "Cognitive Model Supported Tactical Training Simulation". In *Conference on Behavioral Representation in Modeling and Simulation* (BRIMS) (pp. 313-320).

Ericsson, K. A., Krampe, R. Th. and Tesch-Römer, C. (1993). "The Role of Deliberate Practice in the Acquisition of Expert Performance". *Psychological Review*, 100(3), 363-406.

Gonsalves, J. D. (1997). *The Tactical Decision Game: An invaluable training tool for developing junior leaders* [Web Page]. URL: http://call.army.mil/products/etc_bull/armor/mayjun97/ article1.htm.

Klein, G. (1998). *The source of power: how people make decisions*. Cambridge, MA: MIT Press.

Norling, E. (2003). "Capturing the Quake Player: Using a BDI Agent to Model Human Behaviour". In *Proceedings of the Second International Joint Conference on Autonomous Agents and Multiagent Systems*, pages 1080-1081, 2003.

Schmitt, J. F. (1994). *Mastering tactics: A tactical decision games workbook*. Quantico, VA: Marine Corps Association.

Stokes, A. F., Kemper, K. and Kite, K. (1997). "Aeronautical decision making: cue recognition and expertise under time pressure". In C. E. Zsambok and G. Klein (eds), *Naturalistic Decision Making* (pp. 183-196). Mahwah: Lawrence Erlbaum Associates.

Zachary, W., Ryder, J., Santarelli, T. and Weiland, M. (2000). "Applications for executable cognitive models: A case study approach". In *Proceedings of the Human Factors and Ergonomics Society 44th Annual Meeting*. Santa Monica, CA: Human Factors and Ergonomics Society.

PART 5
Teams and Complex Decision Making

Chapter 21

Why Training Team Decision Making is Not as Easy as You Think: Guiding Principles and Needs

Eduardo Salas, Joseph Guthrie and Shawn Burke

Introduction

Team Decision Making (TDM) is a field that has emerged out of the military and organisational need for using teams to perform complex, interdependent, dynamic and ambiguous tasks. Most teams in the military and in some industries perform in environments where there is no right answer, where there is a need for continuous information seeking, and where the consequences for errors are costly. Given these constraints it becomes paramount that teams make optimal decisions. Teams must use all available information and resources to ensure the decisions they make are safe, efficient and ultimately lead to successful task completion. Unfortunately, this does not always happen; some teams derail and fail in achieving their goals. This is partly due to the complexity inherent in team dynamics and the environments in which they operate. So, it is not easy to be consistently effective in team decision making. We must turn our teams of experts into expert decision making teams. And training is one way to accomplish this.

TDM training systems have been developed to "meld individual skills into a team capability that enables members to coordinate and adapt their expertise" (Kozlowski, 1998, p. 122). They usually begin with classroom training for individual knowledge acquisition, followed by training on how to coordinate each individual's expertise with other experts on the team (Kozlowski, 1998). Teams are then given the opportunity to practise TDM skills in simulations. However, as simulations are often not available or cost-effective, teams are regularly assigned real tasks to perform before they have truly learned to coordinate their decision making efforts. Within these "real world" environments little is known about the most effective way to design and deliver TDM systems, and specifically no precise theoretically sound, empirically validated guidelines for designing TDM training systems currently exist.

Why training TDM is not as easy as you think

Why is this the case when much is known about training (Salas, Cannon-Bowers and Smith-Jentsch, 2001), when much is known about team effectiveness (Salas, Stagl and Burke, 2004) and decision making processes have been widely studied (Salas and Klein, 2001)? One reason is that researchers and practitioners from these three fields seldom communicate with each other, nor do they frequently collaborate or engage in systematic research aimed at understanding TDM and its requisite training requirements. Therefore, in order for the field of TDM training to move forward, a paradigm shift must occur. We need a new perspective to leverage all we know from these distinct but related research foci. We elaborate next on why the shift is needed and how it can be brought about.

Why a paradigm shift in TDM training is needed

What does the literature tell us?

Not much. TDM research is in its infancy. The field has not yet provided a precise definition or benchmark for effective TDM. There are a number of reasons for this. First, although there are many theories of decision making in general, few theories are specific to TDM and TDM performance. Few theories explicitly link team-level constructs to decision making processes. Consequently, in the absence of theory there are no guiding, testable hypotheses. In the absence of guiding hypotheses, little (to none) empirical work has focused on uncovering what competencies translate to effective TDM performance.

Of the scant research that has been conducted on TDM, the methodologies used are often times subjective, obtrusive, labour-intensive and weak at best. More importantly, most research has been conducted as laboratory studies, with naïve participants performing contrived tasks. Based on what is known about naturalistic decision making, it is doubtful that much of this work will generalise to team decision making in the field. Clearly what is needed is research that focuses on TDM "in the wild". In sum, the literature does not help much at this point.

What do the experts in the field say?

Experts in the field have their own viewpoints on what TDM training systems lack. Experts often indicate that training systems are not realistic because they only train "X", when in the real world teams must consider "X, Y and Z". In addition, because of the lack of specificity in the training tasks, trainees sometimes are not sure what they are supposed to have learned. Furthermore, TDM training systems do not offer enough tools and exercises to adequately train team coordination skills. Nor are they designed with the specific learning objectives and skills critical to TDM performance at the forefront. In sum, those who are tasked with TDM often find that fielded simulations and training systems are not designed to support or develop the skills needed.

What do the instructors say?

Instructors of TDM training systems also have concerns. Many of these concerns are similar to those of the experts. Instructors also note that the scenarios required in the use of simulations take too long to create and in the end are unrealistic. The lack of realism in the training scenarios causes the trainees to doubt the validity of such systems and consequently the TDM training system as a whole is often not taken seriously. In addition, instructors often do not see the need to measure objectively what researchers and theoreticians would deem critical aspects of performance, either because they would rather go by "gut feeling" or because they claim to "already know" how well the team is performing. An additional component to the above conundrum may be the fact that most systems do not make it easy for instructors to gather information to later be fed back to the team. In sum, instructors want a training system that is easy to implement and use, has face validity, and that provides objective and diagnostic performance data that can be easily translated into feedback for the teams in training.

Given the lack of theoretical guidance and the apparent need both from those charged with TDM as well as the instructors, something drastic needs to happen … a paradigm shift. A paradigm shift requires better science, robust methods and precise practice. A key component in moving towards these goals is for those involved in team, training and decision making (research as well as application) to better communicate, coordinate and integrate their findings. For example, practitioners in all three fields need to understand what is known about each domain and what is not known. If not, TDM training, for example, may fall prey to several myths that still prevail in training (see Table 21.1). The prevalence of these myths has often resulted in ineffective training delivery due to poor design and/or methods.

Table 21.1 Prevailing myths in training

Myths	Reality
SMEs can articulate training needs and design the training	Not so! Must have partnership with learning experts
All you need is to expose trainees to the environment	Not Enough! Must have guided practice
Practice, Practice, Practice	Not enough! Give performance feedback
The more we know about the context, the better for training	Take simplistic views of training
Engineering and technology will solve the problems	Not quite; Learning is a behavioural/ cognitive event
Training is an event	No, it requires continuous learning

What a new paradigm requires

Based on the deficits of current TDM training literature and state of the practice, the following sections will briefly discuss the requirements of a new paradigm and provide a set of guiding principles for team, training and decision making researchers to use in the coordinated development of better TDM training systems. These guidelines (see Table 21.2) can be delineated into three major categories, those dealing with: clarification of TDM construct, learning methodologies, and meaningful assessment of dynamic TDM performance. Each of these categories will be elaborated upon.

Table 21.2 TDM guidelines

1:	TDM must be theoretically based
2:	TDM must adopt a systems approach – before, during and after
3:	TDM training must be learner-centred
4:	TDM must provide relevant information; demonstrate effective decision making performance; create opportunities to practise; diagnose performance; receive feedback and remediate
5:	TDM must guide the effective behaviours/cognitions and these need to be reinforced
6:	TDM must clarify expectations, early
7:	TDM must set a climate for learning
8:	TDM must encourage participation/feedback (constructive) by team members – self-correct
9:	TDM must create scenarios that provide opportunities to practise TDM performance – 'The scenario is the curriculum'
10:	TDM must create events (or set of related) that allow diagnosis of performance – dynamic assessment

Integrating theoretically rooted descriptions of TDM performance

We need better theories of TDM performance in order to combine individual skills to form a team of competent decision makers. These theories should be based on what we know about team effectiveness and TDM performance. They must be parsimonious, coherent and integrated. Not an easy task, but much needed.

Principle 1: The design of TDM training must be theoretically based.

In developing theories for TDM, researchers must look to the approach taken by naturalistic decision making (NDM) theorists in developing models and frameworks based on how decision making occurs in real world situations (Lipshitz et al., 2001).

In addition, they must also look to experts in training to design systems based on what is known about how they perform "in the wild". This is also difficult, but a must.

Principle 2: Those charged with creating TDM training must take a systems approach and consider how the environment prior to, during, and after training may impact its success.

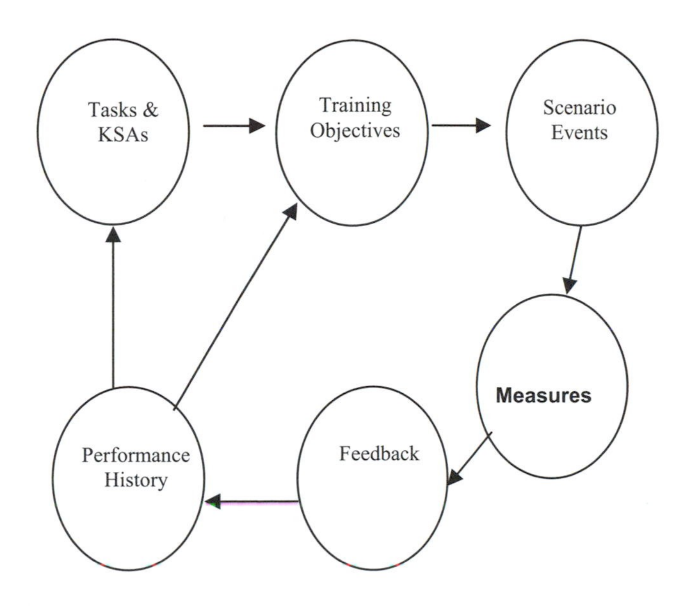

Figure 21.1 Development cycle of SBT
(Adapted from Cannon-Bowers, Burns, Salas and Pruitt, 1998.)

Integrating a robust learning methodology

One method that has proven very effective in creating real world scenarios that allow for specific training objectives to be accomplished is scenario-based training (SBT) or event-based training (EBAT; see Figure 21.1) (Prince et al., 1993; Oser et al., 1999). SBT begins with a job/task analysis to define a skill inventory, which is then translated into training and learning objectives.

Principle 3: TDM training must provide information deemed relevant to the problem at hand.

Principle 4: TDM training must clarify expectations early by setting clear learning objectives.

The development of actual training scenarios in SBT is based on identified learning objectives. Events are embedded within each training scenario to allow the trainees to practise the specific learning objectives targeted in the training system (Oser et al., 1999). The embedded events, specific to each training objective, vary in difficulty and occur at multiple times throughout the scenario.

Principle 5: Create training scenarios that provide opportunities to practise TDM; "scenario is the curriculum".

Principle 6: Practise using the targeted TDM competencies through the use of guided practice in the form of a priori defined embedded events.

SBT explicitly links learning objectives, exercise events, performance measures, and the associated feedback (see Figure 21.1), but also allows for standardised measurement and training and reduces the workload for those in charge of observation and collection of performance data. Because performance measures are coupled with defined events, the observer does not have to observe every instance of behaviour in order to collect performance data.

Principle 7: TDM training must create opportunities for practice of targeted behaviours and ease the burden on those observers that must diagnose performance.

Integrating the ability to assess meaningfully dynamic performance

After the events and the associated scenarios are created for SBT, measurement instruments need to be developed that will be used to assess task performance during each a priori defined event. Within TDM training and real world contexts an unfulfilled need is the development of assessment instruments that will allow near real time assessment.

Principle 8: Measurement tools must facilitate dynamic assessment.

While dynamic assessment is important for developing accurate mental models, measurement tools should also allow the assessment of both process and outcome level feedback. While outcome level feedback informs one as to whether the team was successful in its decision, process feedback is diagnostic in that it conveys the important lesson of how the team reached the outcome. Moreover, this feedback should be tied to learning objectives. Measurement instruments using an SBT approach facilitate this need in that performance measures are developed based on identified learning objectives and scenario events.

Principle 9: Measurement tools must capture process as well as outcome feedback.

The development of such measurement tools requires a multi-disciplinary approach that receives input from team, training and decision making researchers so that all critical aspects of TDM performance can be assessed. By developing standardised measurement tools that target specific learning objectives and team processes, researchers will be able to go beyond providing practice and feedback for TDM training. Then, researchers will also be able to develop and provide teams with the diagnostic tools to explain the processes necessary to successfully complete tasks involving decision making.

Principle 10: TDM training must encourage team participation in delivering constructive feedback that promotes team self-correction.

Conclusions

As this chapter indicates, a paradigm shift is needed in TDM in order for significant progress to be made in the design and delivery of TDM training systems. There is much to be done in order for this shift to happen. Team, training and decision making researchers must communicate and work together to produce better theories, methods and a better understanding of what we know about TDM. In addition, the research conducted needs to be a combination of laboratory, simulations and "in the wild" studies that provide systematic assessments and evaluations of TDM performance. We hope that this chapter motivates the beginning of that shift. This community has the tools (and willingness) necessary to make it happen.

References

Kozlowski, S. W. J. (1998). "Training and developing adaptive teams: Theory, principles and research". In J. A. Cannon Bowers and E. Salas (eds), *Making decisions under stress: Implications for individual and team training* (p. 115-154).Washington, D.C.: American Psychological Association.

Lipshitz, R., Klein, G., Orasanu, J. and Salas, E. (2001). "Focus article: Taking stock of naturalistic decision making". *Journal of Behavioral Decision Making*, 14, 331-352.

Oser, R. L., Cannon-Bowers, J. A., Salas E. and Dwyer, D. J. (1999). "Enhancing human performance in technology-rich environments: Guidelines for scenario-based training". In E. Salas (ed.), *Human/technology interaction in complex systems* (Vol. 9, 175-202). Greenwich, CT: JAI Press.

Prince, C., Oser, R., Salas, E. and Woodruff, W. (1993). "Increasing hits and reducing misses in CRM/LOS scenarios: Guidelines for simulator scenario development". *International Journal of Aviation Psychology*, 3(1), 69-82.

Salas, E., Cannon-Bowers, J. A. and Smith-Jentsch, K. A. (2001). "Team training". In W. Karwowski (ed.), *International encyclopedia of ergonomics and human factors* (Vol. 2. pp. 1391-1393). London: Taylor & Francis.

Salas, E. and Klein, G. (2001). "Expertise and naturalistic decision making: An overview". In E. Salas and G. Klein (eds), *Linking expertise and naturalistic decision making* (pp. 3-33). Mahwah, NJ: Lawrence Erlbaum Associates.

Salas, E., Stagl, K. and Burke, C. S. (2004). "25 years of team effectiveness in organizations: Research themes and emerging needs". In C. L. Cooper and I. T. Robertson (eds), *International review of industrial and organizational psychology*. New York: John Wiley & Sons.

Chapter 22

The Migration of Authority in Tactical Decision Making

Sidney Dekker and Nalini Suparamaniam

Why is coordination in disaster relief work difficult?

What makes coordination in international disaster relief work difficult, and why does new technology not necessarily help? This was the compound question at the outset of the research project we describe here. The question is based on an assumption: coordination difficulties in international disaster relief work are primarily about "getting in touch with one another". Dynes (1989), for example, highlighted how collection and distribution of information can help relief workers explore and exploit available action alternatives. This would mean that anything that helps people in the field (team members, team leaders, representatives of aid organisations) "get in touch" better or quicker (or that lets them get in touch at all in the first place) should improve their work. It does not (Suparamaniam, 2003). This assumption, that the provision of access (either to people or computer systems) is key to solving coordination problems, is not unique to this field (for example, Dekker and Hollnagel, 1999).

For three years, we endeavoured to understand the coordination difficulties as experienced by relief workers in the field and inside mother organisations at home. We participated in formal and informal meetings at all levels of disaster relief organisations, attended local and international training exercises and disaster simulations, and reviewed considerable archival material. We interviewed over 150 relief workers and managers, and ended up with thousands of pages of field notes and transcripts. We wanted to make authentic contact with the perspectives and experiences of those involved with the work of our interest; we needed a way to view their world from the inside-out; a way to see the daily pressures, trade-offs and constraints through the eyes of those confronted by them. The more we learned, the more we had to find out: besides raw data, informant statements were a consistent encouragement to go deeper into the world of disaster relief work to discover how people relied on social means of constructing action and meaning. Our analysis ended up being cyclical: more findings would demand more analysis, which would prompt further findings. There was a constant interplay between data, analysis and theory. Where existing theory was lacking, we generated and tested new concepts; where data was confounding, we turned to theory to map, compare and contrast (cf. Strauss, 1987).

Our initial findings on coordination difficulties were not about new technology at all. Instead, our efforts told us rather quickly that the military command-and-control model, described and critiqued by Dynes (1989) as unfit for disaster relief, is alive and well. There is still no evidence of a greater awareness of the value of "auftragstaktik" (or management by objectives, see also Rosenberg, 1998), which would delegate considerable executive authority and degrees of freedom to low-level team leaders as long as an agreed goal or set of goals is met. Despite Dynes' suggestions to shift from "command" to "coordination", and from "control" to "cooperation", and more recent research into the same types of problems (for example, Snook, 2000), there is little evidence of an increased understanding that disaster relief work is not the same as responding to military contingencies. As a result, the organisation of relief work is often still based on false assumptions and misdirected efforts. Much time is spent on documenting and emphasising authority relationships; it is assumed that decision making and authority should be centralised; plans are over-specified which leads to incomplete knowledge and immediate drift from formal plans and procedures once workers enter the field; and communication is assumed to be downward only: the top of the structure knows what must be done (Dynes, 1989; Suparamaniam, 2003).

Our research justifies Dynes' (1989) scepticism of these assumptions. Those at the top do not typically know what to do. In fact, one of our central observations is a dissociation of knowledge and authority – creating a paradox of power. People in disaster relief either have the knowledge to know what to do (because they are there, locally, in the field, but they lack the authority to decide on implementation). Or people have the authority to do it (but then lack the knowledge). Knowledge and authority are rarely located in the same actor. Interestingly, efforts to coordinate the one with knowledge and the one with authority do not seem to solve much of this central dilemma.

Coordination is difficult but often successful

Deference to protocol, procedure and hierarchy is partly a result of the background of many disaster relief workers (indeed, military) and is something that will even increase with shifts in military roles from fighting wars to dealing with disasters. But we also found another powerful ingredient in the hysteresis. The implementation of unplanned action and use of unplanned resources (certainly from elsewhere) is almost never unproblematic, despite compelling local convictions that such help may be critical. Disaster relief work is also about spending and controlling separate national budgets; about protecting or bolstering national reputations; about making political statements or investing in diplomatic capital. Such higher-level constraints on decision making (that is, sensitivity to political, financial, or diplomatic implications of decisions), the subtleties of which may elude local team leaders, demand bureaucratic accountability (cf. Vaughan, 1996) and centralisation (Mintzberg, 1979).

But our research is not just a story about the relentless hysteresis of complex systems; it does not just confirm the basic irrationality of organisations (for example, March and Olsen, 1979), or coincide with sociological characterisations of the immoral calculation by hierarchies and power structures (Vaughan, 1996), nor does it just re-invent the vaunted introduction and subsequent veiled discard of highly praised technologies that end up not helping at all (Woods, Johannesen, Cook and Sarter, 1994). We found that people, especially team leaders in the field, often make their missions work anyway in spite of the adversity, the difficulty and the countervailing pressures of procedure and protocol.

The interesting question, then, is not just why coordination in international disaster relief work is difficult. In this respect, the data we gathered from field missions, exercises and simulations (Suparamaniam, 2003) can be reconciled with existing theories on contextual constraints and difficulties surrounding real decision making, such as surprises (for example, Rochlin et al, 1987; Weick and Sutcliffe, 2001), dynamics, and uncertainty (Orasanu and Connolly, 1993; Zsambok and Klein, 1997). Demands for making decisions and taking action (and thus for coordination) go up with the tempo and criticality of operations (see Woods et al., 2002). The interesting question, really, is how international disaster relief missions' outcomes are often successful, despite the odds that are overwhelmingly stacked against local actors. Attention to such a question is consistent with interest in sources of "robustness" (Woods and Cook, 2002) or "grace" of human performance under severe pressure (see also Weick, 1987; Rochlin, 1999). What role does interpersonal coordination play in creating such success; in exhibiting such robustness? Naturalistic decision theory sheds light on the local, individual mechanisms that determine successful decision performance under limited resources and uncertainty, adding the concept of "team mind" (Klein, 1998) as one of the interpersonal sources available to decision makers. This overlaps with notions of distributed cognition (for example, Hollan et al., 2000), where decision makers draw on and integrate resources outside their own minds – including other people – to perform successfully. Both approaches, however, see the constitution of the team (or distributed cognitive architecture) as essentially non-problematic. The process of forming the distributed architecture is less interesting than modelling how the architecture is used once it has been formed.

Rochlin (1989) comes closer to the formative mechanisms in his description of informal organisational networking as a strategy to avoid crises on Naval Aircraft Carriers. It covers the spontaneous, informal creation of teams whose composition is made to map onto the functional demands of the problem-to-be-solved. In Rochlin's observations, organisational hierarchies agreed with the relevance of such informal work, and indeed, ad hoc teams' operational success appeared possible in large part because of their unproblematic relationship with formal organisational and command structures. Also, the teams were co-located (on the same ship); members (though differing in rank and functional specialisation) shared a common Navy indoctrination and participated in teams willingly. International disaster relief work, in contrast, is governed by less coherent, more distributed and occasionally competitive formal hierarchies that hardly know, understand or acknowledge the existence of such informal local networking – let alone condone or encourage it. Informal teams do not often share a common indoctrination (or even mother tongue), members are not necessarily co-located (they may be in the same country), and may be recruited against their explicit will or against their better political judgement. Finally, informal teams in international disaster relief work do not form to avoid crises, but to deal with crises that have already happened.

Renegotiation of authority

The separation of authority and knowledge in international disaster relief work creates a tension, an imbalance. Imbalance creates pressure for change. Such change can go two ways. Either knowledge goes where authority resides, or authority goes where knowledge resides. We found out that getting knowledge to the seat of authority is difficult. Team leaders often do not even know who has formal authority to decide over the particular problem at hand: they do not know whom to ask. And even if they (think they) do, their request may be forwarded to other levels, participants or agencies. The dispersed nature of relief work (different countries, governments, head offices, mother organisations), and mechanisms of bureaucratic accountability that tend to shift issues higher and higher before decisions are made, make the location of authority for every particular problem unstable except that such authority does not lie at field level. Planners may have assumed authority on part of organisations or people who in fact have no formal mandate at all (Dynes, 1989). The trajectories along which authority travels before action is finally taken can sometimes be profoundly puzzling.

The opposite way, of getting authority to where knowledge is, is relied on much more often. Compelled to act eventually (or immediately), local team leaders frequently take charge even where no formal authority is mandated to them. This typically occurs through a process of mutual adjustment that involves a "flattening" or apparent disregard of formal hierarchy, not unlike that described in contingency theory (for example, Mintzberg, 1978). "Taking charge", however, hides a number of different adjustments that are made in such situations. Indeed, knowledge of local problem demands is necessary but not sufficient as a basis for "taking charge". Local availability of resources such as equipment (for example, trucks, food, medical supplies, tents), personnel (manpower), or expertise (functional specialisation) is a major determinant for the direction in which authority eventually moves. Knowing what to do is one thing; being able to carry the actions out may be quite another. Knowledge and resources to act on that knowledge may not be co-located in the same team leader either: further negotiations with other team leaders may be necessary to coordinate the understanding of problem demands with the delivery of resources to deal with those demands. In these cases, partial authority for implementing actions and directing operations can shift to the one who actually has the resources, away somewhat from the one who had initial knowledge of what should be done. Such negotiations, however, typically play out at field level and appear possible largely by processes of mutual adjustment without much constraint from higher-order goals or imperatives. While solving local problems, the migration of authority also creates problems. These problems are often related to the lack of preparation of team leaders for the possibilities of this imbalance (between knowledge and authority); the emphasis on procedure and protocol, and the affective way in which authority is renegotiated. We discuss these problems below. Finally we return to one of the questions that inspired our research: why does new technology not always help? Here we have to turn to issues of trust (building it through technical media may be difficult) and data overload (more technology means more data, not necessarily all of it meaningful).

Training and preparation, procedure and protocol

Moving authority to the location of knowledge and resources is an adjustment that is hardly ever officially acknowledged. It is not trained for, nor are the reasons or practical need for such migrations of authority officially discussed in the preparation of team leaders. Training rarely touches on the possibility of conflict between demands for local action and global interests held by stakeholders further up in hierarchies. In fact, the professional indoctrination of those who become team leaders (for example, rescue services or military personnel) as well as the training that prepares them for field work, stresses allegiance to distant supervisors and their higher-order goals (cf. Shattuck and Woods, 2000). The superordination of larger, global concerns makes sense from the perspective of those tasked with organising relief work. Without political backing, without the requisite diplomatic leverage, without financial resources, there would be no basis for relief work in the first place, so sensitivity to these aspects (and to not squandering them) is understandable.

Training also emphasises adherence to formal procedure and protocol (cf. Snook, 2000). Such preparation initially makes sense because team leaders may not know their team members, and team leaders may themselves have little experience in the field. It ensures a measure of order and predictability, which in turn can generate a form of "common ground": a stable basis on which to form expectations about the actions, intentions and competencies of other participants (see Dekker, 2000). Commercial aviation similarly relies on procedure and protocol since crewmembers, especially in larger carriers, rarely know each other. Knowing what the other will and can do is based predominantly on procedure, protocol, *a priori* role divisions (pilot-flying and pilot-not-flying) and formal command structure (Captain-First Officer relationship). In international disaster relief work, however (and quite to the contrary of commercial flying), unpredictability is quite common. Reliance on procedures and formal hierarchical protocol quickly becomes brittle in the face of novelty and surprise (Dekker, 2003). Indeed, while team leaders and team members appear to use procedures and protocol as a dominant resource for action in the beginning of missions, increasing experience with one another, and with the typical problems will make them less reliant on pre-specified guidance. Practical experience (both with each other, and with typical problems) gradually supplants "the book" and the official command structure as a resource for action.

Different ways to renegotiate authority

Authority is renegotiated in a number of ways and styles, some of which go down better with relief participants and organisational members than others. In some circumstances, renegotiations would work without hurting feelings (so to say) and sometimes it would not. Two broad avenues were covered where there was a certain gracefulness associated with the renegotiation, or such a grace could be lacking, or at least judged to be lacking by the aggrieved party. Another broad dimension was the degree of mutuality where both parties are equally involved or asserted where it is conducted by one party without much involvement from the other. When authority is asserted, there is little involvement from the party who has to surrender authority. Such assertions, however, may still happen with a certain grace (see Figure 22.1).

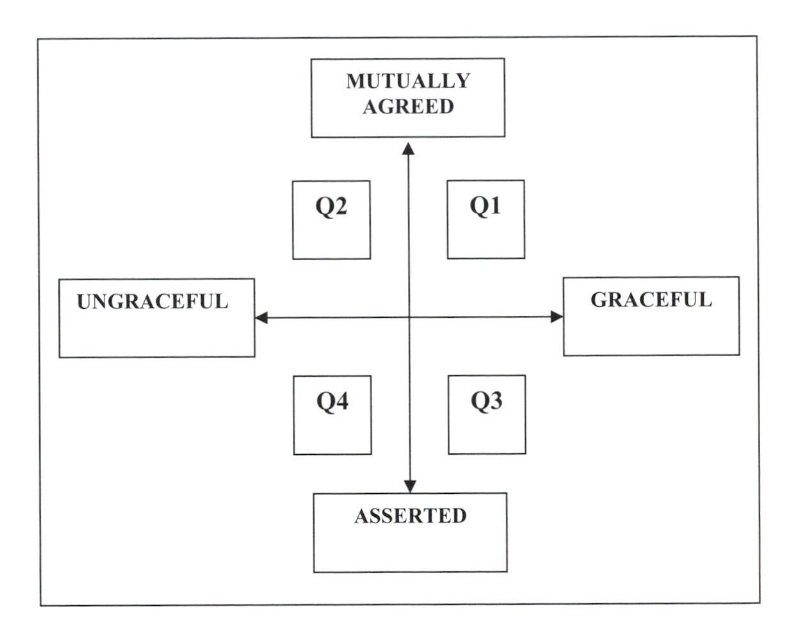

Figure 22.1 Two independent dimensions along which rescue workers renegotiate authority

Note: Processes of renegotiation can be either mutually agreed or asserted, and either graceful or ungraceful. We identified four possible quadrants where rescue workers could place themselves, with Quadrant 1 being the most desireable.

Themes of common ground and experience are consistent with findings from other domains where authority sometimes has to migrate to manage challenging or safety critical situations where prior protocols do not provide support. It is also clear that training, experience and common ground are factors that lack grace and assertiveness in renegotiations of authority. In other words, the more training, the more experience, and the more common ground, the easier it becomes for people to renegotiate authority, to let authority go, or to assume it when necessary. Training is good, and prior protocols are good, but even these are subsumed by experience and common ground. If people know the typical challenges, and if they know the people they are dealing with (and know how these people will react), then renegotiating authority becomes much easier. While no amount of training can substitute for such an accumulation of experience and common ground, it still begs the interesting question of how we can "cheat" the build up of experience, by making people experience these types of situations in advance of actual work. This same question dominates naturalistic decision making research: you cannot make people experienced – they have to do that themselves (see for example, Zsambok and Klein, 1997). But people can be trained by borrowing from their practical future and giving them as much "experience" before they actually start accruing real experience. Recent contributions (for example, Klein, 2002) are providing a direction for progress in this regard.

New technology: capabilities and complexities

In order to renegotiate authority, or sometimes even to realise that authority needs to be renegotiated, people need to coordinate. They need to coordinate across distances and across organisations, teams and countries. Communications technology (for example, e-mail, faxes, telephones) assists as the medium for these interactions. Here the question was on how technology augmented coordination. Our data pointed to a paradox: a mismatch in which technology both augmented coordination but also caused difficulties that interfered with coordination. Technology and human work in other work domains (for example, Dekker and Hollnagel, 1999) often show a similar tension between delivering benefits and creating problems at the same time. Technology is often introduced on the back of promises of quantitative benefits and with the idea that there is a "technological fix" for whatever problems there are with coordination.

Communication technology that helps relief workers coordinate cannot in itself not deal with issues of hierarchy, goal conflicts or authority dislocation. Indeed, technology can speed up coordination but may in fact create data overload. It can help create common ground, but trust in each other (a critical component of teamwork) requires more than contact through a radio or other technological aide. Building trust through technical media may be difficult and exacerbate data overload. More technology easily means more data – not necessarily all of it meaningful. The idea of information that can or cannot be "trusted" is consistent with meaning in the sense in which ecological psychologists use it (for example, Zsambok and Klein, 1997; Vicente, 1999; Flach, 2000). Meaning lies not in the data but in the relationships of data to interests and expectations (Suparamaniam, 2003). In addition to trust in meaning, another aspect of trust via technology included the distance of the person giving commands or information. Our informants were distrustful of data that was transmitted via the various technology aids and had difficulties communicating during coordination.

Finally, getting the one with knowledge in touch with the one with authority does not resolve the likely tension between the two agendas driving each party (one confronted with immediate, local goals; the other governed by larger, global concerns) (see Shattuck and Woods, 2000). Nor does it address the fundamental problem of underspecification (provided communication technology works in the first place): formal hierarchies' understanding of what is going on locally is limited to what can be pushed through the narrow channel of an e-mail, radio message, telephone conversation, or fax. This is not quite the same as "being there".

Conclusions

So why is coordination in international disaster relief work difficult? Perhaps the most foundational reason is that coordination in international disaster relief work is not just coordination. It is *renegotiation*. Coordination is often renegotiation of authority – of who has the say in what gets done. Coordination is often renegotiation of earlier established work rules, procedures. Coordination is often renegotiation of earlier established structures, hierarchies, and formalities. It is renegotiation, in other words, of the ground rules of what makes international disaster relief organisations into what they are: distributed, hierarchical, rule-bound, and politically constrained. Coordination is difficult because it can come down to the renegotiation of the very things that members higher up in the organisational hierarchies hold dear: reputations, political interests, and financial commitments. Coordination is not difficult because people cannot get in touch with each other (in fact, coordination is sometimes difficult precisely because people *can* get in touch with one another, creating clutter and data overload). In the final analysis, coordination in international disaster relief work is difficult because it is inevitably about more than the problem at hand. It is about hierarchies, political constraints, structure, and rules. And it is about experience, local knowledge and the acute sense of having to act, having to do something that disaster relief workers in the field are confronted with when met by fellow humans in need.

Summarising, our key findings are as follows:

- Much of international disaster relief is still planned and organised according to military-style command and control. There is an emphasis on centralised authority, adherence to procedure and protocol and over specification of plans.
- This has to do not only with the background of those involved in international disaster relief work, but with the need to be sensitive to higher-order goals (political, financial, diplomatic) that need to be satisfied to make relief work

possible in the first place. This demands bureaucratic accountability and resists formal downward delegation of authority.

- The result is a paradox of power: knowledge of what to do and authority to do it rarely coincide in the same person. This imbalance continually creates pressure for change.
- Confronted with acute problems to solve, local team leaders engage in a renegotiation of authority. Because of the imbalance, authority migrates: mostly to where the resources are that can meet the problem demands identified by the team leader.
- As time in the field accrues, team member experience with each other and with typical problems gets to supplant procedure and protocol as resources for action.
- Authority can be renegotiated in a number of ways and styles, some of which go down better with relief participants and organisational members than others.
- New communication technology that helps relief workers coordinate cannot in itself not deal with issues of hierarchy, goal conflicts or authority dislocation. New technology can even add problems in the form of data overload and a lack of trust.

References

Dekker, S. W. A. (2000). "Crew situation awareness in high-tech settings: Tactics for research into an ill-defined phenomenon". *Journal of Transportation Human Factors*, 2(1), 49-61.

Dekker, S. (2003). "Failure to adapt or adaptations that fail: Contrasting models on procedures and safety". *Applied Ergonomics*, 34(3), 233-238.

Dekker, S. W. A. and Hollnagel, E. (1999). *Coping with complexity*. Aldershot: Ashgate.

Dynes, R. (1989). "Emergency planning: False assumptions and inappropriate analogies". In *Proceedings of the World Bank Workshop on Risk Management and Safety Control, 1989*, Karlstad: Rescue Services Board.

Flach, J. M. (2000). "Discovering situated meaning: An ecological approach to task analysis". In J. M. Shraagen, S. F. Chipman and V. J. Shalin (eds), *Cognitive task analysis* (pp. 87-100). Mahwah, NJ: Lawrence Erlbaum Associates.

Hollan, J., Hutchins, E. and Kirsh, D. (2000). "Distributed cognition: toward a new foundation for human-computer interaction research". *ACM Transactions on Computer-Human Interaction*, 7, 174-196.

Klein, G. A. (1998). *Sources of power: How people make decisions*. Cambridge, MA: MIT Press.

Klein, G. A. (2002). *Intuition at work: Why developing your gut instincts will make you better at what you do*. New York, NY: Doubleday.

March, J. G. and Olsen, J. P. (1979). *Ambiguity and choice in organisations*. Bergen, Norway: Universitetsförlaget.

Mintzberg, H. (1979). *The structuring of organisations*. Englewood Cliffs, NJ: Prentice-Hall, Inc.

Orasanu, J. and Connolly, T. (1993). "The re-invention of decision-making". In G. A. Klein, J. Orasanu, R. Calderwood and C. Zsambok (eds), *Decision making in action: Models and methods* (pp. 3-20). Norwood, NJ: Ablex.

Rochlin, G. I. (1989). "Informal organisational networking as a crisis-avoidance strategy: US naval flight operations as a case study". *Industrial Crisis Quarterly*, 3, 159-176.

Rochlin, G. I. (1999). "Safe operation as a social construct". *Ergonomics*, 42(11), 1549-1560.

Rochlin, G. I., LaPorte, T. R., and Roberts, K. H. (1987). "The self-designing high reliability organization: Aircraft carrier flight operations at sea". *Naval War College Review*, Autumn 1987.

Rosenberg, T. (1998). "Risk and quality management for safety at a local level". Stockholm, Sweden: Royal Institute of Technology.

Shattuck, L. G. and Woods, D. D. (1997). "Communication of intent in distributed supervisory control systems". In *Proceedings of the Human Factors and Ergonomics Society 41st Annual Meeting* (pp. 259–268). Albuquerque, NM.

Snook, S. A. (2000). *Friendly fire: The accidental shootdown of U.S. blackhawks over Northern Iraq*. Princeton, New Jersey: Princeton University Press.

Strauss, A. L. (1987). *Qualitative analysis for social scientist*. Cambridge: Cambridge University Press.

Suparamaniam, N. (2003). *Renegotiation of authority.* Dissertation No. 830. Linköping Institute of Technology, Linköping, Sweden: Unitryck Publications.

Vaughan, D. (1996). *The Challenger launch decision: Risky technology, culture and deviance at NASA*. Chicago, IL: University of Chicago Press.

Vicente, K. (1999). *Cognitive work analysis: Toward safe, productive, and healthy computer-based work*. Mahwah, NJ: Lawrence Erlbaum Associates.

Weick, K. E. (1987). "Organisational culture as a source of high reliability". *California Management Review*, 29(2), 112-127.

Weick, K. E. and Sutcliffe, K. M. (2001). *Managing the unexpected: assuring high performance in an age of complexity.* San Francisco, CA: Jossey-Bass.

Woods, D. D., Johannesen, L. J., Cook, R. I., and Sarter, N. B. (1994). *Behind human error: Cognitive systems, computers and hindsight*. Dayton, OH: CSERIAC.

Woods, D. D., and Shatluck, L. G. (2000). "Distant supervision: Local action given the potential for surprise". *Cognition Technology and Work*, 2(4), 242-245.

Woods, D. D. and Cook, R. I. (2002). "Nine steps to move forward from error". *Cognition, Technology and Work*, 4, 137-144.

Woods, D. D., Patterson, E. S. and Roth, E. M. (2002). "Can we ever escape from data overload? A cognitive systems diagnosis". *Cognition, Technology and Work*, 4, 22-36.

Zsambok, C. and Klein, G. A. (eds) (1997). *Naturalistic decision making*. Mahwah, NJ: Lawrence Erlbaum Associates.

Chapter 23

The Analysis of Team Decision Making Architectures

Richard Breton and Robert Rousseau

Introduction

Command and control (C^2) decision cycle has been traditionally represented with Boyd's OODA loop (Observe-Orient-Decide-Act). The OODA loop provides a simple and valid representation of own and enemy decision cycle. With its cyclic representation, it captures the continuous aspect of C^2 leading to the assumption that the advantage of the battlefield will go to the one executing the loop better and faster than its opponent. The OODA loop also raises the impact of two important factors, uncertainty and time pressure, in the decision performance. The goal with the OODA phases is to reduce as much as possible the uncertainty in the situation within the time constraints in order to select the most appropriate course of action.

Over the years, the loop has been the object of many critics. The most recurrent one concerns its simplistic representation. Taken as a decision making model, the OODA loop offers a level of cognitive granularity too low to influence the identification of design requirements needed for the development of support systems. Hence, this representation is inappropriate from a cognitive engineering perspective. Also, it suggests a uni-directional sequence of events between the four phases. Such sequence of events cannot properly represent the dynamic properties of the decision making task. It does not illustrate the iterations required for optimal performance within and between the OODA phases. Finally, it can be difficult to consider the team dimension in the classical version of the OODA loop.

To overcome these problems, alternatives have been developed. For instance, Fadok, Boyd and Warden (1995) proposed an extended version of the classical OODA loop by increasing considerably the level of cognitive granularity of the Orient phase. Breton and Bossé (2002) presented a version of the loop in which an iterative process between the Observe and Orient phases allows the reduction of the level of uncertainty in function of the time constraints in the situation. Smith (2002) proposed a version of the OODA loop adapted for Net-Centric Warfare and Effect-Based Operations. Finally, Bryant (2003) suggested a cognitive model for C^2 process modelling, the CECA model, to overcome problems with the classical version of the OODA loop.

The number of alternatives proposed to describe the C^2 decision process can be taken as an indication that one generic model cannot successfully tackle all the dimensions of C^2. It should be more fruitful to develop models that address adequately one specific dimension. However, while the classical OODA loop has been the object of many critics, its simple and schematic representation offers important benefits that make the loop still referred to in military documents and accepted from the military community. Hence, for a matter of simplicity and acceptance, any model representing C^2 should keep explicit the four processes included in the classical OODA loop. This chapter addresses the team dimension in C^2 environments by proposing an approach to develop a Team Decision Making (TDM) model, the Team-OODA (T-OODA), which takes its roots in Boyd's OODA loop. The T-OODA is based on a modelling approach using a modular structure for each phase of the loop. This approach, called the Modular-OODA (M-OODA), has been proposed by Rousseau and Breton (2004).

The team dimension in C^2 environments

With the technological evolution, C^2 environments are more and more complex and time-stressed. In order to cope with these factors, most decision making tasks require team efforts. Within teams, different persons are gathering their efforts and expertise to cope with the complexity of the environment and to execute the task in regard to the time constraints imposed in the situation.

Within any given team, there is an inter-dependency factor among individual efforts in the task execution. Each team member has to consider the presence of other actors. Hence, it is critical, to reach an optimal performance, that the efforts of these different actors be efficiently coordinated. This inter-dependency between team members raises important questions concerning the training programmes and support systems used to execute the task. Systems need to allow and support collaboration and coordination among team members and training programmes need to address the factors that make a team work optimally.

Historically, the modelling of Individual Decision Making (IDM) tasks has positively influenced support systems design and training programmes development. It follows that appropriate modelling of the TDM task could contribute to the understanding of the team aspect in the task execution. It should also influence the development of training programmes for team and the design of team adapted support systems. However, the development of a general TDM model should be difficult if not useless. The major problem lies in the fact that the execution of a given decision making task should be different from one type of organisation to another. For instance, a TDM task should be executed differently in a C^2 environment in comparison with the one made by a jury. It results that TDM and its related sequence of processes could be different from one type of organisation to another. In that sense, it could be difficult to adequately represent all the situation-dependent executions of TDM, with one generic model. This single generic TDM model could be too general and then useless in providing valuable insights required to develop training programmes, systems design and team reorganisation requirements. Such generic TDM would suffer from the same limitations as the classical OODA loop model.

It results that the identification of requirements to develop training programmes and support systems should follow from the development of situation-dependent TDM models. However, without any modelling approach, it should be time-consuming and very costly in terms of efforts to develop TDM models for every situation. In this chapter, we propose a modelling approach that provides guidelines and principles from which different TDM models representing different situations can be developed.

The aim of the present chapter is to describe this approach that is based on the following elements: 1) a basic modular architecture for decision making, making it compatible with teamwork requirements; and 2) Team Functioning Elements (TFE) supporting the TDM modular architecture.

The modelling approach

The modelling approach, based on the M-OODA architecture (Rousseau and Breton, 2004), includes concepts and guidelines from which models are derived. A given TDM model is composed of different modules that represent the Cognitive Functions (CF) from which the decision making task is executed. The modules are assembled in order to represent the sequence of execution of the cognitive processes (for example, perceive) sustaining the cognitive functions (for example, data gathering). The next section describes these concepts and guidelines.

The basic module in the decision making module

The basic module is the core of the model. This module is used to represent the components and properties of the CF included in the decision making task. We propose that each CF is structured in an input-process-state-control system as defined by the simple module presented in Figure 23.1. A module is composed of a module name, three basic components (Process, State and Control), two feedback loops (internal and external) and an input/output described as follows:

Module name: It corresponds to the particular CF included in the decision making task. The name of the module reflects the general or ultimate goal (for example, Data Gathering or Action Selection) of the module.

Inputs: In most cases, inputs are outputs from other preceding modules. It can obviously be information or data selected from the environment.

Process: It is the core active component. It is a goal-directed action applied on an input that produces a state. Its properties depend on the nature of the goal. The process is given a generic name that is closely related to the action included in the module name (for example, perceive, sense, understand, select, act and plan).

State: The state is the result of the process activity. It is a structured representation with properties depending on the nature of the process from which it originates and of the input that was fed to the process. The general properties of state are also defined according to the module goal.

Control: The criteria-based control component is a flow control function gating the delivery of the output to other modules and enabling iterations of the process within the module. Control can interrupt, iterate the process or exercise no gating function depending on the mode of operation required. It can accept a given level of state quality depending on task-goal criteria. Since they are goal-related, the control criteria should be different from one module to another.

Output: The output is the current status of the state resulting from the process that reaches an acceptable level of quality based on the criteria-based control component. The resulting output becomes the input for a subsequent module. Outputs with a high level of familiarity can initiate an automatic process in subsequent modules.

Feedback loops: Feedback loops are implemented within (internal) and between modules (external). They enable control and information requests, amongst other functions. Modules are interconnected in such a way as to represent the execution of the decision making task.

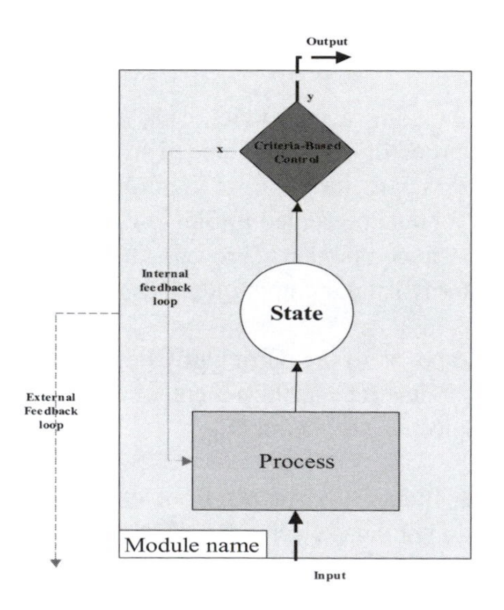

Figure 23.1 The basic model

Team Functioning Elements (TFE)

The TFEs are activities supporting the interactions within the team between team members, units of tasks and the different supporting tools. They are associated with the quality and efficiency of teamwork. The set of TFE selected for the TM models is taken from the NATO RTA IST-019 TG006 (NATO, 2004). It includes Human Communication (HC), Tool Communication (TC), Coordination (Co), Task Allocation (TA), Task Balancing (TB) and Information Distribution (ID).

Human communication: Communication involves the active exchange of information between team members. It seems that the label "communication" is generally used, in literature, to represent the activity of information exchange.

Tool communication: The TC element concerns, in addition to the exchange of information, the concepts of interoperability, data exchange and information exchange protocols and standards between tools.

Coordination: This element concerns the activities between team members. It is related to the merging, in a logical and coordinated manner, of the actions of different individuals to execute adequately a task.

Task allocation: The TA element is related to the role and responsibilities of every team member in the execution of the task.

Task balancing: The TB element is related to the reallocation of subtasks among the team members in such a way that workload is better distributed. This balancing process happens during the execution of the subtasks and then implies that certain decision making processes should be able to be reallocated.

Information distribution: To execute a task, appropriate information sources must be available. It must also be distributed to agent (team members, tools or both) owning the required skills. Then, ID concerns the three relationships between the information sources and the tasks, tools and team member elements.

This list is compatible with those provided by Dickinson and McIntyre (1997) and Smith-Jentsch, Johnston and Payne (1998). However, the benefit of the NATO list is that it addresses not only the human dimension of teamwork but also the repercussions on tools required to execute the task and the organisation of the work.

Modelling principles and rules

A first step in the development of a TDM model is to identify the major processes, functions or subtasks included in the task. Generally, in every IDM model (for example, Naturalistic models (Klein, 1993; Lipshitz, 1993; Beach, 1997)), four general and basic phases can be identified: 1) information has to be picked up from the environment; 2) this information must be understood; 3) a course of action must be selected according to the understanding of the information; 4) the action must be implemented. Thus, these very basic cognitive functions should necessarily be part of any TDM models and they could be labelled as follows: 1) Data Gathering (DG); 2) Situation Understanding (SU); 3) Action Selection (AS); and 4) Action Implementation (AI). These phases are in accordance with the ones included in the Boyd's OODA loop that commonly represents the activities in C^2 environments.

The second step is to represent the sequence of execution of these phases with modules that are based on the basic one presented in Figure 23.1. In these modules, the agents (human or automaton) executing the phases are represented by a specific process box. In fact, these different boxes illustrate the contribution of each agent within the team. It may result, from the work of each agent (represented by a process box), in a different state controlled by a different control component.

The third step is to connect the process boxes or states together with appropriate TFE in order to represent the team architecture.

The type of TDM in operation in a given environment will determine the way agents are organised across and within the modules. That will lead to TDM models with different architectures. In order to provide examples of the approach, the next section illustrates the modelling of three different types of team organisations.

Examples of TDM modules

In order to show the modelling capabilities of the approach, three types, varying on the importance of time pressure and the complexity of the information, are defined. These types are labelled: Autocratic (A), Deliberative (D) and Cooperative (C). The D (low time pressure-low complexity) and A (high time pressure-high complexity) present opposite conditions. The C type is characterised by low time pressure-high complexity conditions. For each type, a fictive situation, in which the TDM task is held, is illustrated in order to represent adequately the situation and its constraints.

Example 1: The autocratic type (A)

An autocratic type can be adopted in highly complex and time-pressed situations. A particularity of this type is that the final decision becomes the responsibility of a team leader. The role of team members is to provide information reports to this team leader in order to support him/her in the selection of the best decision. His/her selection leads to the implementation of the set of actions.

There are two main benefits related to the adoption of this particular type. First, it brings together the required expertises in order to process different sources of information found in the environment. This processing leads to the constitution of the information reports sent to the team leader.

Secondly, debates over the selection of the final decision occurring within a team can be very time-consuming. By leaving the responsibility of the final decision to a single expert, the time to take it can be considerably reduced. To demonstrate the modelling of an autocratic team decision making type, let us take a fictive situation characterised as follows:

- The environment is characterised by different types of data requiring specific and specialised sensors to be processed. In this situation, three different sensors are used to cover different sources of data.
- Each sensor produces a specific representation of the data gathered.
- Different expertises are required to interpret the representations produced by the sensors.
- Experts develop an interpretation of the situation based on the information considered in their analysis. The information is provided by a sensor covering a fraction of the whole picture. Then, it results in a partial interpretation of the situation since it is based on partial information.
- The interpretation of each expert must be sent, at the same moment, to a team leader in charge of the final decision.
- The synchronisation of the report sending is important since a non-significant piece of information in one report could become crucial when put in conjunction with another included in a different report produced by a different expert.
- The team leader has the authority of distributing the workload among team members (human experts or sensors) if it is judged necessary.

Example 2: The deliberative type (D)

As the importance of time pressure and the complexity of the information in the situation decrease, a deliberative mode can be adopted. This mode is very beneficial since it allows the selection of the optimal decision. Then, this type can be adopted when the conditions are favourable (no time constraint, no specific skills required to cope with complex or particular information). In this type, all team members have access to the same information and they have all the time they need for the deliberation process. The goal of this deliberation process is a consensus over an understanding of the situation and the selection of the optimal decision. Thus, the decision making process is ended only with the identification of this optimal solution, whatever the time required to reach that level. The best example of this situation is described by the decision taken by a jury. To demonstrate the modelling of the deliberative team decision making type, let us take a fictive situation characterised as follows:

- The data gathered from the environment does not need specific expertises or skills to be processed. Consequently, a unique and generic sensor processes the data and builds a unique world representation that is made available to all team members.
- Team members develop an interpretation of this unique source of information.

Then, they develop a mental model of the whole situation. Even if no specific expertise is required, their interpretation is obviously based on their subjective judgement.
- Since, there is no time constraint in the situation, team members are encouraged to share information, opinions or ideas on the interpretation of the situation.
- The fusion of all team members' interpretations leads to a unique shared interpretation from which a decision is taken. All team members must agree on the interpretation, whatever the time required to reach that agreement.
- After agreeing on the interpretation of the situation, the team members must agree on the selection of a course of action.

Example 3: The cooperative type (C)

The last example used to illustrate our modelling approach is typical of a situation characterised with the absence of time constraints, but the need for different skills to cope with different sources of information. In this situation, the reason for having a team is to gather different experts required by the presence of specific sources of information. However, since there is no time constraint, all team members can be involved in the comprehension of the problem, the generation of alternatives and the selection of the solution. Then, a cooperative mode can be adopted. The fictive situation used to demonstrate the modelling approach is characterised as follows:

- The environment is characterised by different types of data requiring specific and specialised sensors to be processed. In this situation, three different sensors are used to cover different sources of data.
- Each sensor produces a specific representation of the data gathered. Different expertises are required to interpret the representations produced by the sensors.
- The information is provided by a sensor covering a fraction of the whole picture. Experts develop an interpretation of the situation based on this partial information. However, each expert can have access to other sources of information if desired.
- Since, there is no time constraint in the situation, team members are encouraged to share information, opinions or ideas on the interpretation of the situation.
- The fusion of all team members' interpretations leads to a unique shared interpretation from which a decision is taken. All team members must agree on the interpretation, whatever the time required to reach that agreement.
- After agreeing on the interpretation of the situation, the team members must agree on the selection of a course of action.

Figures 23.2-23.4 show the results of applying the modelling principles for these different types. For the sake of simplification, three different agents, defined as humans, are included in the team. The AI is also kept at the individual level since it concerns more the implementation of the selected course of action instead of its selection itself.

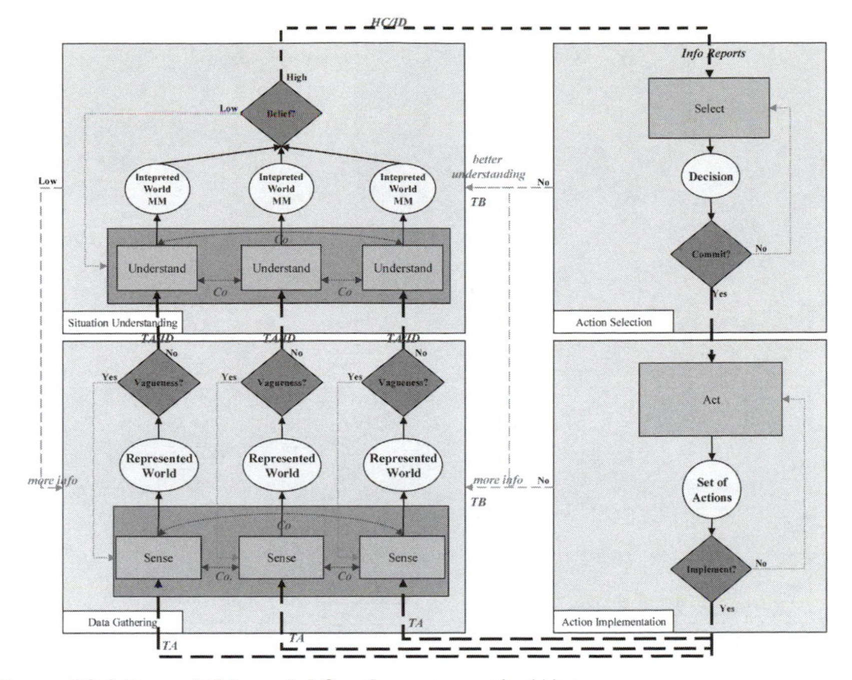

Figure 23.2 Team DM model for the autocratic (A) type

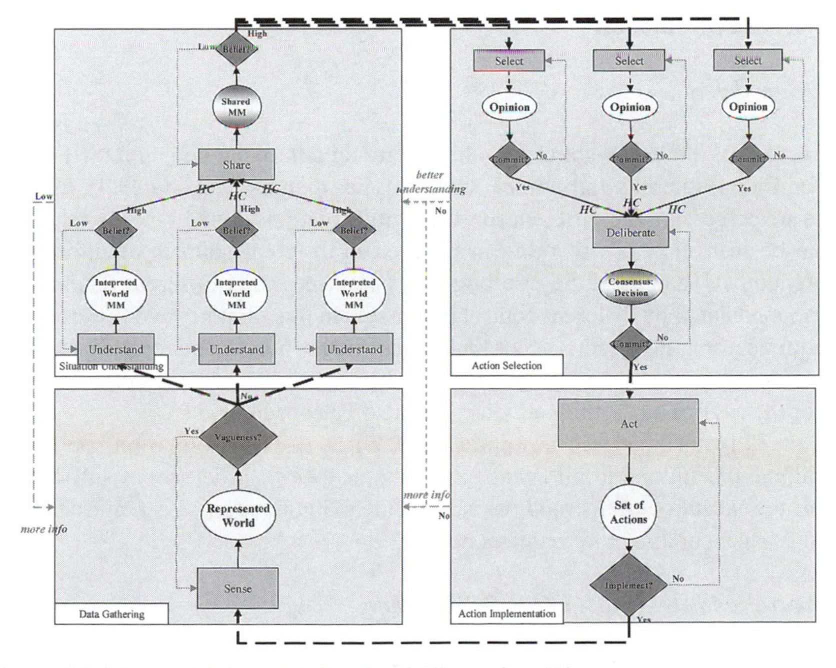

Figure 23.3 Team DM model for the deliberative (D) type

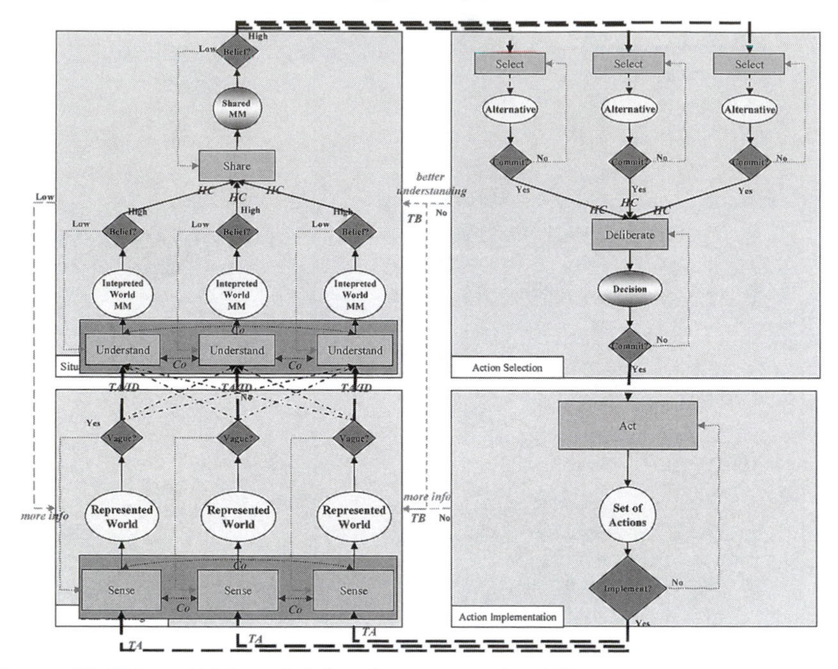

Figure 23.4 Team DM model for the cooperative (C) type

Analysis of TDM models

The DG module

The analysis of the model starts with the first module (left and down on the Figures) labelled Data Gathering. In both the A and C type (Figures 23.2 and 23.4), different agents are processing the different sources of information found in the environment. As can be seen, they are working in parallel on different sources of information (different inputs) and then, the resulting states are necessarily different. These states are also modulated by different control processes. In this module, two different TFE are required. First, the agents, according to their skills, must be adequately allocated to the processing of a specific type of information (TA). Secondly, in order to cover most of the environment, their actions must also be coordinated (Co).

In the D type (illustrated in Figure 23.3), since all the information sources are general and transmitted to all agents, a unique or generic processor is sufficient to support the execution of the module. No specific skill or expertise is required. Then, the single-agent architecture requires no TFE.

The interaction between the DG and SU modules

From the DG module in the A and C types, three different states are produced. These states become inputs for the SU modules. It is critical that these states are adequately distributed to experts owning appropriate skills. However, as is illustrated in the C type by the dotted arrows, all agents can have access to all the different information sources. Then, TA and ID are important elements in the transition between the DG and SU modules. Such elements are not required in the D type since only one generic state is produced and generally distributed to all agents. In this type, all team members have access to this state.

The SU module

The SU module goal is to generate a state defined as an interpretation of the world. In all our examples, this module is under the responsibility of a team. In both the A and C types, team members process different sources of information and consequently, it should result in different interpreting states. However, since there is no time constraint in the C type and agents have access to other information sources (as complementary information), it may be possible that the resulting states in the C type are more complete than the ones produced in the A type. Nevertheless, to ensure an efficient coverage of the information, agents' actions must be coordinated (Co). Coordination is not required in the D type since all team members have access to a unique source of information. Problems related to uncovered or overlapping information are not really critical for this type of TDM.

The A type is used to cope with time constraints. To do so, agents are working in parallel and produce independently a set of interpreted worlds (states) based on the inputs considered. In the D and C types, the low time pressure allows team members to interact together and produce a single state equivalent to a shared mental model of the world. Hence, the HC element is essential in such sharing activities. In our examples, agents are humans. In the case where agents would be automatons, TC would simply replace HC to connect these agents.

The interaction between the SU and AS modules

In the A type, the final decision is under the responsibility of a team leader. The role of other team members is to provide to this team leader information reports based on their interpretation of the situation. Obviously, the HC element is important in the transmission protocol of these reports. A rule is included in the control component that makes possible the synchronisation of all the different reports (resulting states) sent to the next module. Timing is important since the team leader must build an interpretation of the overall situation based on the different information reports. This interpretation development requires a certain fusion of the information. Parts of information reports can be seen as non-significant but being in conjunction with other non-significant parts contained in other reports, these useless parts can become critical ones. If information reports are transmitted with too important delays, it could become difficult for the team leader to execute this fusion process.

In the D and C types, such HC and ID elements are not required since the same team members are in charge of the processing of both the SU and AS modules. They would be required if different members were responsible for these two successive modules.

The AS module

The goal of the AS module is to select a course of action (COA) based on the result of the DG and SU modules. In the A type, since this module is under the responsibility of a team leader, no TFE are required. In the D and C types, assuming that the team members are the same as in the SU module, they interact (HC), within the AS module, in order to reach a consensus (shared state) over a potential COA. The Coordination element is not required since they all debate from the same information that is the shared state build in the SU module.

The AI module

The AI module goal is to implement the selected COA. We have deliberately not considered that module in our modelling effort. Thus, the AI module is modelled with single-agent architecture even though it is likely that it will involve teamwork.

The role of the TB element in the models

The TB element plays a major role in the team organisation in order to cope with the complexity of the situation and the important time constraints. TB allows a readjustment process in which part of some tasks can be reassigned to other less loaded or more appropriate agents. In the A and C types, such reallocation is initiated from the AS and AI modules and concerns mostly the DG and the SU ones. For instance, in order to get more complete and accurate information to select an appropriate COA in the AS module, a request can be sent in the DG module to gather more information or to the SU module to clarify the interpretation provided. In case of the A type, the team leader may decide to reassign the workload in both these modules. In the D type, such a reallocation element is not required since the DG module is under the responsibility of a single agent and all the team members involved in the SU module consider the exact same pool of information.

Discussion

There are some benefits related to the application of the modelling approach. A first one concerns the need in TFE to represent the interaction between the agents part of the team. One may claim that the more an organisation structure requires teamwork, the more TFE will be involved. Consequently, this modelling approach can provide the means to evaluate the importance of the team dimension in any given team organisation. Tables 23.1-23.3 show matrices that represent the inclusion of the different TFE according to the situation. Table 23.1 represents the A type. As can be seen, most of the TFE are present. However, there are no TFE involved within the last two modules (AS and AI). In the D type (Table 23.2), only the HC element is critical. In the C type (Table 23.3), the majority of the TFE come into play.

Table 23.1 TFE involved in the autocratic type

	DG	*SU*	*AS*	*AI*
DG	*TA, Co*	*TA, ID*	*TB*	*TB*
SU		*Co*	*Co, HC,TB*	*TB*
AS				
AI				

Table 23.2 TFE involved in the deliberative type

	DG	*SU*	*AS*	*AI*
DG				
SU		*HC*		
AS			*HC*	
AI				

Table 23.3 TFE involved in the cooperative type

	DG	*SU*	*AS*	*AI*
DG	*TA, Co*	*TA, ID*	*TB*	*TB*
SU		*Co, HC*	*TB*	*TB*
AS			*HC*	
AI				

Note that the TC element is not present in these matrices. It is related to our choice to consider agents as humans in our TDM models. TC could have been required in these types if agents executing the modules had been designed as tools. The identification of the TFE for each potential organisation may have important repercussions on the training programmes and the design of support systems. For instance, for teams adopting an autocratic or a cooperative mode, the communication skills must be trained. Problems and factors affecting the communication of such difference in language or culture, personality traits and leadership must be considered. Tools should be designed to favour the communication amongst team members. Communication protocols should be developed. The training programmes should also address issues related to the coordination of the actions of the different agents involved in the task execution. Support systems should favour such coordination.

A second benefit lies in the comparison between the different team organisations in terms of the importance of the team aspect. Situations are constantly evolving. Conditions and constraints are changing as the time elapses. It may evolve from low time pressure conditions to urgent ones or from very complex to relatively simple ones. A good team should be one that can rapidly adapt its structure based on the conditions prevalent in the situation.

The modelling approach offers a means to measure the benefits-costs in terms of the TFE requirement related to the switch from one organisation to another. For instance, for a team that switches from a deliberative mode to an autocratic one, it is essential that this team be able to allocate adequately parts of the task to the appropriate agent (TA), possess the means to coordinate their actions (Co), have protocols to distribute accordingly the information to proper agents (ID) and have the means to reallocate parts of the task in case of overload or for answering a request (TB).

Training programmes should be developed to support the switch from one mode to another by considering the different TFE affected by the change. Technological systems introduced to support the module execution should also be compatible with the organisational change. The modelling approach offers a means to identify team aspects that should be trained or supported depending on the organisation required in the situation.

Conclusions

Depending on the situation in which a TDM task occurs, the way to execute it should be different from one organisation to another. This fact prevents the development of a generic TDM model that would efficiently represent all the task executions. Such a generic TDM model should be too general to support the understanding process of the task and then to identify valuable training and design requirements. Consequently, situation-dependent models should be developed. However, this modelling effort could be time-consuming and very costly in terms of efforts. To overcome these problems, we have proposed a modelling approach based on a modular architecture.

The modular architecture offers an interesting approach to derive different TDM models and shows the variable importance of TFE within these models. It represents the interactions between team members in the execution of phases in a TDM situation. It can be used to identify design, training and reorganisation requirements in regard to the role and importance of the TFE.

An interesting aspect of the modular approach is the representation of agents with an "input-process-state" perspective. Added to the identification of the interactions between the agents in the process execution, the modular approach can be used to identify which inputs are required, who should be in charge of their processing and what is the desired state.

Principles and guidelines used to develop the TDM models are part of a modelling approach including the M-OODA loop (Rousseau and Breton, 2004. As mentioned above, different models may be required to cover the various aspects of C^2. This chapter proposes a modelling approach to cover the team dimension. Breton and Rousseau (2005), using the M-OODA approach, have developed a model that increases the level of cognitive granularity of the loop. This model, called the C-OODA, should support the identification of design requirements for support systems. All these models take their roots in the classical OODA loop and then have the benefits related to Boyd's model.

In the future, efforts should now be devoted to developing generic building blocks or module architectures based on the input-process-state-control systems that represent the most common team architecture. These building blocks could be assembled in different ways to represent specific team organisations.

References

Beach, L. R. (1997). *The psychology of decision making: People in organizations*. Thousands Oaks, CA: Sage Publications.

Breton, R. and Bossé, E. (2002). "The cognitive costs and benefits of automation". In *Proceedings of NATO RTO-HFM Symposium: The role of humans in intelligent and automated systems*. Warsaw, Poland, 7-9 October 2002.

Breton, R. and Rousseau, R. (2005). "The C-OODA: A Cognitive Version of the OODA Loop to Represent C2 Activities". In *Proceedings of the 10th International Command and Control Research and Technology Symposium*. Washington DC., USA.

Bryant, D. (2003). "Critique, Explore, Compare, and Adapt (CECA): A new model for command decision making". Defence Research & Development Canada – Toronto, TR-2003-105, July 2003, 49 pages.

Dickinson, T. L. and McIntyre, R. M. (1997). "A Conceptual Framework for Teamwork Measurement". In M. T. Brannick, E. Salas and C. Prince (eds), *Team performance assessment and measurement* (pp. 19-43). Mahwah, NJ: Lawrence Erlbaum Associates.

Fadok, D. S., Boyd, J. and Warden, J. (1995). "Air power's quest for strategic paralysis". Maxwell Air Force Base, AL: Air University Press (AD-A291621).

Klein, G. A. (1993). "A recognition-primed decision (RPD) model of rapid decision making". In G. Klein, J. Orasanu, R. Calderwood and C. E. Zsambok (eds), *Decision making in action: Models and methods* (pp. 138-147). Norwood, NJ: Ablex.

Lipshitz, R. (1993). "Decision Making as Argument-Driven Action". In G. Klein, A. Orasanu, R. Calderwood and C. Zsambok (eds), *Decision making in action: Models and methods* (pp. 172-181). Norwood, NJ: Ablex.

NATO RTA IST-019, TG-006. "Modeling of organization and decision architectures". December 2004, 249 pages.

Rousseau, R. and Breton, R. (2004). "The M-OODA: A model incorporating control functions and teamwork in the OODA loop". In *Proceedings of the 2004 Command and Control Research and Technology Symposium*. San Diego, USA.

Smith, E. A. (2002). "Effect-based operations. Applying network centric warfare in peace, crisis, and war". DoD Command and Control Research Program.

Smith-Jentsch, K. A., Johnston, J. H. and Payne, S. C. (1998). "Measuring team-related expertise in complex environments". In J. A. Cannon-Bowers and E. Salas (eds), *Making decisions under stress: Implications for individual and team training* (pp. 61-87). Washington, DC: American Psychological Association.

Chapter 24

Surgical Team Self-Review: Enhancing Organisational Learning in the Royal Cornwall Hospital Trust

Simon Henderson, Matt Mills, Adrian Hobbs, Alan Bleakley, James Boyden and Linda Walsh

Introduction

Although error in medicine and its impact on patient safety has been publicly debated only for the last five years, it has been discussed in the medical literature for at least the last 15 years. In 1994, Leape challenged the ideal that "if physicians and nurses could be properly trained and motivated, then they would make no mistakes" (Leape, 1994). He called for "a culture in which errors and deviations are regarded not as human failures, but as opportunities to improve the system"; this would need "grassroots participation to identify and develop system modifications" and a fundamental cultural change. Leape drew an analogy to the aviation model where system design and training absorbs errors and failures.

In 1995, Gaba and Howard made the distinction between the operating room and flight crews (19?5). Cockpit flight crews share the same mental model and are socialised into the same aviation culture. In contrast the operating theatre has at least three "crews" of professionals: surgeons, anaesthetists and nurses. None can do the other's job; each has their own mental model for the surgical procedure and each identifies with their own subculture. Gaba believed that sharing mental models (situation awareness) is an integral feature of expert performance. He proposed investigating the role of situation awareness in medical domains in real and simulated work environments.

In 1999, the American Institute of Medicine published a report *To Err is Human*, which excited media attention due to its graphic analogy of the results of medical error equating to the crashing of several fully loaded Jumbo Jets each month (Kohn, Corrigan and Donaldson, 1999). However, its real message was that the medical community needed training systems to reduce error and learn from mistakes.

The UK Governmental report in 2000 into learning from adverse events within the National Health Service (NHS) stated that:

> Too often in the past we have witnessed tragedies which could have been avoided had the lessons of past experience been properly learned. … Most distressing of all, such failures often have a familiar ring, displaying strong similarities to incidents which have occurred before and in some cases almost exactly replicating them. (Chief Medical Officer, 2000)

In its recommendations, the report stated that:

> ...the NHS should encourage a reporting culture amongst its staff which is generally free of blame for the individual reporting error or mistakes, *and encourage staff to look critically at their own actions and those of their teams*. (Emphasis added). (Chief Medical Officer, 2000)

In response to this recommendation, QinetiQ has been working with the Royal Cornwall Hospital (RCH), Truro, to develop and trial a series of tools that enable surgical teams to review and enhance performance. The initiative is one of three strands of a patient safety programme. The other two strands are the introduction of Team Resource Management education and training (analogous to the aviation Crew Resource Management programmes) and Close-Call Reporting (analogous to the aviation near-miss reporting systems).

Currently, surgical teams are often formed from a collection of individuals who do not regularly work together. Prior to conducting a surgical procedure this team does not usually engage in any pre-briefing activity (indeed, the first time the team meets is often "over the body").

> ...I've known times when we've started the day and three or four of us are thinking completely differently about what we think the plan is. Sooner or later you realise this, normally you think it's funny, but sometimes it can cause all sorts of nightmares. We should all know the plan for the day and be singing from the same hymn sheet from the start. (Operating theatre team member)

This quote would not have been heard 10-15 years ago, where greater staffing levels meant that team members from the same "crew" could easily cover absentees. The repertoire and complexity of operations was also considerably less than today. However, if an error was being made the rigid hierarchy impeded its capture, and once it occurred the investigation would take place within that "crew" group (culture) according to their professional customs, or informally in the coffee room. The wider team's professional and organisational learning would therefore be lost to the detriment of future patient safety.

With modern work practices, such as flexible working, working-hours directives and staff shortages, stable teams are now the exception rather than the rule. Similarly, referential, coercive leadership is no longer acceptable and a more supportive, facilitative leadership, recognising that knowledge is distributed around the team, is needed. Despite this lack of team continuity briefings rarely occur, which often leads to delay, incorrect equipment being readied, wastage of time and resources and the potential to compromise patient safety. Coupled with this, the team does not usually perform any post-procedure review (debrief) in order to assess performance and identify lessons to be learned. This means that the individual team members, the team as a whole, and indeed the entire organisation, is not learning from the experience of the individuals who are working in the system. Further, such information is not captured and shared across the team, department and organisation. Thus, the same issues, problems, and errors are likely to recur in different teams at different times in the future.

It is against this background that Team Self-Review (TSR) was introduced into RCH to improve team members' non-technical skills, to help them build effective *ad hoc* teams and to improve the patient safety margin.

What is Team Self-Review?

TSR is a process that enables a team to consider its actions (past, present and future) and identify means for promoting and maintaining high quality teamwork and levels of performance, via the generation of issues, reflection and implementation of actions.

Two key techniques comprise the TSR process:

- The TSR pre-brief
- The TSR debrief

The pre-brief is held at the beginning of a team "session" (that is, an operating list, case, training exercise or other event) and the debrief is held at the end of the session. The pre-brief aims to review and clarify the plan for the session *("What do we need to know? What are we going to do? What should we do if...?")* and the debrief aims to assess and review team performance in the session that has just occurred *("How well did we do? What should we do differently in future?")*. Both techniques are characterised by four defining features:

- They are *structured* team discussions.
- They provide an opportunity for every team member to *have a voice* in the team.
- They provide an opportunity to *project and reflect* as a team.
- They provide an opportunity to *learn* as a team; and to learn across teams through the use of TSR logs.

The TSR pre-brief

Many teams experience problems, confusion and communication breakdowns because team members do not have a clear understanding from the outset of the plan and what is expected of them. To help reduce the likelihood of these difficulties occurring, the pre-brief is designed to:

- Enable clarity of direction from the outset
- Facilitate and strengthen coordination within the team and with other dependents outside of the team

- Increase risk and hazard awareness and improve problem spotting and error trapping
- Consider and develop contingency and mitigation plans and actions for problem areas (that is, where things could go wrong, what to do if they do)
- Allow team members to raise queries/concerns and clear any misunderstandings
- Encourage a team culture of open and honest communication and assertive questioning
- Increase sense of team identity (that is, make everybody feel part of the team with a valid role, perspective and opinion).

The TSR pre-brief takes only five to ten minutes to conduct. The full scripted structure is presented on a double-sided A4 laminated sheet. Handy credit card-sized versions with the high level headings and topics have also been produced and have proved to be popular (see Figure 24.1).

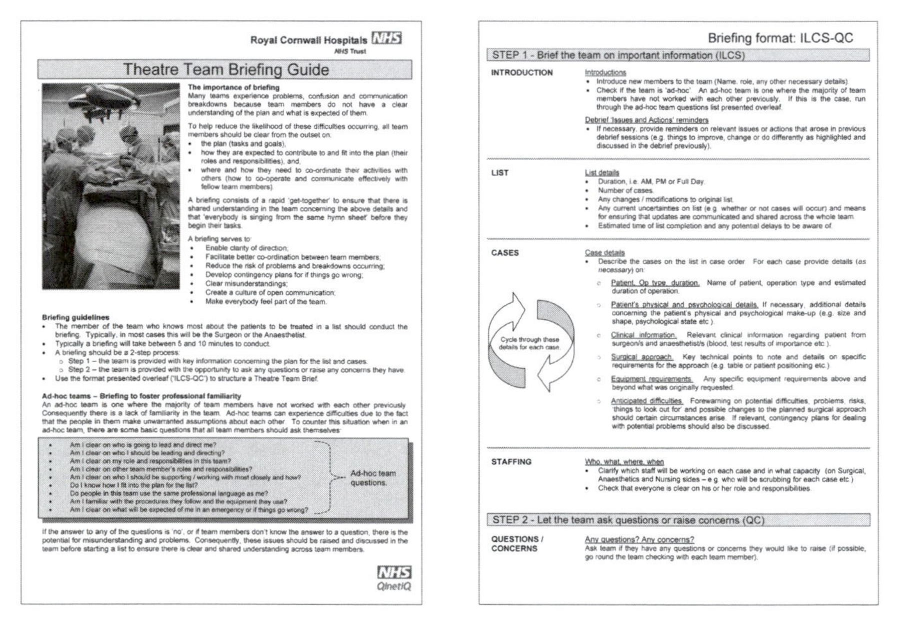

Royal Cornwall Hospitals NHS
NHS Trust

Theatre Team Briefing Guide

The importance of briefing
Many teams experience problems, confusion and communication breakdowns because team members do not have a clear understanding of the plan and what is expected of them.

To help reduce the likelihood of these difficulties occurring, all team members should be clear from the outset on:

- the plan (tasks and goals),
- how they are expected to contribute to and fit into the plan (their roles and responsibilities), and,
- where and how they need to co-ordinate their activities with others (how to co-operate and communicate effectively with fellow team members).

A briefing consists of a rapid 'get-together' to ensure that there is shared understanding in the team concerning the above details and that 'everybody is singing from the same hymn sheet' before they begin their tasks.

A briefing serves to:

- Enable clarity of direction;
- Facilitate better co-ordination between team members;
- Reduce the risk of problems and breakdowns occurring;
- Develop contingency plans for if things go wrong;
- Clear misunderstandings;
- Create a culture of open communication;
- Make everybody feel part of the team.

Briefing guidelines

- The member of the team who knows most about the patients to be treated in a list should conduct the briefing. Typically, in most cases this will be the Surgeon or the Anaesthetist.
- Typically a briefing will take between 5 and 10 minutes to conduct.
- A briefing should be a 2-step process:
  - Step 1 – the team is provided with key information concerning the plan for the list and cases.
  - Step 2 – the team is provided with the opportunity to ask any questions or raise any concerns they have.
- Use the format presented overleaf ('ILCS-QC') to structure a Theatre Team Brief.

Ad-hoc teams – Briefing to foster professional familiarity
An ad-hoc team is one where the majority of team members have not worked with each other previously. Consequently there is a lack of familiarity in the team. Ad-hoc teams can experience difficulties due to the fact that the people in them make unwarranted assumptions about each other. To counter this situation when in an ad-hoc team, there are some basic questions that all team members should ask themselves:

- Am I clear on who is going to lead and direct me?
- Am I clear on who I should be leading and directing?
- Am I clear on my role and responsibilities in this team?
- Am I clear on other team member's roles and responsibilities?
- Am I clear on who I should be supporting / working with most closely and how?
- Do I know how I fit into the plan for the list?
- Do people in this team use the same professional language as me?
- Am I familiar with the procedures they follow and the equipment they use?
- Am I clear on what will be expected of me in an emergency or if things go wrong?

Ad-hoc team questions.

If the answer to any of the questions is 'no', or if team members don't know the answer to a question, there is the potential for misunderstanding and problems. Consequently, these issues should be raised and discussed in the team before starting a list to ensure there is clear and shared understanding across team members.

NHS
QinetiQ

Briefing format: ILCS-QC

STEP 1 - Brief the team on important information (ILCS)

INTRODUCTION

Introductions

- Introduce new members to the team (Name, role, any other necessary details).
- Check if the team is 'ad-hoc'. An ad-hoc team is one where the majority of team members have not worked with each other previously. If this is the case, run through the ad-hoc team questions list presented overleaf.

Debrief 'Issues and Actions' reminders

- If necessary, provide reminders on relevant issues or actions that arose in previous debrief sessions (e.g. things to improve, change or do differently as highlighted and discussed in the debrief previously).

LIST

List details

- Duration, i.e. AM, PM or Full Day.
- Number of cases.
- Any changes / modifications to original list.
- Any current uncertainties on list (e.g. whether or not cases will occur) and means for ensuring that updates are communicated and shared across the whole team.
- Estimated time of list completion and any potential delays to be aware of.

CASES

Case details

- Describe the cases on the list in case order. For each case provide details (*as necessary*) on:
  - Patient, Op type, duration. Name of patient, operation type and estimated duration of operation.
  - Patient's physical and psychological details. If necessary, additional details concerning the patient's physical and psychological make-up (e.g. size and shape, psychological state etc.).
  - Clinical information. Relevant clinical information regarding patient from surgeon/s and anaesthetist/s (blood, test results of importance etc.).
  - Surgical approach. Key technical points to note and details on specific requirements for the approach (e.g. table or patient positioning etc.)
  - Equipment requirements. Any specific equipment requirements above and beyond what was originally requested.
  - Anticipated difficulties. Forewarning on potential difficulties, problems, risks, 'things to look out for' and possible changes to the planned surgical approach should certain circumstances arise. If relevant, contingency plans for dealing with potential problems should also be discussed.

STAFFING

Who, what, where, when

- Clarify which staff will be working on each case and in what capacity (on Surgical, Anaesthetics and Nursing sides – e.g. who will be scrubbing for each case etc.)
- Check that everyone is clear on his or her role and responsibilities.

STEP 2 - Let the team ask questions or raise concerns (QC)

QUESTIONS / CONCERNS

Any questions? Any concerns?
Ask team if they have any questions or concerns they would like to raise (if possible, go round the team checking with each team member).

Figure 24.1 TSR pre-brief

The TSR debrief

The TSR debrief provides an opportunity and structure for the team to talk about teamwork and performance during the previous session. It provides an opportunity for every team member, irrespective of rank or status, to address concerns or questions.

Further, it enables the team to identify, implement and monitor behavioural changes to enhance performance.

A variety of TSR debriefing techniques were developed (15 in total) which map to different environmental constraints and pressures (for example, time available, familiarity of staff, critique depth required). In this way the surgical team is effectively provided with a "tool bag of techniques" from which to choose a TSR debrief appropriate for a particular context.

A teamwork model was developed specifically for surgical teams, to provide the topics for review. The model uses an analogy of "team health". There are five key areas of team health with each comprising three health dimensions. Figure 24.2 presents the health areas and dimensions.

1. Team Management	Planning and organisation	Leadership and direction	Inter-team working
2. Team Thinking	Shared understanding	Thinking ahead	Decision making
3. Team Behaviour	Information passed	Communication style	Monitoring and support
4. Team Climate	Recognition and respect	Managing disagreements	Stress and flashpoints
5. Team Safety	Safe practice	Equipment use	Low-energy and fatigue

Figure 24.2 Dimensions of team health

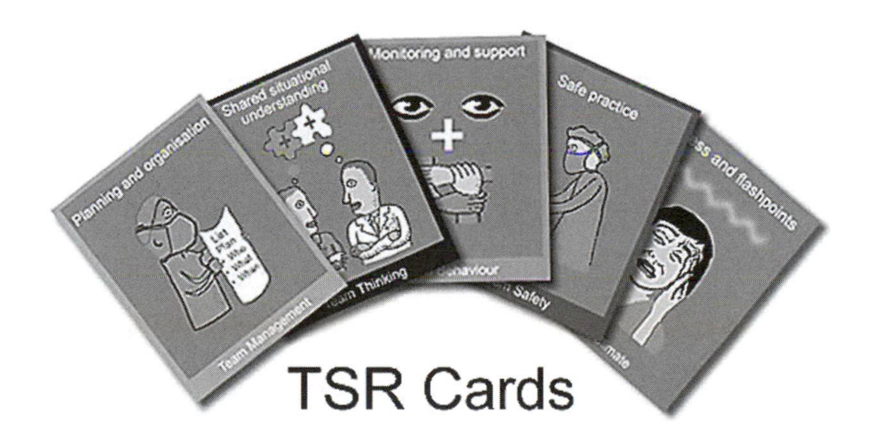

Figure 24.3 TSR cards

Each debriefing technique presents different ways of structuring a review session based upon all, or a chosen subset, of these health dimensions. A number of the techniques make use of a set of TSR cards, purpose-built for facilitating debriefs (see Figure 24.3).

The TSR debrief is designed to facilitate two key processes:

- Identifying lessons to be learned
- Generating actions to implement the lessons.

In this way lessons *identified* are actually converted into lessons *learned* via modifications in team behaviour. The TSR debrief is also concerned with packaging these outputs in such a way that they can be shared outside of the team; thus enabling teams working in the same department to learn vicariously from each other. Consequently, a series of recording sheets has been designed together with recommendations for managing the collection, feedback and exploitation of the issues and actions generated.

The RCH experience

RCH staff played an active role in the requirements elicitation, design and trial of the TSR tools. Workshops, observational studies, focus groups and interviews were conducted with staff to design, review and iteratively improve the tools. The TSR tools were initially tested over 21 lists with staff in June and July 2003, exposing over 80 relevant teamwork, performance and safety-related lessons, issues and actions; such issues would normally have remained tacit without the debrief having taken place. Figure 24.4 shows the spread of issues raised by the initial self-reviews performed. The issues raised were coded using the dimensions from the team health model.

Figure 24.4 depicts good teamwork as a positive and issues that need improvement as a negative. From the table it is readily apparent that the top issue needing improvement concerns the management or leadership of the team. The next issue requiring addressing is "shared situational understanding", analogous to the situation awareness of Gaba and Howard. The third issue was inter-team working (between the operating theatres and the ward). Pre-briefing could help address all of these issues.

Equipment use issues highlighted the need for continual staff training and refresher training on new, complex items of medical equipment.

Staff reactions

Staff at RCH have reacted positively to the concept of Team Self-Review, although there was an initial Hawthorne effect evident when QinetiQ were facilitating the reviews. Initially staff felt quite self-conscious being together outside the actual operating theatre and reflecting on the day's events. The use of the TSR cards has facilitated the initiation of a discussion and ensured that everyone has said something, even if they later reflect that they could have said more.

One student operating department practitioner wrote:

> It was a good way for the team to get together to discuss the positive and negative aspects of the day. Each person had their say and was listened to. ... On reflection I could have put more input into the discussion and offered a student slant on issues that were raised.

Reflection – as an individual and also as a team – is one of the key aims of the process.

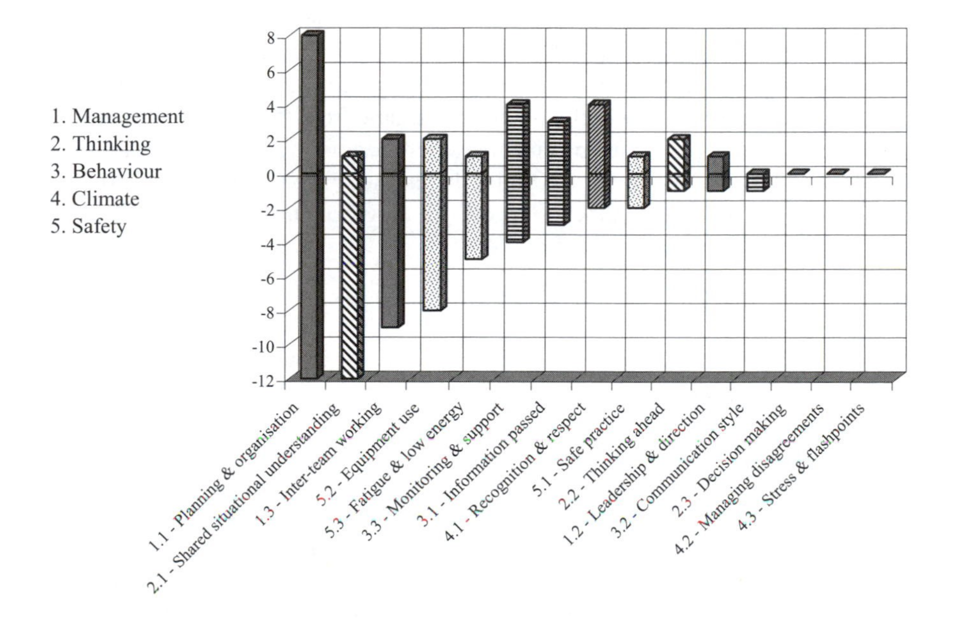

Figure 24.4 TSR outcomes

Narrative excerpts from the TSR logs illustrate this, as shown in Figure 24.4:

- Reflection – "we realised that the MRSA protocols are different at other hospitals"
- Empowerment – "the runner would in future speak up as he was clearer on the protocol"
- The need for briefing – "we could have done with a bit more information on the third case"; "the regular staff assumed it had been done and did not check with the new staff"
- Traditional hierarchies invoking powerlessness – "the surgeons shoes were dirty and he went everywhere with them – nothing was done about it".

The last comment came from a Team Self-Review without the surgeon. At RCH, the surgeons are the group most likely to absent themselves from the TSR. Reasons include time pressure, continuing patient care and the fear of criticism. Traditionally, they were perceived as the coercive team leader; some still try this style of leadership – possibly they feel unable to adopt a more social style. Team Self-Review would help them gain this and help build the non-technical skills of the rest of that day's team.

Team Self-Review issues are changing as time goes on. Full details of the changes at RCH are not reported here; however, team management is a recurring theme, highlighting the need for briefing. Safe practice and shared situational understanding are being more frequently discussed as team members feel able to voice concerns and realise that they are not all sharing the same mental model for the procedure. Teams that are regularly self-reviewing are also reporting more positive team behaviours.

Pre-briefing was rare. Hopefully the team pressure from the TSRs will persuade more surgeons to pre-brief. Nursing and anaesthetic teams already pre-brief for their sub-speciality, but it is the surgeon who initially has the "big picture" and pre-briefing would ensure that all team members start off "singing from the same hymn sheet".

Current status of TSR

After the initial assessment period, RCH continued to employ TSR on a regular basis, eventually gaining sufficient confidence in the approach to roll it out to a second theatre block a year later.

In 2005, the UK National Audit Office (NAO) conducted a follow-up review of organisational learning in the NHS to see if things had improved since the Government's 2001 White Paper. The review included an evaluation of the impact of the TSR system on patient safety at RCH. The audit cited an internal review conducted by the hospital, which concluded that "staff in the theatre complex exposed to team self-review showed statistically significant improvement in teamwork climate and some improvement in the safety climate (using the Safety Attitude Questionnaire, a reliable and formally validated research tool used in over 350 hospitals worldwide) than staff who were not offered debriefs or did not attend."

The NAO also funded further research by the Trust to evaluate the long-term impacts and benefits of Team Self-Review. The Staff Attitude Questionnaire data identified that:

- Briefing and debriefing had a positive impact on non-technical skills and patient safety
- Pre-session briefing was important for safety and effective team management (those interviewed reported that the process improved teambuilding and communication, and enhanced preparation and anticipation of potential problems for theatre lists)
- Debriefing was valued as a process by which the teams could learn from

problems encountered during lists and plan how care could be improved in the future.

The National Patient Safety Agency is currently considering plans to make the TSR system available to hospitals nationwide.

Acknowledgements

The authors would like to acknowledge the support and enthusiasm of all theatre staff at the Royal Cornwall Hospital who contributed to the success of the project, and also the NPSA for their vision in sponsoring and supporting such innovative, safety-oriented research.

References

Chief Medical Officer (2000). *An organisation with a memory.* Report of an expert group on learning from adverse events in the NHS, chaired by the Chief Medical Officer. London: The Stationery Office.

Gaba, D. M. and Howard, S. K. (1995). "Situation awareness in anaesthesiology". *Human Factors*, 37(1), 20-31.

Kohn, L., Corrigan, J. and Donaldson, M. (eds) (1999). *To Err is Human: Building a safer health system*. Washington: National Academy Press.

Leape, L. L. (1994). "Error in medicine". *Journal of American Medical Association*, 272(23), 1851-1868.

National Audit Office (2005). "A safer place for patients: Learning to improve patient safety". Report by the Comptroller and Auditor General. HC 456 Session 2005-2006. November 2005.

PART 6
Assessment and Measurement

Chapter 25

A Method for Need Analysis before Decision Making Based on Ecological Psychology

Thierry Morineau

Introduction

Research in natural decision making has highlighted the importance of situation awareness in the quality of decision making (Endsley, 1995). In the same way, the initial stage of need elicitation in the design process has been underlined as highly critical for the quality of final product. Yet, several methods are available to assist needs analysis in the process of design. The main known method is the functional analysis of need (Norme, 1991). The goal of this method is to describe needs by a set of functions and to avoid expressing the needs directly into some final solutions. Nevertheless, this approach of need definition is not always easy to perform, because of the lack of a concrete dimension that seems necessary for designers in order to estimate constraints upon object functioning (Darses, Détienne and Visser, 2001). Some methods attempt to take into account the concrete dimension in need identification. The use of scenarios can allow an accurate projection of need in context and can be realised with the help of text and schemas describing an object. However, some critical questions rapidly arise as to which scenario to select, and what are the essential aspects of a scenario? (Hertzum, 2003).

When developing a new method of interface design, Vicente and Rasmussen (1990) proposed a frame of description of work domain in which the system would be embedded. This description is based on two orthogonal scales: the Part-Whole decomposition hierarchy, describing the product as a system composed of subsystems; and the Abstraction Hierarchy, formulating the product as a set of Means-Ends ordered functions. For instance, this tool has been used to develop a ship for the US Marines (Bisantz et al., 2003).

The main advantages of this method are to handle the entire abstract and concrete dimensions of functions for object description and to be user-oriented. However, some drawbacks can jeopardise the use of this method. First, a guide for effective use of this method does not exist. Secondly, this method imposes a mode of idea structuring that can be viewed as a constraint by the subject (Lind, 1998). And thirdly, we can interrogate the concept of "function" and its significance in an ecological perspective. Is a set of functions sufficient to describe an object and its integration in a context? Is a set of functions sufficient to describe a need? To give some elements of response, we suggest an investigation of the ecological nature of need.

Ecological nature of need

The root of the notion of "need" has been found to be below the symbolic cognitive processes, like reasoning or working memory process. The need comes from psycho-physiological mechanisms related to the organism's adaptation to his ecological environment, in which she or he looks for resources. Towards its motivational role, need intervenes as a pre-conceptual influence in the formulation of goals for action (Clancey, 2002). This pre-conceptual aspect of need implies that needs are mainly in a state of latency within an organism, without symbolisation, and that need involves an articulation with some features of the ecological niche. It means that direct elicitation of needs is not evident and that some aspects of the environment have to be taken into account. Some evidence concerning the sub-symbolic aspect of need and its ecological dimension has been shown in ecological psychology. The concept of affordance proposed by Gibson (1979) corresponds to a strong articulation between the adaptive needs of an organism and some pieces of information picked up. In another way, Barker (1968) developed an ecological study of live places of human beings. He shows some relationship (synomorphy relation) linking environmental features and stable extra-individual behaviour patterns. For instance, a schoolroom is composed of desks and chairs, which are placed in correspondence with teacher and children behaviours. These places are qualified as "behaviour settings". In this way, Barker notes that common sense leads people to define behaviour settings with physical attributes of the place and extra-individual behaviour patterns. For instance, a route would be defined as a way (physical attribute) to travel or carry on merchandises (extra-individual behaviour patterns).

Also, the concepts of affordance and behaviour settings allow us to envisage an analysis of need based on a relationship between three entities: the functional features of the environment, which we call "resources"; the adaptive needs of individuals, called "needs"; and the characteristics of concerned individuals, called "population". According to the works of Rasmussen and Vicente, we will consider need as based on a hierarchical structure of abstraction. With this framework, we have engaged the building of a method of need analysis in the context of decision making in a municipality.

Method of need analysis based on a study in a municipality

Our study has been completed in a little tourist town in South-Brittany (France), with one thousand inhabitants. The municipality had to cope with an urban problem where young inhabitants tend to quit the town, whereas an older population comes to set up second homes. In this context, decision makers had the purpose of using some landscape in the centre of the town to develop a politic of urban and service development. At the moment of our intervention, the decision makers looked for a definition of their needs, whilst they would meet a property developer presenting a specific proposal of urbanisation. Note that the decision team had to deal with multiple parameters (financial, political, economic, demographic…) that led the project to be particularly complex. In order to help the decision makers, we proposed to intervene in the course of need identification through four stages of study.

First stage: Access to the ecological elements of the need

In accordance with our point of view on the cognitive status of need as a sub-symbolic process, we suggested to elicit the needs of the subjects by non-directed interview from which mental representation should emerge without a pre-defined frame of ideas coming from the observer. Moreover, verbal communication is a common mode of exchange for a manager in the context of her/his work (Kuo, 1998).

We interviewed the Mayor and the Deputy Mayor before and after the strategic meeting with the property developer. We also participated in this meeting. During this meeting, the property developer proposed a very finalised option of urbanisation of the area including 25 small houses, which could be used as second homes. This proposal consisted of building a private housing estate that is usually attractive for tourists and the retired. But *a priori*, this commercial offer was in contradiction to the political choices of the decision makers.

Second stage: Ecological element coding

The coding scheme applied to the recorded interview consisted of three kinds of tags corresponding to the three ecological elements defining a need: Resource, Need, and Population. Note that we included another tag representing the other cognitive elements verbalised by the decision maker and relevant for the project management (called "Other"). We considered that a mental representation was composed of a set of viewpoints. A sentence, a part of a sentence or several sentences can represent a viewpoint. A set of tagged and related elements defines a viewpoint (called P – see Table 25.1 for an example).

Table 25.1 Example of coding (Mayor, first interview)

<P> Effectively, the Municipal Council thinks **<Need Settlement= 'building the sector'**>we must build this sector**</Need><Resource landed='fields in city centre'>**which is finally a sum of fields in the city centre**</Res></P >** **<P><Need project='a suiting project'>**but also we wish that the project suits us **</ Need></P>**

In our study, we calculated the percentage of the different tags in the interviews with the two decision makers, before and after the meeting (Figure 25.1). This analysis allows us to highlight first the small part of Mayor's discourse allocated to the Resource, Need, and Population definition.

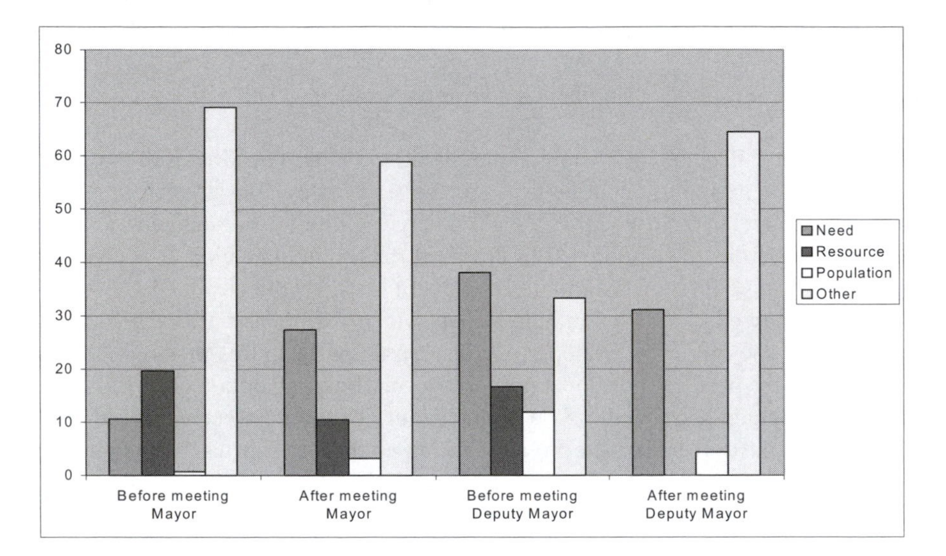

Figure 25.1 Percentages of elements coded in each interview

In the first interview, almost 70 per cent of the elements are not concerned with ecological aspects of the need (about 60 per cent in the second Mayor's interview). The semantic content of these other elements focuses on the stages of the management of any project that could be engaged. Concerning the Deputy Mayor, we observe another pattern of frequency in which the ecological elements firstly appear as the most salient in the discourse. This political decision maker underlines the importance of the population's need study to select a fitting urban project. After the meeting with the property developer, his discourse was divided into two extreme points: respectively the need corresponding to about 30 per cent of the tags and the constraints and limits that the property developer's project would imply and which corresponds to the tag "other" (about 60 per cent of the tags in the second interview).

Third stage: Map of need and measures on the level of need specification

Having all the tags of the ecological elements of need, we structured them by making a hierarchical organisation for each kind of tag (a branching from general to particular elements). Especially, we focused on the "Need" tag for the purpose of presenting the level of specification of need expressed by each decision maker at one moment in the course of the decision making (see Appendix).

After this hierarchical classification of "Need" tag in each verbal corpus, we elaborated a scale of specification level in order to allow quantification and so, comparison of the different specification levels in each interview. We put together all the occurrences of "Need" tag in all the interviews collected with the decision team and ordered them as the function of general/particular scale. The tags that were more general received the value "1", the next tags received the value "2" and so on. After that, we assigned the value of specification to each node in each branching. This scale, based on the levels of specification found in the decision team, allowed us to situate the level of specification of one decision maker at one moment as the function of all the team. Also, two indicators can be measured. The first one is the level of depth in specification, which corresponds to the level of specification obtained by each branch of the "need" branching. The second indicator is the level of coherence for each branch. For each link between two nodes on the branch, we calculated the number of specification level lacking. The higher this number is, the more the coherence in the specification of the need is low relative to what the decision team can specify. Figure 25.2 shows the depth and the coherence of specification obtained by each branch of Mayor (called "M" in the figure) and Deputy Mayor's Need (called "A") maps.

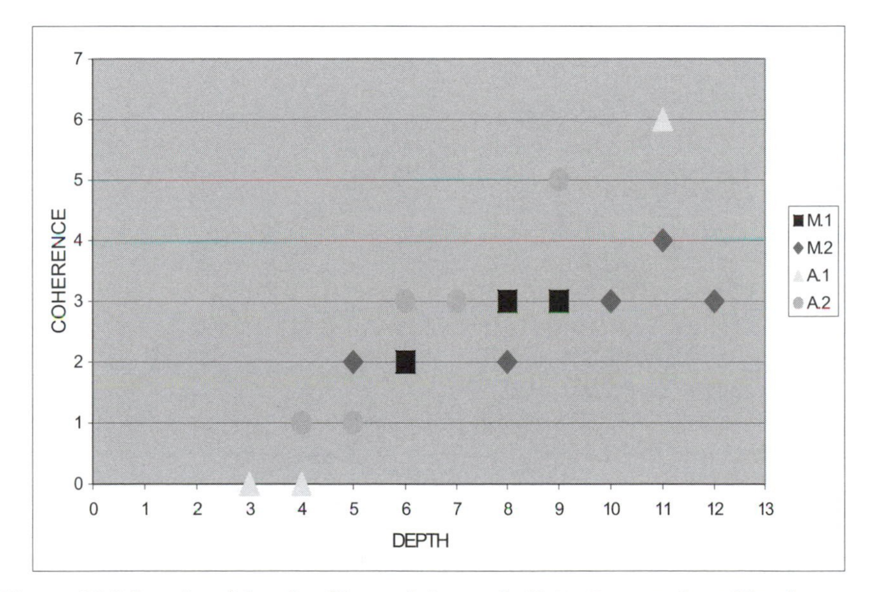

Figure 25.2 Levels of depth of branch in need elicitation produced by the decision makers

Note: In y: number of lacking nodes and in x: number of nodes in the branching.

Figure 25.2 shows the combined representation of levels of depth and coherence for the two decision makers. We observe that in his ecological elements elicitation, the Mayor is generally more coherent (number of lacking nodes reduces) and has more depth (number of nodes by branch more large) than the Deputy Mayor.

Fourth stage: Deepening of need expression

The maps of ecological elements (resource, need, population) and the indicators of depth and coherence in specification serve to be presented to the decision makers in order to help them improve the quality of their mental representation of needs. The possibility of assessing the levels of coherence and depth relative to the other decision makers in the team can support communication and collaborative work in the team.

In our study, the presentation of these issues provided by our analysis led the decision makers to delay their decision in order to refine their project of urbanisation. In the future, we wish to automate the coding of interview with the use of XML tag and to implement our method in a system supporting need analysis in decision making.

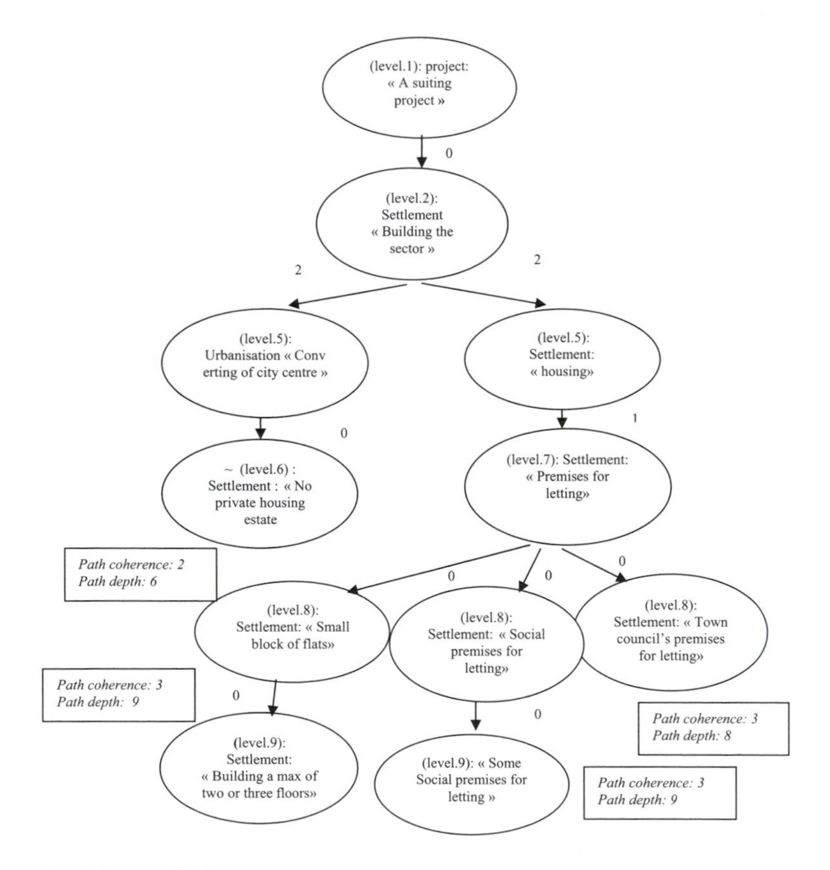

Appendix Figure 25.1

Acknowledgements

The author thanks Delphine Monier, Emmanuel Frénod, and the OGGAM association for their help. We thank the members of the Municipal Council for their participation.

References

Barker, R. G. (1968). *Ecological psychology*. Stanford: Stanford University Press.

Bisantz, A. M., Roth, E., Brickman, B., Gosbee, L. L., Hettinger, L. and McKinney, J. (2003). "Integrating cognitive analyses in a large-scale system design process". *International Journal of Human-Computer Studies*, 58, 177-206.

Clancey, W. J. (2002). "Simulating activities: relating motives, deliberation and attentive coordination". *Cognitive Systems Research*, 3, 471-499.

Darses, F., Détienne, F. and Visser, W. (2001). "Assister la conception: perspectives pour la psychologie ergonomique". In *Proceedings of Congrès EPIQUE de la Société Française de Psychologie*. Nantes, France (pp. 11-20).

Endsley, M. R. (1995). "Toward a theory of situation awareness in dynamic systems". *Human Factors*, 37, 32-64.

Gibson, J. J. (1979). *The ecological approach to visual perception.* London: Lawrence Erlbaum Associates.

Hertzum, M. (2003). "Making use of scenarios: A field study of conceptual design". *International Journal of Human-Computer Studies*, 58, 215-239.

Kuo, F-Y. (1998). "Managerial intuition and the development of executive support systems". *Decision Support Systems*, 24, 89-103.

Lind, M. (1998). "Making sense of the abstraction hierarchy". In *Proceedings of the CSAPC.*Villeneuve d'Ascq, France (pp. 195-200).

Norme NF X50-151 (1991). *Analyse de la valeur, analyse fonctionnelle*. Paris: AFNOR.

Vicente, K. J. and Rasmussen, J. (1990). "The ecology of human-machine systems II: mediating 'direct perception' in complex work domains". *Ecological Psychology*, 2, 207-249.

Chapter 26

Operational Net Assessment: A Canadian Human Factors Analysis

Philip Farrell

Introduction

The world is changing rapidly and military organisations all around the world are compelled to self-reflect and look for transformational ways of doing business. Effects-Based Operations (EBO) as opposed to Threat-Based Operations during the Cold War is a transformational concept where the intent is not to match the adversary's capability one for one, but to ensure that national and international aims are achieved and maintained. Multinational Forces must work together using Diplomatic, Information, Military, and Economic (DIME) instruments of power to achieve the strategic aim and associated desired effects.

Joint Forces Command in the United States (US) is investigating EBO with coalition partners Australia (AU), Canada (CA), Germany (GE), United Kingdom (UK), and NATO (North Atlantic Treaty Organisation). The multinational Limited Objective Experiment 2 (LOE 2), conducted in 2002, explored the Operational Net Assessment (ONA) process, and how different information sharing agreements might influence that process.

The ONA represents a model of an adverse system organised under Political, Military, Economic, Social, Information, and Infrastructure (PMESII) disciplines. The global concept is that given a model of an adverse system, one can analyse the model and predict possible actions and resources required to achieve desired effects.

The LOE 2 objectives were to:

1. Explore the ONA process of EBO
2. Investigate new Multi-National Information Sharing (MNIS) strategies.

The Coalition Headquarters was staffed with military Planners and civilian PMESII experts (or System-of-Systems Analysts: SOSAs) from the six partners. The multinational SOSA teams built the ONA database collaboratively by obtaining, interpreting, and storing information in the database. Both the Planners and SOSA used MNIS agreements between countries to release information, which increased the complexity of the experiment.

LOE 2 incorporated both exploratory (objective 1) and hypothesis-testing (objective 2) experimentation into one event (Stenbit, Wells and Alberts, 2002), which presented many challenges in its design and execution. The coalition experiment team took one year to plan the event. Close to 70 players participated in the event, and they played distributed in their respective nations. Participants used Groove™ to collaborate with each other. Groove™ provided text and voice chat, document preview and editing, sketchpad, and other collaboration tools.

After the experiment, the US compiled a comprehensive report of the event (J9 Joint Experimentation Analysis Division, 2003). Canada took the opportunity to report on the Human Factors (HF) results and discuss them from a Canadian perspective. This chapter includes excerpts from the Canadian report (Farrell, 2003a).

The Canadian report explores the HF issues related to team information sharing, including Workspace Design, Human-Computer Interaction (HCI), Distributed Planning, Team Dynamics, Problem Solving, Cultural Issues, as well as Multiple Agent Interaction, and Situation Awareness (SA) and confidence (reported in Lichacz and Farrell, 2005). Farrell (2003a) takes each issue in turn, briefly looks at the literature on the issue, collates and analyses, primarily, observation data, discusses the results, and makes recommendations for future HF research areas required to advance Effects-Based Operations.

Workspace design

Few studies examine virtual workspaces and their impact on human performance and workload (Johns and Blake, 2001). During the event, some participants were co-located in the same room (physical workspace) while others communicated with distributed team members using text and audio chat (virtual workspace). The key HF issue was how did the participants use physical and virtual spaces during their collaborations?

The time per visit, time per user, and visits per user were calculated from the number of workspace participants, time spent in a workspace, and the number of times the workspace was visited by participants (see Table 26.1). Unfortunately, Groove™ made no distinction between analysts and participants, and thus the data were confounded. Nevertheless, assuming that the analysts' intervention was minimal and consistent in all the spaces, the data were examined for trends.

The "Space" column indicates the virtual workspaces where various teams could meet and exchange information. For example, the coalition space (COA) for Planners and SOSAs had the highest time per visit (65 seconds). Users spent 6 minutes and 33 seconds on average per day in COA. CA participants spent only 17 seconds per visit in their national space, but had five visits per user (that is, shorter times but more visits). Users from countries that had multi- (ML), tri- (TL), and bi-lateral (BL1 and BL2) Information Sharing Agreements spent even less time in these spaces, leading one to believe that they were redundant, and most of the collaboration could be done in the COA space.

Sixty out of 718 observations were made about the usefulness of the workspaces, and only nine of those were favourable. A typical comment was, "this multiple space thing is becoming a problem".

The design considerations that come from the results are: 1) face-to-face interaction is preferred whenever possible; 2) COA and national spaces are the minimum requirement for virtual spaces; but 3) redundant communication modes (for example, Internet Protocol phone) is recommended.

Table 26.1 Time per visit and user averaged over eight days

Space	Time/visit	Time/user	Visits/user
COA	0:1:05	0:6:33	6
US	**0:0:44**	**0:3:51**	**5**
SOSA	0:0:44	0:1:38	2
ML	0:0:23	0:1:12	3
Planner	0:0:39	0:1:40	3
GE	**0:0:46**	**0:2:38**	**3**
NATO	**0:0:34**	**0:2:18**	**4**
UK	**0:0:38**	**0:1:36**	**3**
AU	**0:0:39**	**0:2:05**	**3**
CA	**0:0:17**	**0:1:24**	**5**
TL	0:0:19	0:0:53	3
BL2	0:0:22	0:1:05	3
BL1	0:0:14	0:0:33	2

Human-computer interaction

Human-Computer Interaction aspects include display design (Mullet and Sano, 1995), cognitive compatibility, and alternative control technologies to name a few, but MNLOE 2 produced no additional/new HCI insights.

Thirteen per cent of all HF observations were concerned with the Groove™ interface design. For example: "*– need to build trust among coalition members – how do we do this when we are limited to voice and [text] chat*". Participants liked the audio and text capability to communicate. However, audio chat degraded rapidly as more players joined a space, thus frustrating the users. Furthermore, there was little customisation that addressed the challenges for building complex ONA databases.

Excessive time delays as well as overloading the Groove™ audio chat generated frustration amongst players and observers alike. At one point, the IP phone became the primary means of communication. It was quickly recognised that this would render the experiment invalid and so Groove™ audio chat was reinstated the following day. A sample comment from the observation data was: "*audio failed*". Although Groove™ was designed for network collaboration and information sharing, it was not designed for collaboration with 70 people in 2002.

In conclusion, Groove™ seems to be adequate for the tasks it was designed for, but perhaps not for this LOE, based on the frustration expressed by the users. The technology and interface requirements for ONA development should be addressed in future experimentation, particularly a tool that reflects the ONA process and business rules, which would aid staff members in performing their tasks.

Distributed planning

LOE 2 explored the concept of distributed effects-based planning. Several models exist for individual planning (Powers, 1973), but few for distributed planning (Myers, Jarvis and Lee, 2002). On one hand, multiple cooperating nations may generate more robust plans than a single nation. On the other hand, a single nation may produce a better plan than the coalition due to restrictions imposed on information sharing agreements.

LOE 2 had the potential to explore coalition versus national planning. However, the vignettes and injects did not push Planners to engage in distributed planning. That is, the scenario was in peacetime, there were no strategic objectives identified, and there was no need for military action. There might have been a need for Diplomatic, Information or Economic action, but again there were no clear strategic objectives.

Table 26.2 Total number (%) of ONA access/changes

	Accessed	Changed
Nodes	3376 (57%)	306 (58%)
Effects	1468 (25%)	124 (24%)
Actions	678 (11%)	56 (11%)
Resources	428 (7%)	36 (7%)

Table 26.2 provides an indication of ONA activity. Given that Nodes and Effects are primarily SOSA activities, players accessed Nodes 57 per cent of the time and Effects 25 per cent of the time. Given that Actions and Resources are primarily Planner activities, players accessed Actions 11 per cent of the time and Resources 7 per cent of the time. That is, there was 4.5 times more SOSA than planning activity in the ONA database. Forty-two comments were made about distributed planning. For example:

- US planners and SOSA cells completed a discussion of injects and actions to take. The process was orderly with tasks assigned to certain planners – such as

requesting releases from the FDO [Foreign Disclosure Officer] and tracking actions taken on injects on a log form. This followed SOSA concern that important decisions be captured and not lost in the course of play.
- UK team members are not happy at way the coalition discussion and interactions are proceeding today. Too many different opinions between teams within nations as well as teams between nations.

Half of these comments also dealt with teams and nations unable to come to a decision about actions and resources for inclusion in the ONA database. That is, actions could not be identified until desired effects were articulated, keeping in mind that a desired effect may be realised through the natural unfolding of events. In pre-crisis, the Planners' role might be to determine the effects that will result without any military intervention. If undesired effects are discovered during the ONA process, then an opposing desired effect must be identified and Planners can begin to plan a course of action that might achieve these desired effects.

In summary, LOE 2 did not have incentives for participants to remain fully engaged in information sharing and distributed planning because: 1) it was pre-crisis and by definition there is no need for action; 2) the desired effects may occur with the normal unfolding of time; and 3) national will and coalition intent needed to be clearly articulated so as to justify actions and resources. It is hoped that these issues will be resolved before and during Multi-National Experiment 3, which investigates Effects-Based Planning.

Team dynamics

It is expected that team dynamics such as personality types, cognitive consistency, social perception, and crowd behaviour (Gleitman, 1981) would be evident during LOE 2. Common Intent (CI; has aspects of cognitive consistency) is a key social psychology factor that the coalition needs to have in order to build an effective ONA. Team members may develop CI over time from individual interpretations of the strategic objectives and individual expectations of the final product. Their interpretations and expectations are based on the individual's experience, training, values and culture. The degree to which these interpretations and expectations are common amongst team members will have an impact on goal achievement.

The team leader is another important factor towards the achievement of the objective. He or she, as well as all staff members, needs to have the right balance of responsibility, authority and competency (Pigeau and McCann, 1995). Whenever there is an impasse, the team looks to the team leader to resolve it in a manner that increases their confidence in the leadership and that the goal will be achieved. A leader who demonstrates leadership competencies will motivate team members to take on personal authority and intrinsic responsibility towards building common intent and achieving the objective.

LOE 2 comprised a matrix of teams: national teams, SOSA and Planner teams, and the Coalition team. First, some teams did not have a leader, such as the Planners' team. Secondly, an individual may be a member of more than one team and may have conflicting objectives and allegiances. Thirdly, some teams were distributed and some were co-located. One can hypothesise that those teams with an identified leader, clear de-conflicted goals, and face-to-face interaction would perform better than any of the other teams that lack one or more of these facets.

The experiment was not explicitly designed to test this hypothesis but there was enough evidence recorded to report trends as shown by the following excerpts from the observation data:

Common intent

- Comment on US discussion – appears that they do not confer on a position for the US before entering text chat. Need to have a coherent US line as all other nations do to minimise confusion.

Co-location

- Until we see a quantum leap in data retention/storage techniques and machine/human interfaces, we'll remain more comfortable collaborating face to face.
- Doubts about the usefulness and effectiveness of the distributed teamwork – creative teams need face to face work conditions.

Leadership

- Too many different opinions between teams within nations as well as teams between nations.

Also, 101 out of 718 comments (14 per cent) dealt with teams. Sixty-four of those comments were under the category of "user-user breakdown" – that is, there was some type of miscommunication or misunderstanding between two or more people. Some of these comments are associated with other HF aspects studied herein, and Table 26.3 summarises these associations.

Data showed that common intent, co-location and leadership are critical elements for teams, and need to be addressed with respect to team performance. Anecdotally, the national teams seemed to fair better than any of the other teams, primarily due to communication delays. Even within the CA national team, a few days were required to clearly define roles and responsibilities. Once leadership and intent was clearly established, the CA team worked effectively, and better than the SOSA, Planner, or Coalition teams in which the Canadians were involved.

Table 26.3 Team comments associated with itself and other issues

Other Issue	# of Comments	Sample Comment
Team	74	Vivid group work – good team spirit. 16 Foreign injects treated
Distributed Planning	11	The team is in the process of updating database with nodes and resources. The process seems slow and time consuming. What if decisions were needed quickly or time played a crucial role in taking imitative in a crisis
Workspace Design	7	Team members have a preference in printing all relevant documents…
Communication medium	4	Groove limitations that resulted in broken audio forced many US players out of the COA space thus limiting team interaction

Problem solving

Although problem solving was identified as a critical HF issue for LOE 2, there was no situation that challenged the participants to have innovative solutions. For example, the national teams were given information that required a decision to share the information within a given agreement, change the agreement, delay the release of information, or retain the information. Since the information was either unclassified or did not expose any Canadian vulnerabilities, nor did it affect the ONA development, the default decision was to release information using the suggested information sharing agreement. In one or two cases, a different decision was made – just to make the game play interesting.

The observation data yielded 26 comments that referred primarily to technical problems or experimental problems – not problem solving with respect to the teams and to the objectives. There was no data collected that could shed some insight into problem solving. If problem solving is a priority for the next experiment, then care must be taken in the experimental design to encourage problem solving behaviours.

Cultural issues

The Cultural hypothesis for LOE 2 was that a coalition ONA would be better than any one national ONA due to the cultural diversity resident in the coalition. In contrast, the Human Factors analysis wanted to capture the cultural issues that inhibit information sharing and the ONA process.

Cultural issues are linked to values, and values may clash within a multinational coalition. The mere fact that a coalition exists means that some values are aligned, and this must be the starting point for collaboration. Iraqi Freedom is an example where France, Germany and Canada had differing core values from the US and UK who formed a large part of the US-led coalition. In LOE 2, several observations were made on culture indicating that there will still be cultural and value differences even after nations agree to work together, as follows:

- Messages and communication between COA countries tend to get distorted. Intent and context of messages are not understood. The first is a language problem with non-native English countries. Secondly, English-speaking countries have different ways of saying things causing slight morphing of messages and misinterpretation.
- US30 asks GE to concur with Canadian comments but GE assumes they are asking them to begin presentation of their nodes. The mistake was not necessarily because of language barriers but subtleties of language are obviously lost between nations.

Language is a strong driver of cultural issues. In this experiment, GE was the only non-native English-speaking nation. However, there was still confusion due to differences in English accents, and the use and meaning of English words.

Military culture seemed to unify the coalition. It became the *de facto* way for briefs, debriefs, communication protocols, and so on. This was a common language that most participants understood and felt comfortable with. However, there was a strong civilian element to contend with, that is, most of the SOSAs were civilian government employees. Interestingly, they were amenable to receiving commander's guidance, particularly if it made sense in achieving the overall goals.

- Civilian vs Military culture has been observed. The default is to adopt a civilian model for meetings–- that is collaboration – consensus – asking for volunteers.

As with problem solving, observers were asked to record any issues related to culture, but there was nothing in the design of the experiment that would expose any cultural differences.

Multiple agent interaction

Related to team dynamics is an emerging Human Factors topic called "multiple agent interaction", that is, the interaction between intelligent agents. An intelligent agent acts on the world (its immediate sphere of influence), processes data from the world, adapts to changes in the world, and is constrained by the world. An agent may be animate or inanimate, human or machine, real or virtual.

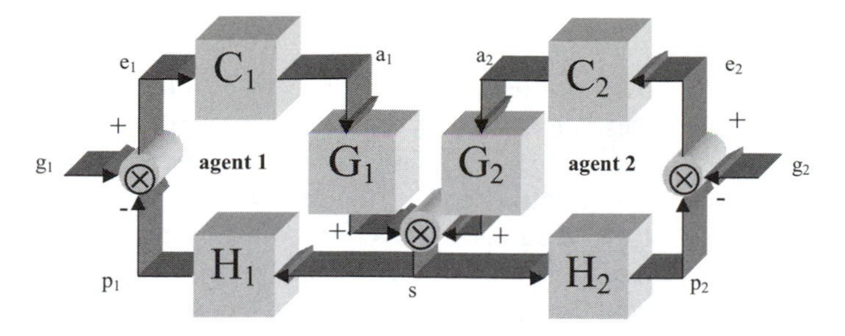

Figure 26.1 Dual agent

A mathematical analysis was conducted on the stability of interacting multiple agents (Farrell, 2003b). The main result of this study showed that the interaction tends to be stable when agents receive as much information as possible but act on separate parts of the world, as shown in block diagram form in Figure 26.1. That is, each agent acts (a_1 and a_2) on separate world states (G_1 and G_2) thus producing sensory information (s) that is transformed by both H_1 and H_2 into the agents' perceptions (p_1 and p_2). The perceptions are compared to their respective goals (g_1 and g_2) and a perceptual error is generated (e_1 and e_2). The error becomes the catalyst for making decisions (C_1 and C_2) to act, thus closing the loop. This framework is based on Perceptual Control Theory (Powers, 1973).

MN LOE 02 provided data for testing the following Multiple Agent Interaction model hypotheses:

- *Hypothesis 1*: Effective interaction has clear and distinct goals.
  Result: *"Trouble also in articulating the AU goals of the exercise."* Participants did not know what their goals were, and so the interaction was ineffective.
- *Hypothesis 2*: Effective interaction has clear feedback.
  Result: Participants were never sure of how the ONA was progressing and when the ONA was completed.
- *Hypothesis 3*: Effective interaction requires a means of influencing world states.
  Result: Each player had access to Groove™ and the Internet to influence information sharing and the ONA database (world states).
- *Hypothesis 4*: An agent with great influence will reach their goal, sometimes at the expense of other agents' goals.
  Result: *"US are still trying to push military options over other options."* True collaboration was not evident. On the other hand, sometimes nothing gets accomplished without forceful leadership.

Information sharing

Situation Awareness (SA) was used to investigate the multinational information sharing (MNIS) hypothesis that a future coalition agreement should yield better SA than the current multilateral agreements. The SA data were analysed within a 6 (partners) x 2 (SOSAs vs. Planners) x 2 (current vs. future MNIS) x 4 (days) x 3 (surveys/day) analysis of variance (ANOVA) with repeated measures on the last three variables. Overall, there was a mean SA score of 67.75 per cent. SA was significantly higher during the current MNIS (70 per cent vs. 65 per cent), $F(1,34) = 12.70$, $Mse = 349$, $p < .001$. There was a significant interaction between MNIS and country, $F(5,34) = 3.70$, $Mse = 349$, $p < .009$. Post hoc analyses using Tukey's Honestly Significant Difference revealed that GE had a significantly lower SA during future MNIS ($p<.002$). Analysis of the participants' confidence ratings in their SA responses revealed that participants were over-confident in their SA responses.

Conclusions

This chapter contains excerpts from the LOE 2 HF analyses from the Canadian perspective, and the full report is available from the author.

The Workspace Design results showed that participants acclimatised themselves to working in virtual spaces within eight days. Human-Computer Interaction was inadequate primarily because the tool was not designed to handle the large number of players. LOE 2 did not have incentives for participants to remain fully engaged in information sharing and Distributed Planning. Anecdotally, the national Team Dynamics seemed to be more effective than the distributed teams primarily due to technical issues associated with the collaboration tools. The experimental design did not challenge the players to use unusual Problem solving techniques. Cultural Issues will occur even within coalitions whose core language and values are the same. The Multiple Agent Interaction model is a powerful method for exploring goal achievement and interaction effectiveness. Information Sharing produced an SA result of about 68 per cent, and participants were over-confident in their responses.

Overall, more research is required in each of the HF issues. The continuing challenge will be to control for as many variables as possible in order to achieve a repeatable result. However, the experimentation designs are growing in complexity. It is recommended that most of these HF issues are studied separately from the main event if possible, and then the results can flow into the design and implementation of subsequent large events.

Acknowledgements

I would like to thank US JFCOM and the multinational team that planned and executed LOE 2. Their series of experimentation continues to provide a rich venue for investigating HF issues related to EBO.

References

Farrell, P. S. E. (2003a). "Limited Objective Experiment 02: Canadian Human Factors Report". A report prepared for Canadian Forces Experimentation Centre. DRDC-Toronto TR-2003-166, Toronto, Canada: Defence R&D Canada (60 pp.).

Farrell, P. S. E., (2003b). "Human factors implications from operator-agent interaction analysis". In *Proceedings of 49th Annual General Meeting and Conference of the Canadian Aeronautics and Space Institute.*

Gleitman, H. (1981). *Psychology*. New York: W. W. Norton & Company, Inc.

J9 Joint Experimentation Analysis Division (2003). "Multinational Limited Objective Experiment II (MN LOE II) Final Report". *US Joint Forces Command*, prepared with contributions from Australia, Canada, Germany and the UK, December 2003.

Johns, C. and Blake, E. (2001). "Cognitive maps in virtual environments: facilitation of learning through the use of innate spatial abilities". In *Proceedings of the 1st Conference on computer graphics, virtual reality and visualisation* (pp. 125-129). New York: ACM Press.

Leggatt, A. P. (2002). "Limited Objective Experiment 02 UK Proposal for Human Factors Measures". Draft document version 4.

Lichacz, F. M. J. and Farrell, P. S. E. (2005). "The calibration of situation awareness and confidence within a multinational operational net assessment". *Military Psychology*, 17(4), 247-268.

Mullet, K. and Sano, D. (1995). *Designing visual interfaces*. Mountain View, CA: Sun MicroSystems Inc.

Myers, K. L., Jarvis, P. A. and Lee, T. J. (2002). "Active coordination of distributed human planners". In *Proceedings of the 6th International Conference on AI planning and scheduling*. American Association for Artificial Intelligence.

Pigeau, R. and McCann, C. (1995). "Putting 'command' back into command and control". In *Proceedings of the Command and Control Conference.* Ottawa: Canadian Defence Preparedness Association.

Powers, W. T. (1973). *Behavior: The control of perception.* Hawtorne, New York: Aldine De Gruyter.

Stenbit, J. P., Wells, L. and Alberts, D. S. (2002). *Code of best practice for experimentation.* DoD Command and Control Research Program. Washington, D.C., USA: CCRP Publication Series.

Chapter 27

Psycho-physiological Measures of Situation Awareness

Han Tin French, Elizabeth Clarke, Diane Pomeroy,
Melanie Seymour and C. Richard Clark

Introduction

In the military domain, situation awareness (SA), generally understood to mean "knowing what is going on", is an important concept. The presumption is that in battles, all combat personnel from the lowest to the highest ranks must have SA, without which they may face defeat or make serious mistakes that may result in dire consequences. Land Operations Division of the Defence Science & Technology Organisation (DSTO), Australia, is conducting research on SA, specifically on the methodology for measuring commanders' SA when conducting Command and Control (C2) functions. An ability to measure SA will provide a powerful tool for studying cognitive processes involved in C2 tasks, and for evaluating advanced technology tools developed to support the command processes. This chapter describes an exploratory study to examine the feasibility of using psycho-physiological measures to assess SA.

Techniques for measuring SA

The definition of SA used in this work is that commonly employed in the research community, namely "the perception of the elements in the environment within a volume of time and space, the comprehension of their meaning, and the projection of their status in the near future" (Endsley, 1995). The merits and disadvantages of the four primary methods for measuring SA have been discussed by Pew (1995). In the present work, direct experimental techniques were used. The direct questioning technique based on SAGAT (Situation Awareness Global Assessment Technique) (Endsley and Smolensky, 1998) has earlier been used in a command post exercise involving Brigade Headquarters staff (French and Hutchinson, 2002). SAGAT involves freezing the activities at random times during which participants are questioned. The probes cover three levels of SA: perception of elements; understanding the elements in the current context; and projection to the near future. A limitation of SAGAT is its intrusiveness. Psycho-physiological data can be obtained continuously while the person focuses on the tasks at hand, thus removing the need for interference. The relationship between psycho-physiological measures with SA and other cognitive domains such as mental workload and fatigue have been studied for many years. A relatively recent review (Wilson, 2000) summarised the status in this research area and suggested five areas for testing the utility of psycho-physiological measures. In the review it was mentioned that in a 1994 study, EEG (electroencephalogram) showed promise as an indirect measure of SA in air-to-ground simulation tasks.

Experimental method

Essentially the experiment consisted of having a group of participants play a computer game that had been designed to stimulate their SA. The players, who were given specific tasks as part of the scenario, had to gain and maintain SA to achieve them. EEG, EOG (electrooculogram) and respiration rate data were collected during game time. SA questions were administered five times during natural breaks in the game, and at its conclusion.

Computer game

Military command post exercises, even though they may be tightly controlled and scripted, are not suited for experiments involving EEG. Computer simulations can be used as a test-bed for studying SA. A range of commercial off-the-shelf computer games, some of which have been used for military training (Ford, Barlow and Lewis, 2003), are available. Operation Flashpoint, produced by Codemasters, was chosen because it allows the users to develop and script new scenarios. With the expectation that not enough military personnel would be available to act as participants in the experiment, a game narrative was developed that would not require specialised military knowledge. Within the scenario, the player was given the task of rescuing a scientist as the overall mission. The game was divided into several sections during which the player was given more immediate tasks. Information was provided to the player by visual cues and text messages on the screen, as well as audio signals and messages through a pair of head-phones. The game also provided virtual tools such as a digital map, a pair of binoculars and a compass.

SA levels

Central to the experiment was the ability to code the stimuli corresponding to the three SA levels. Initially event-related electrical potential (ERP) analyses were considered. The operationalisation of the three SA levels, namely (i) the detection of stimuli in an environment, (ii) the recognition of the significance of environmental stimuli, and (iii) the expectancy of change in the environment, map very well to measures that can be obtained from ERPs, in the form of the P1-N1-P2 complex, the N2-P3 slow wave complex, and the Contingent Negative Variation (CNV) respectively (Coles and Rugg, 1995). The three indices are obtained conventionally through the averaging of a sufficient number of trials of scalp EEG time-locked to the onset of relevant events. Unfortunately the timing of the onset of the events in the game could not be obtained to the degree of precision required for ERP analysis from the computer game selected. Analysis was therefore conducted on raw EEG data.

Following the completion of the data collection phase, the EEG records were time-marked with SA level 1, level 2 and level 3 events. SA level 1 events were those in which stimuli presented to the player did not have any direct relevance to the task at hand. Level 2 events were stimuli that had relevance to the immediate task, such as the sound of a truck, when the task was to look for the truck. These were differentiated from level 3 events, which had relevance to the overall mission, for example, messages relating to the long-term aim of rescuing the scientist.

Data collection

Twelve male participants experienced at playing computer games participated in the study. Detailed information about the scenario and tasks were provided well in advance of the experiment. Prior to game play the players went through a training session to familiarise themselves with the control keys used in the game.

A SYNAMPS system (Neuroscan, Sterling, VA, USA) was used to record continuous EEG data from 19 tin scalp electrodes positioned according to the 10:20 system. EOG were recorded from above and below the right eye and from the outer canthus of each eye. Electrode impedances were less than 10 kΩ. EEG and EOG were amplified 1000 times (DC-100 Hz) at a 500 Hz sampling rate.

Analyses

Power spectra

The EEG data was corrected for eye movements using the Scan software (version 4.2) eye movement reduction algorithm. Task EEG data was epoched for 2048 ms prior to each SA event time-marker (Pre epochs), and for 2048 ms following each time marker (Post epochs). Each epoch was baselined and visually inspected for excessive electrical noise, and any epoch with such artefact was rejected.

For each participant, the data was averaged in the frequency domain using Fast Fourier Transformation (FFT). Power values were obtained for delta (0-4 Hz), theta (4-8 Hz), all alpha (8-12 Hz), low alpha (8-10 Hz), high alpha (10-12 Hz), beta (12-30 Hz), gamma 1 (35-44 Hz), gamma 2 (45-54 Hz), gamma 3 (55-64 Hz) and gamma 4 (65-74 Hz) frequency bands. Paired t-tests were performed to compare differences between Pre and Post values of spectral power for each frequency and electrode combination, Pre and Post values of spectral power across each SA level, and Post and Rest values of spectral power for each of the particular frequency and electrode combinations. To control for the number of comparisons, alpha level was set to 0.001.

Discriminant analyses

For each participant, the power spectral density was calculated in $\mu V^2/Hz$ for single EEG trials by using FFT with a Hanning window and 1024 point block size. Data were then collapsed across individuals to determine whether it was possible to identify a common way to classify the three SA levels. As discriminant analysis is sensitive to outliers, any trial containing an outlier greater than three standard deviations from the mean was removed from the data set, prior to analyses being carried out in SPSS 11.0 for Windows, using the stepwise method with Wilks Lambda and leave-one-out classification options.

Results

Behavioural

The game, which took about one hour to complete, appeared to have succeeded in engaging the players in the task, with mean SAGAT scores ranging from 36 per cent to 76 per cent. Some players found it difficult at times to navigate in the virtual terrain; it became necessary to help orient the players to progress with the game.

The means (and standard deviations) of the number of SA events per participant were 16.1 (3.5), 40.2 (4.2) and 19.5 (2.7) for level 1, 2 and 3 respectively.

Power spectra

Significant differences were found between Pre and Post values of spectral power for SA level 2. Mean theta, gamma 1, gamma 3 and gamma 4 power were significantly greater for the Post than Pre event types at electrodes F4; F3 and F4; FP1, F3 and F7; and F4 respectively. No such differences were found for SA levels 1 and 3.

Discriminant analysis

Group level

Table 27.1 Summary of group discriminant analysis classifications

SA Level	No. Trials	Classification			% Correct*
		SA1	SA2	SA3	
1	134	26	99	9	19.4 (11.9)
2	282	19	252	11	89.4 (82.6)
3	145	15	106	24	16.6 (7.6)

Note: The number of trials for each SA level is shown, as well as whether trials within each SA level were classified as belonging to SA level 1, 2, or 3.

For each participant, data was normalised by averaging across all sites for each frequency band. Discriminant analysis classification (see Table 27.1) was poor overall (53.8 per cent on training and 46.3 per cent on a validation data set) for all but level 2 SA with many of the levels 1 and 3 SA trials also classified as level 2. Although the overall classification when trying to distinguish between SA levels 1 and 2 (70.2 per cent training; 63.7 per cent validation), 2 and 3 (70.3 per cent training; 62.5 per cent validation), and 1 and 3 (62.0 per cent training; 49.5 per cent validation) was better, these still indicated poor classification of SA levels 1 and 3.

Individual level

Discriminant analyses were carried out on the non-normalised data for individual participants. The results for three representative persons (4, 12, 13) are shown in Table 27.2. In some cases, classification was good (for example, Participant 4). However, for others it was not, with the outcome quite variable. For example, there was difficulty classifying trials related to level 3 SA from Participant 12, whereas level 1 SA was not classified at all for Participant 13. In total, it was possible to distinguish the three SA levels in only three participants.

Table 27.2 Summary of individual discriminant classifications for three representative participants

			Classification (EEG based)				
Person	**SA**	**Trial**	**SA1**	**SA2**	**SA3**	**Sites**	**% Correct***
4	1	24	20	4		10	83.3 (75)
	2	30	2	27	1		90.0 (80)
	3	12			12		100.0 (100)
12	1	17	12	4	1	3	70.8 (71)
	2	35	2	30	3		85.7 (80)
	3	21	4	11	6		28.6 (24)
13	1	21		20	1	3	0.0 (0)
	2	51		43	8		84.3 (84.3)
	3	27		13	14		51.9 (50.9)

Note: The number of trials for each SA level is shown, as well as whether trials within each SA level were classified as belonging to SA level 1, 2, or 3. The number of sites selected by the discriminant analysis routine for use in the classification algorithm is also shown.
* Figure in brackets represents classification rate for a validation data set.

The number of sites involved in classification varied between participants, as did the site location and frequency bands involved. The only commonalities across participants were that frequencies within the alpha and gamma bands were involved in all cases.

Discussion

The results of the study indicate an increased involvement of frontal lobe activity in the theta and gamma bands during the processing of SA level 2 events. Gamma is associated with higher cognitive function, feature binding and decision making (Pulvermueller et al., 1995; Haig et al., 2000). Frontal theta is associated with increased mental load (Gevins et al., 1997), particularly during focused attention (Benham et al., 1995). These same effects were not obtained for SA1 and SA3 events. This is probably related to there being fewer of these events than for level 2, thereby reducing statistical power. Clearly, more level 1 and 3 events were needed.

The "free-play" nature of the game meant that the experimenter did not have absolute control over the events. Depending on how the game unfolded, some of the stimuli might not have been presented. Also, it was difficult to determine the exact time when the player might perceive a visual cue. The appearance of an image on the screen does not equate to player's perception. This was less of a problem with audio cues or messages, but some players may have still not encountered all of the stimuli depending on how they played the game.

Satisfactory classification of each of the three SA levels was only possible at the individual level, and even then only for a few people. The discriminant analyses of individual participants highlight the presence of variability, due possibly to individual differences in physiology or cognitive style.

Overall, the high level of misclassification of SA levels 1 and 3 as level 2 could be due to a variety of factors. This misclassification could reflect the small number of trials for SA levels 1 and 3, an inability of the experimental design to separate out the three levels of SA, or the presence of confounds.

Although muscle activity (electromyography or EMG) generated from facial or neck muscles during EEG collection might be a potential discriminator between the different levels of SA, there was no measure of this in the study. Further, since the frequency of EMG occurs within the same bandwidth as EEG gamma, there could be a confound between these different physiological activities, even though visually identifiable EMG was removed.

Game play involves many decisions. Therefore there is also a potential confound between EEG related to decision making, actions undertaken as a result of the decisions, and that related to SA. Although SA results in decisions being taken and actions executed, no measure was taken during this study of participants' game decisions. Participants received assistance when required during the game, which made it difficult to assess their performance. In future studies it will be valuable to measure not only the participants' SA, but also the quality of their decisions and performance.

In summary, this experiment has provided some support for the view that EEG data may be used to assess SA. A greater number of trials and ability to time-lock EEG data will allow use of ERPs as well as (or instead of) EEG frequency bands. It may be that a particular cortical region is important for either overall SA or the individual SA levels and collection of data from more scalp sites will improve spatial resolution and provide more sites overlying a particular cortical region.

This study illustrates the inherent differences between laboratory research and applied research in environments such as the military. There is often a conflict between methodologies used to address questions of relevance to the military (and other "real world" areas) and scientific techniques. The desire to address holistic issues can result in the use of designs that do not adequately allow investigation of phenomena of interest, as is the case with this study. However, it has allowed an appreciation of the differing cultural views.

Two further studies utilising different experimental designs have been undertaken to address some of the limitations of the current study. Analyses are still being conducted; however, it is appropriate to present a brief summary of findings at this stage. The first of the follow-up studies measured changes in EEG frequencies whilst participants watched a short segment of three different movies. Role playing instructions were used to induce a mental set corresponding to a designated SA level whilst participants were watching a movie clip, and each movie had clearly defined events that could be related to each of the three SA levels. The orders of induced SA levels and movie clips were counter-balanced across the participants. Consistent with findings of the current study, although it was possible to distinguish level 2 SA from levels 1 and 3, the EEG related to SA levels 1 and 3 was comparable. More specifically, inducement of a level 2 SA mental set resulted in increased power within the delta band compared to the other SA levels. As already mentioned, it is possible to map the different SA levels onto specific ERP component, with recognition of the importance of environmental events (level 2 SA) being associable with the ERP known as the P3. The increased delta power associated with level 2 SA in this study is therefore consistent with the proposed relationship between EEG frequency within the delta band and the P3 ERP component (Karakaş, Erzengin and Başar, 2000; Başar-Eroglu et al., 1992) as well as research associating delta activity with decision making (Başar et al., 2001).

The design of the second follow-up study resolved the time-locking issues present in the first and second studies. It enabled the use of the originally proposed mapping of the individual SA levels to a specific ERP measure: SA level 1, the P1-N1-P2 complex; SA level 2, the N2-P3 slow wave complex, and SA level 3, the Contingent Negative Variation (CNV). Data from this study are still being analysed.

References

Başar, E., Başar-Eroglu, C., Karakaş, S. and Schürmann, M. (2001). "Gamma, alpha, delta, and theta oscillations govern cognitive processes". *International Journal of Psychophysiology*, 39, 241-248.Başar-Eroglu, C., Başar, E., Demiralp, T. and Schürmann, M. (1992). "P300-response: Possible psychophysiological correlates in delta and theta frequency channels. A review". *International Journal of Psychophysiology*, 13, 161-179.

Benham, G., Rasey, H. W., Lubar, J. F., Fredrick, J. A. and Zoffuto A. C. (1995). "EEG power-spectral and coherence differences between attentional states during a complex auditory task". *Journal of Neurotherapy*, 2-3(1), 1-10.

Coles, M. G. H. and Rugg, M. D. (1995). "Event-related brain potentials: An introduction". In M. D. Rugg and M. G. H. Coles (eds), *Electrophysiology of mind: Event-related brain potentials and cognition* (pp. 1-26). New York: Oxford University Press.

Endsley, M. R. (1995). "Toward a theory of situation awareness in dynamic systems". *Human Factors*, 37, 32-64.

Endsley, M. R. and Smolensky, M. W. (1998). "Situation awareness in air traffic control: The picture". In M. W. Smolensky and E. S. Stein (eds), *Human factors in air traffic control* (pp. 115-154). London: Academic Press.

Ford, M., Barlow, M. and Lewis, E. (2003). "An initial analysis of the military potential of COTS games". In *Proceedings of SimTect 2003*, Adelaide.

French, H. T. and Hutchinson, A. (2002). "Measurement of situation awareness in a C4ISR experiment". In *Proceedings of the 7th International Command and Control Research and Technology Symposium*. Quebec City, Canada: CCRP.

Gevins, A., Smith, M. E., McEvoy, L. and Yu, D. (1997). "High-resolution EEG mapping of cortical activation related to working memory: Effects of task related difficulty, type of processing, and practice". *Cerebral Cortex*, 7, 734-785.

Haig, A. R., Gordon, E., Wright, J. J., Meares, R. A. and Bahramali, H. (2000). "Synchronous cortical gamma-band activity in task-relevant cognition". *NeuroReport*, 11(4), 669-675.

Karakaş, S., Erzengin, O. U. and Başar, E. (2000). "The genesis of human event-related responses explained through the theory of oscillatory neural assemblies". *Neuroscience Letters*, 285, 45-48.

Pew, R. W. (1995). "The state of situation awareness measurement: Circa 1995". In D. J. Garland and M. R. Endsley (eds), *Experimental analysis and measurement of situation awareness*. Daytona Beach, FL: Embry-Riddle Aeronautical University Press.

Pulvermueller, F., Lutzenberger, W., Preissl, H. and Birbaumer, N. (1995). "Spectral responses in the gamma-band: Physiological signs of higher cognitive processes?" *NeuroReport*, 6(15), 2059-2064.

Wilson, G. F. (2000). "Strategies for psychophysiological assessment of situation awareness". In M. R. Endsley and D. J. Garland (eds), *Situation awareness analysis and measurement* (pp. 175-188). Mahwah, NJ: Lawrence Erlbaum Associates.

Chapter 28

Signal Detection Theory and the Assessment of Situation Awareness

Barry McGuinness

Introduction

Put simply, situation awareness (SA) is "knowing what is going on so I can figure out what to do".[1] Humans are increasingly called upon to make important decisions in complex situations that are often uncertain, unpredictable and stressful. Examples include military combat, air traffic control, complex surgical procedures and civil emergencies. More often than not, poor decisions follow directly from an inadequate, partial or even false understanding of the situation. Having accurate, appropriate and up-to-date SA is therefore crucial for decision makers. For this reason, supporting the SA of individuals and teams has become a high priority factor in human engineering.

SA assessment

We can determine the human effects of new systems by carrying out controlled assessments using realistic tasks, including human-in-the-loop simulations. Experiments like this must be scientifically objective and use valid, robust methods to give meaningful results. Often, they require methods to assess the quality of decision makers' SA. The assessment of SA is far from straightforward, however. This is due partly to the multi-faceted nature of SA itself, and partly to the fundamental difficulty of observing what is happening in another person's mind. These two factors give rise to a range of options for assessing SA.

We could, for example, focus on the subjective aspect of SA by asking the participant to rate his/her own SA, as is the case with instruments like the Situation Awareness Rating Technique (SART; Taylor, 1990) and Crew Awareness Rating Scale (CARS; McGuinness and Foy, 2000). However, this does not tell us about the state of the participant's SA as such; it tells us only how the participant *perceives* their own SA.

Alternatively, we could assess SA by focusing on the mental and behavioural processes supporting SA, such as information acquisition and inference making. The patterns of activity of such processes may be reflected in physical correlates such as brain activity or eye-scanning behaviour. While providing more objective data than

1 Definition offered by a pilot, cited in Adam (1993).

subjective ratings, process measures of SA still fail to provide much insight into a person's actual perception and understanding of reality.

The most direct approach to SA assessment is to focus on the contents of awareness itself, that is, to assess how well the participant's mental representations of the situation fit the facts of the situation, or the "ground truth". One form of this approach is to *probe* the participant's perceptions and interpretations of the situation by eliciting responses to pertinent questions. Asking well-chosen probe questions, responses to which can be readily checked against ground truth, can give considerable insight into the accuracy, completeness and currency of the person's SA.

SA probes

In some respects, SA probes are similar to tests of students' knowledge. There are two basic ways of using probes, which we can call "supply" and "selection". With *supply* probes, the participant must supply the information being asked for. Asking the participant to provide a situation report (sit rep) update is a straightforward form of supply probe. Supply probes can also be in the form of very specific, open questions; for example, "*Which hostile track currently presents the greatest threat*?"

With selection probes, in contrast, the correct information is presented to the participant along with one or more incorrect options; the participant is asked to select the correct one. An example of a selection probe technique is the use of multiple-choice questions. This method is embodied in what is probably the most well established SA probe technique, the Situation Awareness Global Assessment Technique (SAGAT; Endsley, 1995).

In theory, the extent of a person's awareness of a situation is indicated by his or her success in being able to judge the truth or falsity of propositions related to it (Ebel and Frisbie, 1991). With this in mind, an alternative probe technique we have been developing is the use of *true/false probes*. A true/false probe is a description of some aspect of the situation, a description which may or may not in fact be true. The description is presented to a participant who, on the basis of his or her awareness of the situation, indicates whether it is true or false. For example:

Probe statement
"A column of enemy tanks is now leaving the city."

Response
True [ ] *False* [✓]

An intriguing aspect of using true/false probes is that the participants' responses are naturally amenable to analysis in terms of hit rate, miss rate, false alarm rate and correct rejection rate (Figure 28.1), which are the basis of perceptual analysis in Signal Detection Theory.

		Participant's response	
		"TRUE"	"FALSE"
Probe type	TRUE	**HIT**	**MISS** (error)
	FALSE	**FALSE ALARM** (error)	**CORRECT REJECTION**

Figure 28.1 Contingency table showing the four possible outcomes of a true/false probe response, depending on type of probe (true or false) and the response made ("True" or "False")

Signal detection theory

Signal Detection Theory (SDT) is not merely a theory but also a mathematical technique for analysing perceptual performance. It originated as a model of how human observers perform when they must detect ambiguous visual stimuli of a certain type, such as targets on a radar screen (Tanner and Swets, 1954). The theory describes the task as one of distinguishing between specific target stimuli and other, irrelevant stimuli, referred to as *signals* and *noise* respectively. For example, the radar observer's task is to detect meaningful radar 'blips' (signals) whilst ignoring or rejecting all irrelevant stimuli (noise). The theory also posits two important internal factors influencing an observer's performance on a signal detection task:

1. The observer's *sensitivity* in being able to correctly discriminate true signals from non-signals.
2. The observer's response *criterion* (or *bias*) when it comes to ambiguous stimuli. In other words, the observer's strategy for handling those stimuli that require a deliberate judgement.

Thus, SDT recognises that individuals are not merely passive receivers of stimuli or information; when confronted with uncertainty, they also actively engage in the process of deciding whether what they perceive signifies one thing rather than another.

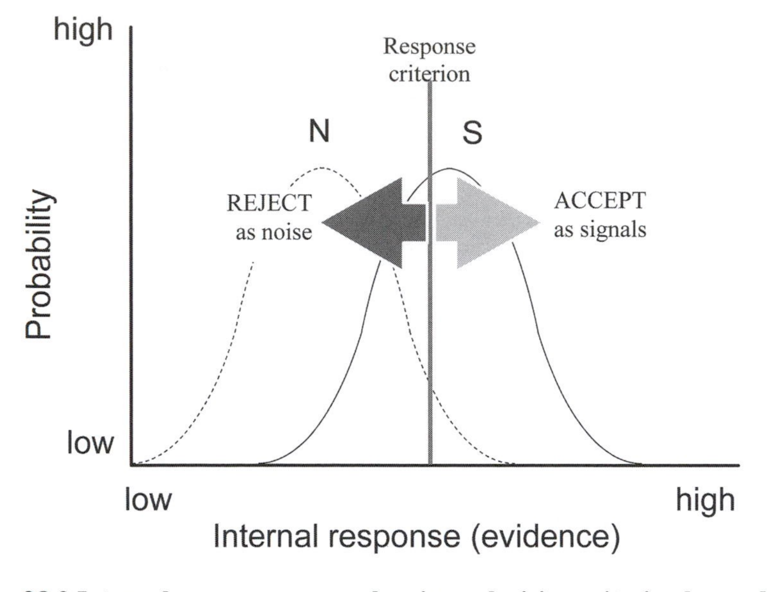

Figure 28.2 Internal response curves showing a decision criterion beyond which the observer will accept stimuli as signals

Finally, SDT provides a graphic model for understanding performance on a signal detection task. The underlying notion is that overt responses to stimuli are mediated by internal responses within the observer. These can be thought of as different levels of neural activation, which may be experienced as a subjective sense of "evidence" for the presence of a signal. Figure 28.2 shows a graph of two hypothetical internal response curves. The curve on the left (N) is the probability distribution for internal responses that would be generated if the observer were presented only with non-signals, that is, noise-only stimuli. The curve on the right (S) is the probability distribution for the strength of internal responses when real signals are present.

Notice that the two distributions can overlap. That is, sometimes the internal responses for non-signals can be as strong as internal responses for real signals. This is the essence of ambiguous stimuli and gives rise to the possibility of discrimination errors. What kinds of errors are made depends to a large extent on where the observer has set his or her *response criterion*. A given observer may, for example, prefer to err on the side of caution by setting a high criterion and rejecting all but the most definite stimuli. This would be termed a "conservative bias". Alternatively, an observer could ensure that no possible signals slip through the net by setting a low criterion and accepting all but the most obviously false stimuli. This would be a "liberal bias".

Relevance of SDT to situation awareness

The early application of SDT to studies of sensory performance is considered a major advance in the understanding of human perception. To some extent, though, these origins obscure the theory's more general applicability. In fact, SDT has been applied to a host of varied decision problems that extend far beyond the detection of sensory signals. Over the decades, SDT methods and measures have been adopted for the study of such diverse real world tasks as military target detection, motorists' detection of hazards, medical diagnosis, and other diagnostic tasks in fields like information retrieval, weather forecasting, survey research, aptitude testing, polygraph lie detection, and vigilance. Another recent extension of SDT has been to the analysis of recognition memory.

In essence, SDT models the ability of an agent to compare and match some given input with one or more known categories. It provides a practical tool for the analysis and understanding of real-life judgements or diagnoses. Can it also be used in this way to assess situation awareness?

We can consider this question by comparing the processes of signal detection (as modelled in SDT) with the process of situation assessment. The task of situation assessment is to arrive at a mentally perceived situation with a full understanding of its implications. This is obviously far more complex and abstract than perceptual signal detection, and involves acquiring information and interpreting that information on the basis of prior knowledge and expectations. Nevertheless, there is a degree of equivalence in the fact that both signal detection and situation assessment involve *discrimination*. Specifically, the "observer" in situation assessment must be able to discriminate between at least the following:

- *Valid versus invalid information*. Invalid information is that which appears to represent the current situation but is in fact erroneous or unreliable. For example, an item of information may be too old to be current, or originate from an untrustworthy source.
- *Valid versus invalid interpretations*. It is important to correlate different items of information to establish a coherent picture of the situation as it actually is. For example, one might interpret a warning light as indicating a system fault, whereas other evidence indicates that the warning light itself is at fault.
- *Valid versus invalid inferences*. In this case, the discrimination is to do with the validity of one's logic rather than one's interpretations. For example, the idea that "the enemy will surrender as soon as they see us coming", an inference based on the assumption that lesser powers are intimidated by our technological supremacy, may not be valid.

A further parallel with signal detection is that the observer must have some criterion for deciding how to respond when uncertain about any of the above discriminations. For instance, a military commander may receive conflicting intelligence about the status of a particular enemy movement, and must then judge which information to accept as valid, and which version of the situation to believe.

There are also parallels between the types of error that can be made in signal detection and situation assessment. Aside from the failure to detect critical stimuli, the possible errors of situation assessment include accepting invalid information and rejecting valid interpretations of the situation. Such errors can arise because of the human vulnerability to *confirmation bias*: the automatic tendency to seek primarily those sources of information that confirm what we already believe to be true. The USS *Vincennes* incident in the Persian Gulf is a case in point: because of an expectancy of imminent air attack, some cues were overlooked and others were completely misread, resulting in the shooting-down of an Iranian passenger jet (Klein, 1998). Confirmation bias can be regarded as a low criterion setting in the observer, that is, an over-willingness to accept stimuli as evidence for a particular situation.

Application to SA probes

How can the SDT framework and analytical techniques be used to assess SA? In human factors research, the cognitive content of SA has typically been assessed using some kind of hit rate, that is, the proportion of probes that are responded to correctly. While this seems an obvious statistic to use in terms of face validity, on its own it is an inadequate index of SA accuracy for two reasons. First, a participant's hits disclose nothing about his or her false perceptions or false beliefs about the situation; nor does it reveal their correct awareness of what is *not* the case (for example, knowing that a certain piece of on-screen information is false). Secondly, hit rate alone fails to provide a full picture of the participant's awareness because, in terms of SDT, it confounds sensitivity and response bias (Swets and Pickett, 1982). Discriminating between participants' sensitivity on the one hand and judgement strategy on the other could therefore be valuable for understanding patterns in people's situation assessments.

Practical example

We have so far applied this technique (dubbed QUASA, for *Quantitative Analysis of Situation Awareness*) to SA probe data obtained from a small number of military trials of varying size. The most significant of these has been a multinational experiment held in February 2003. Led by the US Joint Forces Command, LOE2 (the second in a series of Limited Objective Experiments) involved five nations plus NATO collaborating via the Collaborative Federated BattleLab Network, a secure online environment designed to facilitate allied experimentation. The experiment focused on "operational net assessment" (ONA) (described in Chapter 26), a new process by which coalition analysts and planners work jointly on the development of a shared knowledge base for an emerging crisis situation. This multinational task was used to test collaboration and information sharing across different security domains within a fictional scenario.

One aim of the LOE2 analysis activity was to focus on the human issues of situation awareness and shared awareness. To this end, 58 players (located in five countries) were asked to respond to SA probes at two-hourly intervals in order to provide a measure of their awareness of the current situation. The probes were descriptive statements of elements of the situation of interest compiled both from baseline knowledge in the ONA database and new information that was to be added to it during the experiment.

Equal numbers of true and false probe statements were carefully formulated, with the probe construction process going through several iterations for refinement. First, the probes were shown to a set of independent evaluators (with no other involvement in the experiment and no knowledge of the scenario), who were asked to judge the likelihood of each statement being true or false based purely on the given wording. When inadvertent cues were found, the wording of a probe was altered to make it more neutral. Secondly, the probes were assessed by a German human factors analyst for intelligibility. Probes were then altered if necessary to ensure that the non-native English speakers participating in the experiment would be able to understand them clearly. Finally, each probe was evaluated by a subject matter expert for its operational significance and relevance within the experimental scenario. Only those statements that successfully passed all three tests were used in the final probe set.

Each probe statement was followed by four questions (Figure 28.3). The participants were instructed to complete the questions in silence, without consulting other participants or the database. They were also asked not to discuss the questions after presentation. When the answer to a probe was not known, participants were instructed to make a guess at the true/false response and then indicate they were guessing by marking "very low" on the confidence scale.

The Commander of the [...] Air Force has recently resigned over corruption charges.

1) **Is this** [knowledge] **relevant to your task/role?**
   [ ] yes [ ] maybe [ ] no
2) **Is this statement true or false?**
   [ ] true [ ] false
3) **What is your level of confidence in the true/false response?**
   [ ] very low [ ] low [ ] medium [ ] high [ ] very high
4) **Which countries will mostly answer this probe correctly?**
   [ ] AUS [ ] CA [ ] GE [ ] RE [ ] UK [ ] US

Figure 28.3 Example of an LOE2 probe and the questions asked after each

During the experiment, five probes were presented to all participants every two hours. By the end of the two-week experiment, 45 participants had answered at least 100 probes each.

SDT analyses

Hit rates and false alarm rates were found both for each individual participant and for each of the five national teams, and then used to generate measures of sensitivity (d') and response bias (ß). The team scores are summarised in Figure 28.4, which compares average hit rates versus average false alarm rates (a graph of this type is termed a Receiver Operating Characteristic or ROC graph).

Sensitivity

Given that sensitivity measures respondents' ability to discriminate between correct and incorrect descriptions of the situation, d' can be taken as a quantitative index of SA. This statistic has a minimum value of zero (indicating no SA), whereas values in the range 3-4 are to be regarded as very high. In the LOE2 SA probe data, a wide range of sensitivity scores was found across individuals, with a few showing good discrimination between true and false statements (d' > 2.0) and a few others performing relatively poorly (d' < 0.5). The average level of sensitivity was moderate (d' = 1.0, s.d. = 0.5). In terms of average team-level sensitivity there was little difference between the nations, with team averages for d' ranging from 1.0 to 1.6. In other words, one nation was as good as another on the whole at detecting true versus false statements.

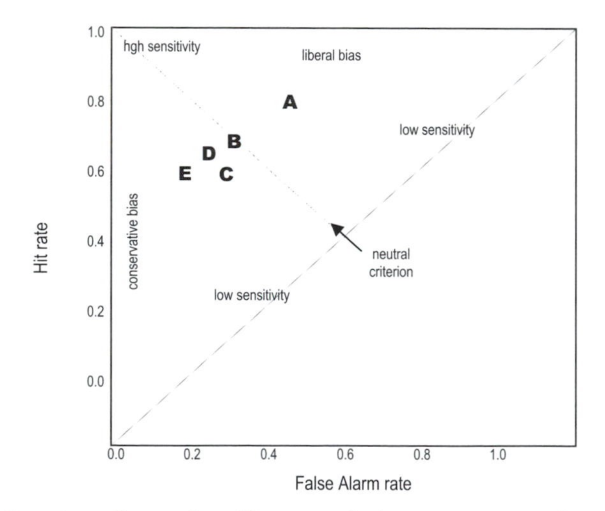

Figure 28.4 Receiver Operating Characteristic curve comparing average hit rates and false alarm rates of the five nation teams

Criterion/bias

With a 50:50 ratio of true and false statements in this experiment, the optimum criterion setting would have been with neutral bias, that is, ß = 1.0. The average across all participants was indeed essentially neutral (ß = 1.1, s.d. = 0.4), but there was a spread of bias scores between individuals, ranging from 0.4 (fairly liberal) to 2.6 (fairly conservative). An interesting difference was found between national teams, with team A alone (ß = 0.7) showing an overall liberal bias (a tendency to accept false statements as true) while other nations were either neutral or relatively conservative in their probe responses. Team A had on average a higher hit rate than the others, but it also had a higher false alarm rate (hence, net sensitivity was about the same).

It was also found that this team had the lowest overall confidence in its probe responses (based on responses to Question 3 after each probe). Subsequent analysis indicated possible reasons for this: Team A had missed out on some initial training and experience due to technical problems, and appeared to be adapting by trusting other nations to provide the information needed.

There is space here to provide only a snapshot of the SDT analyses applied to the SA data from this experiment. The main focus of this discussion is on what this experience has to teach us about the pros and cons of using the QUASA technique.

Evaluation of the technique

True/false probes

One clear disadvantage of true-false probes is that the task of generating them can be labour-intensive both before and during an experiment. The researcher must have at least a reasonable understanding of the domain and the scenario and keep up to speed with the situation as it changes so that appropriate probes can be generated. The need to generate not only appropriate true statements but also false alternatives is a well-known problem with selection-type probes.

Another disadvantage of true/false probes is their binary nature. Through educated guesswork alone, an individual could score a correct response rate of better than the chance level of 50 per cent. (This compares unfavourably with the SAGAT four-alternative format where the chance success level is 25 per cent.)

The main advantage of true-false probes is that they are easy to score objectively and the results obtained are very amenable to analysis in terms of hits, misses, false alarms and correct rejections. In addition, participants generally find them quick and easy to respond to. We also find that applying SDT analysis to true/false SA probe data yields sensitivity and bias statistics that can give insights into participants' SA that would not be available using hit rate alone. We have also found that combining these with subjective self-ratings gives a fuller picture. In fact, it appears to us that combining SA probes with subjective ratings of response confidence and statistically analysing their interrelationship would generally be a highly effective way to assess situation awareness, with or without the addition of SDT analysis.

Outstanding issues

Since entertaining the use of true/false probes and Signal Detection Theory to assess SA, our investigations have identified several key issues:

1. *How appropriate is SDT as a method for the assessment of situation awareness? What are the limitations?* For instance, SDT requires that "signals" are objectively identifiable. That is, we must be able to compare the participant's responses against a known, actual situation. In some cases this is definitely possible; we simply compare participants' perceptions and understanding against the objective "ground truth". There are aspects of SA, however, that involve non-veridical inferences, such as what could happen in the future and what may be done about it. Having no objective referent, these aspects may not be amenable to analysis by SDT. (Note that the same constraint also applies to other probe techniques not using SDT.)
2. *What guidelines are needed to ensure that probes are appropriately constructed?* For instance, we have found that it is important to ensure that probe responses capture participants' actual awareness (or reflect the lack thereof) rather than their ability to make informed guesses and clever judgements. It was found in LOE2, for example, that simply negating a true statement to construct a false statement often inadvertently cued the reader to a likely falsehood. Each probe therefore went through a rigorous process of checks prior to its use in the experiment.
3. *How many probes are needed to give a valid SDT analysis?* There are questions as to the validity of the statistics insofar as validity is dependent upon the number of responses on which the analysis is based, and hence on the number of probes used.
4. *What do the SDT statistics actually mean in terms of participants' cognitive performance?* For example, does a strong conservative bias in probe responses reflect a similar bias in the same participant's assessment of the situation? This remains to be determined through further research.

Conclusion

In conclusion, SDT appears to lend itself to the quantitative analysis of SA using true/false probes, and potentially may be a valuable tool for SA assessment. However, there are some outstanding issues requiring further research.

References

Adam, E. C. (1993), "Fighter cockpits of the future". *Proceedings of the 12th DASC, the 1993 IEEE/AIAA Digital Avionics Systems Conference*, 318-323.

Ebel, R. L. and Frisbie, D. A. (1991). *Essentials of educational measurement* (5th ed.). Englewood Cliffs, NJ: Prentice-Hall.

Endsley, M. R. (1995). "Measurement of situation awareness in dynamic systems". *Human Factors*, 37(1), 65-84.

Klein, G. (1998). *Sources of power: How people make decisions.* Cambridge, MA: The MIT Press.

McGuinness, B. and Foy, L. (2000). "A subjective measure of SA: The Crew Awareness Rating Scale (CARS)". In *Proceedings of Human Performance, Situation Awareness and Automation Conference*, Savannah, Georgia. SA Technologies, Inc.

Swets, J. A. and Pickett, R. M. (1982). *Evaluation of diagnostic systems: Methods from Signal Detection Theory*. New York: Academic Press.

Tanner, W. P., Jr. and Swets, J. A. (1954). "A decision making theory of visual detection". *Psychological Review*, 61, 401-409.

Taylor, R. M. (1990). "Situation Awareness Rating Technique (SART): The development of a tool for aircrew systems design". In *Situation Awareness in Aerospace Operations*, AGARD-CP-478 (pp. 3/1-3/17). Neuilly-sur-Seine, France: NATO-AGARD.

Chapter 29

Psycho-physiological Measurements of Mental Activity, Stress Reactions and Situation Awareness in the Maritime Full Mission Simulator

Thomas Koester

Introduction

The work of the officer on watch on the bridge of a ship can be characterised by its composition of proactive and reactive behaviour and communication. Good performance is characterised by a high amount of proactive behaviour and communication in which future events and actions on the voyage are anticipated and prepared; this is in some parts of the literature called the third level of situation awareness (see Koester and Rabjerg, 2005 for further discussion of the concept of situation awareness). This third level of situation awareness is considered to be essential for the prevention of human error, incidents and accidents and therefore for the overall safety of the vessel (Grech and Horberry, 2002).

The logic of the concept of situation awareness is that anticipation (situation awareness on Level 3) implies perception (situation awareness on Level 1) and comprehension (situation awareness on Level 2) (Endsley, 2000). It is possible on the basis of observations of crew communication among officers on watch on ferries in regular service to find examples of situation awareness on Level 1 and 2 indicated by reactive communication and examples of situation awareness on Level 3 indicated by proactive communication (Koester, 2003).

Changes in the crew behaviour and communication can be observed directly as responses to changes in demand for situation awareness, but the underlying cognitive processes including the situation awareness are not directly observable and measurable.

It is assumed that change in demand for situation awareness generates other crew responses than changes in behaviour and communication, for example, psycho-physiological reactions related to changes in level of stress and mental activity. The measurements of stress levels and levels of mental activity of the officer on watch can therefore be used in an analysis and interpretation of the crew response to variations in demand for situation awareness.

The psycho-physiological reactions can be measured with methods based on the measurement of physiological parameters such as galvanic skin response (GSR), heart rate, electrocardiogram (ECG/EKG) or electric activity in the brain (that is, electroencephalogram or EEG) (Koester and Sørensen, 2003). This chapter describes how variations in stress level and level of mental activity could be analysed and interpreted on the basis of measurements of brain wave activity (EEG) (Koester and Sørensen, 2004).

Hypothesis

It is the hypothesis that situations with a high demand for situation awareness on Level 3 will generate high levels of stress and mental activity and that a low demand for situation awareness will generate equivalent low levels of stress and mental activity. The hypothesis will be tested using two simulated situations in a maritime full mission simulator. One situation has high demand for situation awareness on Level 3 while the other situation has low demand for situation awareness on Level 3. Both situations have demand for situation awareness on Level 1 and 2.

Experimental design

The method of the study is a combination of measurement of electrical activity in the brain (EEG) and observation of crew behaviour and communication and events in a set of simulated voyages in a maritime full mission simulator. The set of simulator measurements included two realistic scenarios in a full mission simulator, both arrivals to Rostock in Germany with the large car and passenger ferry M/S Colour Festival. It was expected that the arrivals would generate changes in demand for situation awareness caused by the changes in situation related to the arrival and to the procedures a short time before arrival. The participants in the simulator experiments were Greek captains with experience from the same route including the arrival to Rostock. The vessel was, although not exactly the same, quite similar in manoeuvring characteristics to the actual vessel usually sailed by the participants on the route.

The voyages are divided into four phases of different length:

1. *Open water* – from start to first VHF-radio call to Rostock Vessel Traffic Service (VTS) Centre
2. *Approach* – from call to Rostock VTS to dredged channel
3. *Channel* – in the dredged channel
4. *Harbour* – inside harbour from passage of breakwater to arrival/end of simulator voyage

The situation with a high demand of situation awareness on Level 3 is a simulated alarm of an engine failure occurring at a moment in the voyage where it requires rescheduling and re-planning of the rest of the voyage. The re-planning is required, because arrival times have to be reported to the Rostock Vessel Traffic Service centre. A simulated fire alarm is used as a control event in the other voyage. Both the alarm for engine failure and the fire alarm have safety critical potential, but the fire alarm has no significant importance for the planning and scheduling of the voyage. It is therefore assumed that the fire alarm has a low demand for situation awareness on Level 3, while the engine alarm, due to the need for re-scheduling, is related to a high demand for situation awareness on Level 3.

The situation with a low demand for situation awareness on Level 3 is the arrival of the vessel in Rostock which is the harbour phase of the voyage. Although this phase of the voyage has the highest amount of alarms and radio communication, is the most difficult task with respect to the manoeuvring of the large vessel alongside the quay in narrow space and requires intense manual control by means of manoeuvre handles, the amount of planning and the time frame for the planning is very limited. Since situation awareness on Level 3 is related to anticipation of future events and since the amount and importance of future events decreases dramatically at the end of the voyage, when the remaining time of the voyage decreases, it is expected that the levels of stress and mental activity will decrease in the harbour phase due to the decrease in demand for situation awareness on Level 3.

EEG spectrum

The electrical activity in the brain (EEG) can, according to Pettersen and Hoffmann (2002), be expounded as a reflection of the mental state or activity of the person. The following frequency bands are used (as shown in Table 29.1).

Table 29.1 Electrical activity in the brain – frequency bands and their associated mental state or activity

Frequency band	Frequency range	Mental state or activity
Delta	0.5-4 Hz	Sleep
Theta	4-8 Hz	Dreams
Alfa	8-13 Hz	Awake and alert
Beta-1	13-20 Hz	Mental activity, cognition, perception, attention
Beta-2	20-36 Hz	Stress, anxiety, fear

Equipment

EEG is measured by means of electrodes placed directly on the skin on the skull after certain predefined principles and standards, for example, the international 10-20 electrode system (Stern, Ray and Quigley, 2001). The signals from the electrodes are amplified and the results are recorded electronically by means of analogue or digital equipment. The NERVUS system, designed for measurement of EEG (and other psycho-physiological reactions such as electrocardiography (ECG/EKG), galvanic skin response (GSR) and so on), was used for data collection in the experiments described in this chapter. The components of the equipment are an electrode cap designed for the international 10-20 system of placement of electrodes on the skull, a NERVUS amplifier with 16 channels and a cable connection to a computer with the NERVUS Monitor software for data capture and analysis. Figure 29.1 illustrates the experimental set-up with the electrode cap and the amplifier (fitted in the belt of the participant) in the maritime full mission simulator. The EEG is measured by a sampling rate of 256 times per second, and the EEG spectrum is measured as 15 seconds averages according to common standards (Stern, Ray and Quigley, 2001; Fisch, 1999).

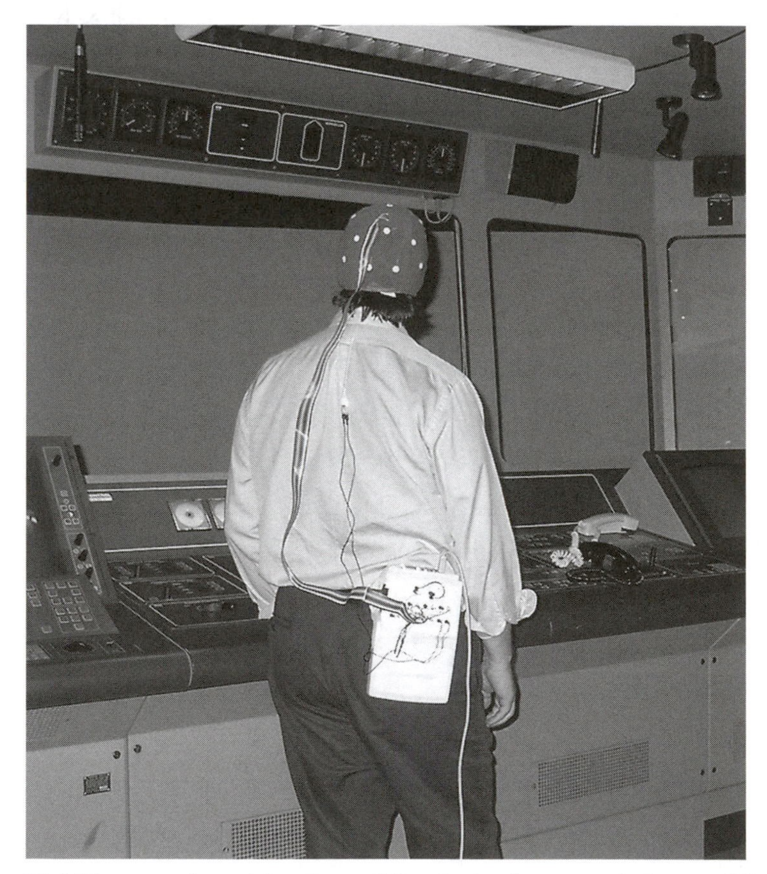

Figure 29.1 The experimental set-up of the electrode cap and the amplifier (fitted in the belt of the participant) in the maritime full mission simulator

Data collection

The measurements of EEG made in the full mission simulator were analysed according to the conceptual meaning of the different frequency bands. The analysis included both the Beta-1 band indicating mental activity and the Beta-2 band indicating level of stress. The average levels of Beta-1 and Beta-2 activity in percent of the total amount of brain wave activity was calculated for each of the four phases of the voyage: Open water – from start to first VHF call to Rostock Vessel Traffic Service centre; Approach – from call to Rostock VTS to dredged channel; Channel – in the dredged channel; and Harbour – inside harbour from passage of breakwater to arrival.

Results and discussion

Findings from the measurements of EEG in the first voyage in the simulator show that although variations in mental activity are found inside each phase, there is no significant difference in mental activity (Beta-1 frequency band) between the four phases of the voyage, when they are compared.

However, the stress level (Beta-2 frequency band) is significantly higher in the approach phase than in open water (*t*-test, $p<0.001$) and harbour (*t*-test, $p<0.05$). The stress level in the approach phase is also higher than in the channel, but this difference is not significant (*t*-test, $p=0.07$). See Figure 29.2.

These results show that although the level of mental activity is unchanged, there is a significant increase in stress level in the approach phase compared to both the open water phase and the harbour phase. The frequency of alarms and radio communication is in total lower in the approach phase than in the harbour phase. This means that the high stress level in the approach phase cannot be explained by the amount of alarms and/or radio communication alone.

However, the high stress level could be explained by a certain event in the approach phase. During this phase an engine failure was simulated. This engine failure required reduced speed for a short period and also some communication between the test person and the engine control room (a person acting as engine officer) and between the test person and Warnemünde VTS Centre with information about time of arrival, changes in schedule due to the engine failure and about weather conditions inside the harbour and so on (a person was acting as VTS officer).

This engine failure increased the demand for situation awareness on Level 3, because arrival times had to be recalculated given the speed reduction, and it also raised demand for situation awareness on Level 1 and 2 in the perception, interpretation and comprehension of alarms and messages from engine crew related to the engine failure. Furthermore, in this phase the test person is changing from reactive mode in open water to proactive mode for preparation for arrival. Situation awareness on Level 3 is in focus because this phase of the voyage requires detailed planning of the arrival, which is still approximately 40 to 80 minutes away in the future.

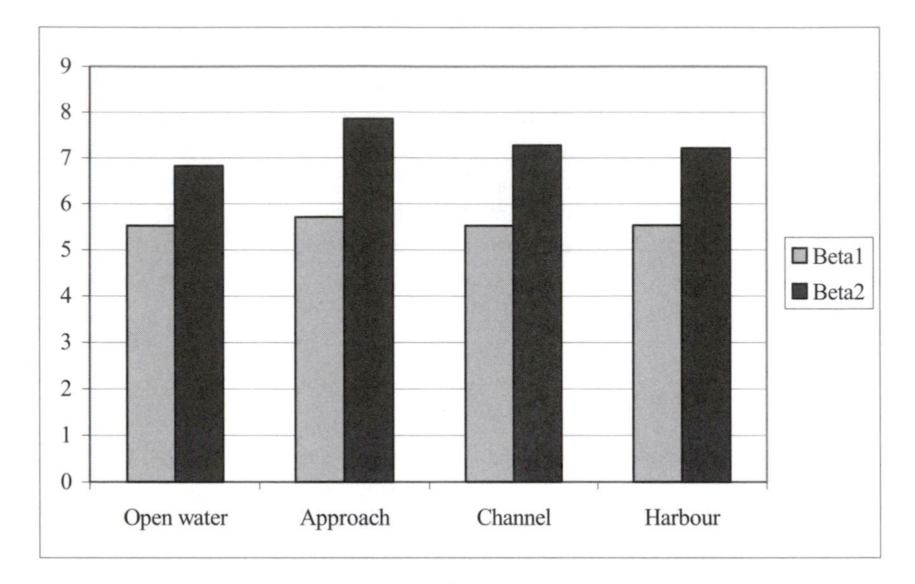

Figure 29.2 Average level of Beta-1 and Beta-2 activity in the four phases of the first simulated voyage

Note: Levels show the average percentage of Beta-1 and Beta-2 activity in the brain.

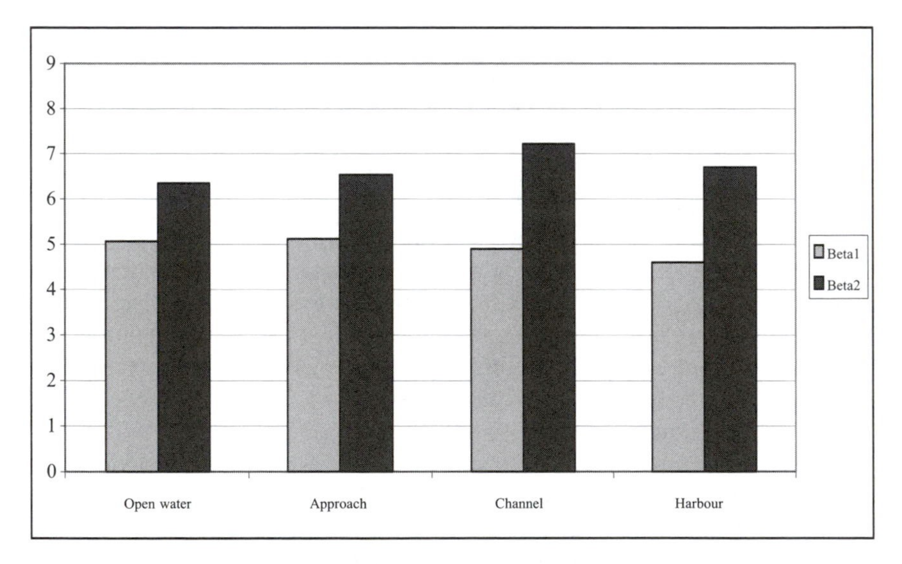

Figure 29.3 Average level of Beta-1 and Beta-2 activity in the four phases of the other simulated voyage

Note: Levels show the average percentage of Beta-1 and Beta-2 activity of the total electrical activity in the brain.

A possible interpretation is therefore that the increase in level of stress is related to the workload demand generated when situation awareness on all three levels, and especially on Level 3, is required from the context.

The other simulated arrival to Rostock illustrates a significant (*t*-test, $p<0.05$) decrease in level of mental activity (Beta-1) in the harbour phase compared to the open water and approach phases of the voyage (see Figure 29.3). The harbour phase is characterised by this relatively low level of mental activity compared to the open water and approach phases, even though the amount of alarms and radio communication is much higher, the task is difficult (manoeuvring a large vessel alongside the quay in narrow space) and requires intense manual control by means of manoeuvre handles. Since this is the last part of the voyage, the anticipation and preparation of future event is assumed to be of less importance. The low level of mental activity could therefore, with respect to the other findings, be explained by a contextually related decrease in demand for situation awareness on Level 3.

Control measurements

A series of on-board measurements were used as control measurements. The on-board measurements included three arrivals on the route between Rødby in Denmark and Puttgarden in Germany. The measurements were made on-board the car and passenger ferry M/S Prins Richard. The participant in the on-board measurements was a voluntary crew member with many years of experience from that exact route and vessel.

The results from on-board measurements on three voyages with a ferry between Rødby in Denmark and Puttgarden in Germany show homogeneity of stress level (Beta-2 activity) in four different phases of the voyage (see Figure 29.4): *Departure* – first 15 to 25 minutes; *Transit* – following 10 to 20 minutes; *Approach* – seven to nine minutes before arrival; and *Arrival* – from five to seven minutes before arrival until time of arrival. Rather high variations in Beta-2 activity for the first 15 to 25 minutes of the 46-minutes-long voyage are seen in all three cases.

The tasks performed in this period were typically related to the passage of the other ferry on the same route and the establishment of a proper situation awareness and overview of the traffic situation on the route ahead. The first 15 to 25 minutes of the voyage was followed in all three voyages by an equivalent period of about 10 to 20 minutes on open water with lower levels of Beta-2 activity.

After this phase of the voyage and seven to nine minutes before the time of arrival, a sudden and significant increase in Beta-2 brain wave activity was found in two out of three voyages (*t*-test, first voyage $p=1.6*10^{-6}$ and second voyage $p=4.2*10^{-4}$). At this moment the vessel was in a position where it was approaching the harbour, but not yet in the harbour basin, and the crew was preparing for arrival. Eventually the level of Beta-2 decreases five to seven minutes before arrival and remains at a rather low level until the time of arrival in the ferry berth (*t*-test, first voyage $p=2.8*10^{-4}$, second voyage $p=0.026$ and third voyage $p=0.0016$). The tasks performed in this last period include manual control of the vessel by means of manoeuvring handles and controls as well as communication with other crew members about tasks related to the vessel's positioning in the ferry berth.

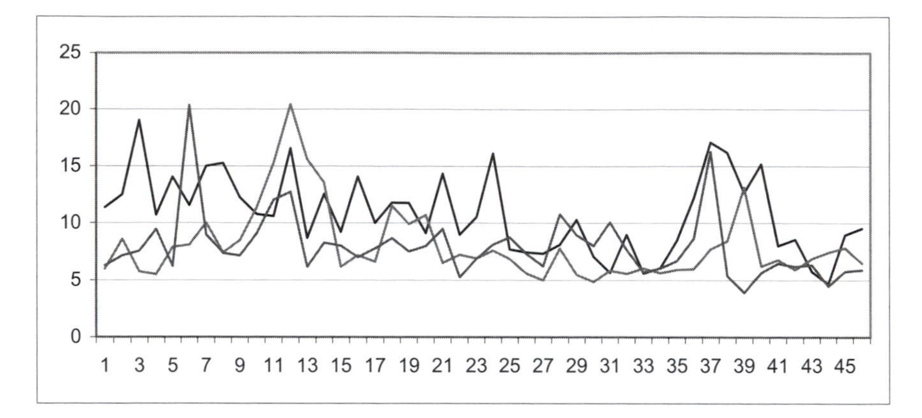

Figure 29.4 Variations in one-minute averages of Beta-2 activity in percent of the total brain wave activity on three voyages with M/S Prins Richard

Note: The x-axis is the time after departure in minutes.

Since Beta-2 activity reflects the level of stress, the increase in the approach phase could be interpreted as a short-term stress reaction related to the preparation for and anticipation of the arrival. The decrease immediately after and five to seven minutes before arrival indicates further that the stress reaction is more likely a result of an increased demand for situation awareness on Level 3 rather than a result of intense manual control of the vessel. Otherwise we would have expected a maintained high stress level until the moment of arrival rather than the actual observed decrease.

In other words, the on-board measurements support the findings from the simulator experiments: that a decrease in level of stress is found when there is a decrease in demand for situation awareness on Level 3.

Conclusions

The results show that a demand for situation awareness on Level 3 can generate an increased level of stress. Further, it is exemplified how decreased demand for situation awareness on Level 3 can generate a decreased level of mental activity even though the overall workload demand and therefore demand for situation awareness on Level 1 and 2 is maintained. The finding from the simulator experiments, that a decrease in demand for situation awareness on Level 3 generates a decrease in the level of mental activity, is supported by on-board measurements of stress level, where a decrease in stress level is found at the time of arrival where the demand for situation awareness on Level 3 decreases dramatically. The conclusion is that the hypothesis is supported by the empirical findings from both the simulator experiments and the on-board measurements. Further, it was shown to be evident that the use of psycho-physiological techniques has great potential in the measurement of mental activity, stress reactions and situation awareness.

Acknowledgements

I would like to thank Scandlines A/S (Denmark) and Superfast Ferries (Greece) as well as volunteer crew members from these shipping companies acting as test persons, Cephalon A/S (Denmark) for valuable support on EEG equipment, Dr. Olle Wikström and staff at the Neurophysiological Department, University Hospital MAS, Malmö (Sweden) for guidance and training in use of the equipment, and the Merchant Marine Academy of Macedonia (Greece) for support and discussion related to the simulator voyages. The work presented in this chapter is partially built on the research project SPIN-HSV co-funded by the European Commission under the 5th Framework Programme for Research and Development.

References

Endsley, M. R. (2000). "Theoretical underpinnings of situation awareness: A critical review". In M. R. Endsley and D. J. Garland (eds), *Situation awareness analysis and measurement* (pp. 3-32). Mahwah, NJ: Lawrence Erlbaum Associates.

Fisch, B. J. (1999). *Fisch & Spehlmann's EEG Primer* (3rd ed.). Amsterdam: Elsevier.

Grech, M. and Horberry, T. (2002). "Human error in maritime operations: Situation awareness and accident reports". In *Proceedings of 5th International Workshop on Human Error, Safety and Systems Development*. Noah's On The Beach, Newcastle, Australia.

Koester, T. (2003). "Situation awareness and situation dependent behaviour adjustment in the maritime work domain". In *Proceedings of the 10th International Conference on Human-Computer Interaction (HCI International 2003)*, Crete, Greece.

Koester, T. and Rabjerg, T. (2005). "A new framework for situation based analysis of the work on board ships". In *Proceedings of Royal Institution of Naval Architects (RINA), Human factors in ship design, safety and operation*. London, UK.

Koester, T. and Sørensen, P. K. (2003). "Human factors assessment". In *Proceedings of MARSIM'03*, Kanazawa, Japan.

Koester, T. and Sørensen, P. K. (2004). "Measurement of stress and mental activity among maritime crew members". In *Proceedings of Human performance, situation awareness, and automation (HPSAA)*, Daytona Beach, Florida.

Pettersen, A.-H. and Hoffmann, E. (2002). "Hjernebølgetræning af DAMP-børn". *Psykolog Nyt*, 23, 3-9.

Stern, R. M., Ray, W. J. and Quigley, K. S. (2001). *Psychophysiological Recording* (2nd ed.). Oxford: Oxford University Press.

Chapter 30

Measures of Attention and Cognitive Effort in Tactical Decision Making

Sandra Marshall

Introduction

The complex activity of tactical decision making has many components, including the focused attention and increased mental effort of those engaged in the decision making. This chapter describes one study from an ongoing collaboration to model the attention and mental effort of officers engaged in team decision making. The collaboration is funded by the Office of Naval Research, and the simulation environment used in the collaboration is the Distributed Dynamic Decision Making Simulation (Kleinman, Young and Higgins, 1996). In this setting, a team of decision makers is given a mission, a set of predefined mission requirements, and information about the team members' own assets. Working together, they must formulate plans of action and execute those plans to accomplish the overall mission objective.

Background

The principal goal of the collaboration is to examine how well teams under different structural organisations adapt to varying mission contexts (Diedrich, Entin, Hutchins, Hocevar, Rubineau and MacMillan, 2003; Entin, Diedrich, Kleinman, Kemple, Hocevar, Rubineau and MacMillan, 2003). In real world situations, the organisational structure of a decision making team may be misaligned with the operational setting in which it is forced to work. In such cases, the team needs to consider whether to modify its organisation or formulate alternative plans for completing its task.

For the study described here, teams were first trained to perform under one of two basic organisational structures and then asked to carry out missions that were closely aligned with these types of organisation. Thus, each team carried out a mission designed to maximise its underlying organisation and then also carried out a mission that was mismatched with its underlying organisation.

The two organisation structures studied here are commonly defined as *functional* and *divisional*. Under the functional organisation, each team member specialised in one or two functions (such as air warfare) and was responsible for those specific functions across the entire scope of the mission. Under the divisional organisation, each team member controlled all assets for one multi-functional platform (such as an aircraft carrier) and was responsible for all functions within a restricted location.

Both organisations require the team members to work together to plan objectives of the mission, but the two schemes require different levels of coordination.

Performance measures

Two useful measures of successful team decision making are overall performance and communication among team members. General results from an experiment involving eight six-person teams of Navy officers have recently been reported by other members of the collaboration. One analysis found that the overall performance of the teams was better when working on missions that matched their training structure than on missions that were mismatched with training (Diedrich et al., 2003). Similarly, overall communications increased significantly in the mismatched cases. A second analysis examined the same data in terms of scenario tempo over time and concluded that the effects on communication and performance emerged early and held across the entire scenario (Entin et al., 2003).

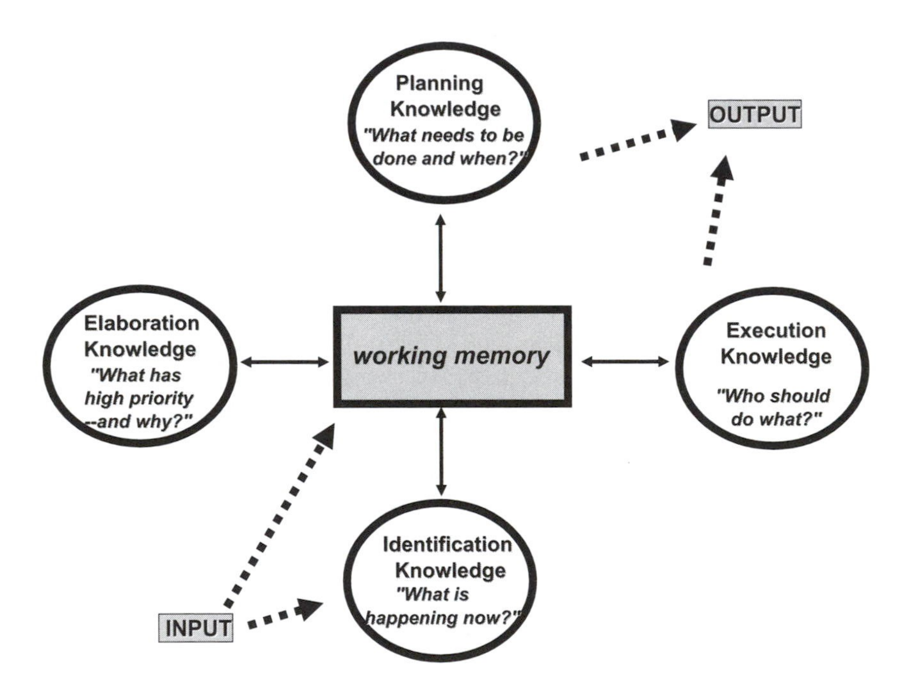

Figure 30.1 The schema model of decision making
Source: Adapted from Marshall (1995) with permission from Cambridge University Press.

Schema model

The schema model employed here is the model first defined by Marshall (1995) for the domain of problem solving and later extended to tactical decision making (Smith and Marshall, 1997). The basic structure of the model is shown in Figure 30.1.

In team decision making, the four knowledge components that make up the schema are often the basic elements of communication among team members. In order to perform as a well-integrated team, the members must keep each other informed about their own assets and actions and must also be alert to the assets and actions of others. This communication takes the form of specific statements that convey the necessary Identification, Elaboration, Planning, and Execution Knowledge. Table 30.1 contains examples of these four types of communications observed in the present study.

Table 30.1 Schema knowledge components

Identification	*"All stations, this is ORANGE. Be aware that we have a mine on port. Over."*
Elaboration	*"GREEN, this is BLUE. My SOF [Special Operations Forces] Team is in range of Naval Base East. Over."*
Planning	*"BLUE, this is BROWN. Can you send one strike down to the bridge to take out enemy ground force? Over."*
Execution	*"BROWN, this is BLUE. I am sending in a TLAM to GSAM 357."*

A closely shared body of schema knowledge should result in strong performance and crisp, decisive communications. However, when the situation does not match the team's expectations, the team members may no longer be able to anticipate the assets and actions of their fellow team members. In that case, either essential shared schema knowledge may be missing or the communication of existing shared knowledge may deteriorate. In such instances, one expects to observe rising cognitive workload in the team members as the mission progresses.

In the present study, it was hypothesised that teams working in situations in which their organisation matched the mission structure would have lower workload than teams working in situations in which the team organisation contrasted with the mission structure. It was also hypothesised that their schema knowledge would be more cohesive for the matched condition than for the mismatched condition.

Cognitive workload

A new metric based on pupil dilation was used to measure cognitive workload. It is well known in the psychological literature that mental effort is accompanied by changes in pupil dilation. A number of laboratory tests have confirmed this result for many areas such as visual search, reading and problem solving (Beatty, 1982). The new technique used here provides continuous recording and analysis of pupil size, allowing estimation of cognitive workload over sustained periods of time across many different task components for a single individual (Marshall, Pleydell-Pearce and Dickson, 2003).

Pupil size is a signal that can be processed like any other signal. The Index of Cognitive Activity is calculated from high-frequency components of this signal as an individual performs a specified task. It is a measure of relative change that reflects the number of times each second that unusual and abrupt increases occur in the amplitude of the pupil signal. The index is patented under US Patent No. 6,090,051 (Marshall, 2000).

As most researchers know, the pupil signal is noisy and hard to analyse. A special challenge in analysing the signal is to separate two reflex responses that often occur simultaneously, the light reflex and the dilation reflex (Loewenfeld, 1993). Two sets of muscles govern pupil dilation, the circular muscles surrounding the pupil and the radial muscles extending outward from the pupil. In the presence of light, the circular muscles typically are activated while the radial muscles are inhibited, causing contraction of the pupil and producing the light reflex. In the presence of a cognitive stimulus, the radial muscles are activated and the circular muscles are inhibited, resulting in the dilation reflex. The index measures the latter reflex.

The index is derived from wavelet analysis (Daubechies, 1988, 1992; Ogden, 1997). Wavelet analysis involves repeated transformations of a signal, with the goal of decomposing the original signal into orthogonal components. A number of commercial applications are available for computing wavelets, such as MATLAB's Wavelet Toolbox.

At the heart of wavelet analysis is a "mother wavelet", a small oscillatory function that decays rapidly to zero in both positive and negative direction, that is, a little wave. There exist a number of different mother wavelets. The index utilises the family of Daubechies' mother wavelets, which produces an orthonormal basis for representing the data. Each element of the basis has compact support. Daubechies' mother wavelets can take a number of different sizes. The size of the wavelet dictates the number of coefficients involved in the analysis, with the wavelet having twice the number of coefficients specified by its size.

Wavelet analysis proceeds iteratively. Using the mother wavelet function, the dilation transformation first extracts the high frequency details from the signal. Next, using a scaling function that is orthogonal to the wavelet function, a second transformation extracts from the signal all information not captured by the wavelet transform. The second extraction yields a smoothed version of the signal. The difference between the smoothed version and the original signal is called the detail set. The detail coefficients are used by the ICA (Index of Cognitive Activity). The index uses the high-frequency details of the signal that are extracted in the initial decomposition. Details of small magnitude are eliminated by a threshold comparison. The index can be calculated across a signal of any length.

Because the focus is on changes in dilation, the wavelet coefficients are then compared with the original signal to identify the coefficients that correspond to significant pupil *increase*. A threshold procedure is applied at this point to retain only those coefficients that correspond to increases in pupil size, that is, dilation coefficients. The threshold can be set to any value and depends upon the unit of measure of the pupil signal.

To measure an individual's response, it is necessary to determine the number of significant dilation coefficients occurring across a task of *n* seconds. Two measures are typically of interest: the average number of such coefficients per second, which provides a ratio of total effort to total time, and the second-by-second trace across the entire task. The average provides an estimate of total task difficulty as gauged by the individual's mental effort. The second-by-second trace shows the variability in effort that occurs during the task and highlights the points at which effort is extremely high or extremely low. Average values are generally more useful for task and group comparisons. Second-by-second traces are valuable for examination of a single individual's performance. In both cases, the index provides the number of abrupt increases in pupil size per second, either averaged across a number of seconds or taken for a single second at a time.

Two of the mechanical aspects in computing the index are selection of a wavelet and determination of the appropriate threshold for detail coefficients. Wavelet size will depend upon the sampling rate and the scale factor used to record pupil diameter. A comparison involving 11 different wavelets and six different criterion thresholds on several data sets suggests that Daubechies' wavelet 8 with a threshold of 4.0 is satisfactory across a wide range of tasks and individuals. These settings are used in the analyses presented in the following sections.

Method

Participants

Forty-eight officers attending the US Naval Postgraduate School participated in the collaborative study. Eight six-member teams were formed, and one member of each team volunteered for the eye-tracking portion of the study.

Procedure

Four teams were trained under functional organisation and four teams were trained under divisional organisation. Details of the training may be found in Daubechies (1988, 1992). All teams were then assessed as they engaged in two simulated warfare scenarios. One scenario was most suited to the divisional organisation, and the other scenario was most suited to the functional organisation, but both could be performed successfully under either organisation.

One member of each team was monitored using SR International's EyeLink II eye tracking system as the team worked through the two scenarios. Following a brief introduction to the apparatus, the individual went through a short calibration procedure and then continued to participate as usual with the rest of the team.

The EyeLink system recorded point of gaze and pupil size at a sampling rate of 250 Hz continuously across the two 35-minute scenarios. These data are analysed in near real time to show the entire gaze history as well as the number of times that important elements of the tactical display are noticed, the specific focus of the officer during critical events, and the frequency with which s/he reviews essential information such as the requirements of the mission or his or her remaining assets and their capabilities.

Data analysis

The two test scenarios were videotaped, with the tapes showing the eye movements of one individual as he or she engaged in the simulation. The tapes also contain all audio communications among the team members. In addition, all eye data were recorded digitally for later analysis.

Two participants were selected for single subject analyses. One participant was trained as a member of a functional team, and the other was trained as a member of a divisional team.

Each team member was assigned a colour code for communication purposes (for example, Blue, Green, Brown, Orange, Purple or Red). The participants studied here were coded Blue (from the Functional team) and Purple (from the Divisional team). Blue's responsibilities under the functional organisation were to destroy Scud missiles and missile launchers and his/her assets were tactical/steerable Tomahawk missiles (TTOMs) and anti-ballistic missiles (ABMs). Purple's responsibilities under the divisional scenario were to manage all assets from a destroyer platform and to conduct ground operations in a specified region of the mission.

The Blue participant described here received functional training that highlighted his use of assets to focus on locating and destroying Scud missiles. To do this successfully, he had to coordinate closely with other team members, especially Purple.

The Purple participant described here received divisional training that highlighted her responsibility to use all assets associated with her ship. This scheme required her to work closely with Red and Orange.

Results

Communications

The first analysis examined the pattern of communication for both teams on both scenarios. Figure 30.2 shows the relative balance of communications among the team members for the matched and mismatched conditions. This figure provides the proportion of communications made by each team member during the two scenarios. For example, the upper left panel shows that Blue made 39 per cent of all communications during this scenario.

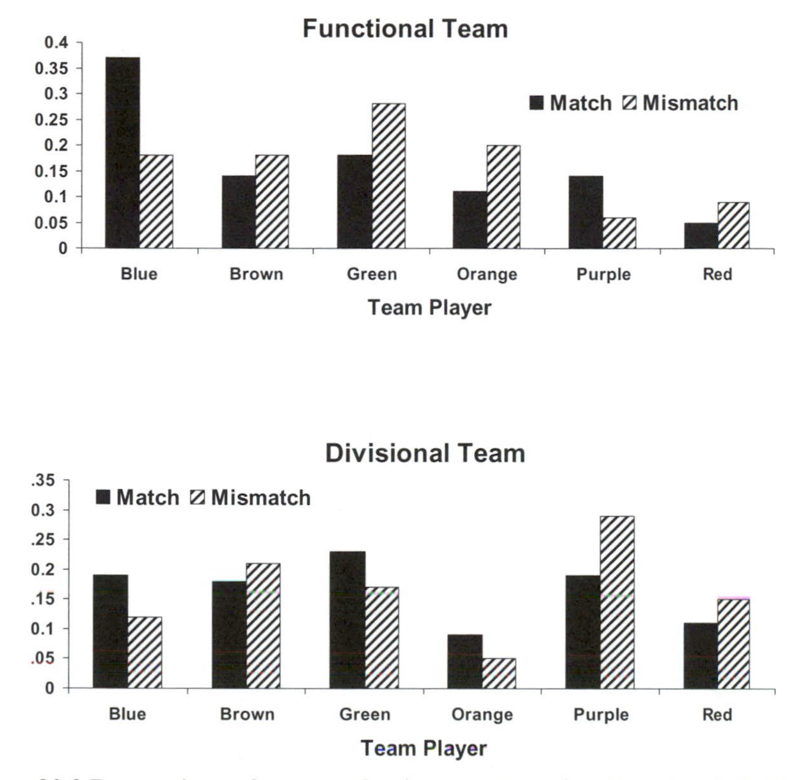

Figure 30.2 Proportions of communications made by functional and divisional team members during matched and mismatched conditions

The top panel shows the functional team during the functional scenario (left) and during the divisional scenario (right). The most striking finding here is the dominance of Blue during the matched condition. As the team engaged in the mismatched scenario condition, the number of communications made by the Blue team member decreased dramatically, while those of Brown, Green and Orange increased.

The bottom panel shows the divisional team during the divisional scenario (left) and the functional scenario (right). Again, there is a shift in the number of communications made by team members. No single team member is dominant for either scenario. For both teams, when the scenarios were not closely aligned with the team's organisation, different team members began talking more often.

The total number of communications for these two teams across the two scenarios differed significantly as shown in Figure 30.3, with χ^2=84.30, with each team issuing fewer communications during its matched condition than its mismatched one. This analysis confirms that it was not the nature of the scenario itself (that is, either functional or divisional) that elicited more communication. The matched condition for each one was approximately the same.

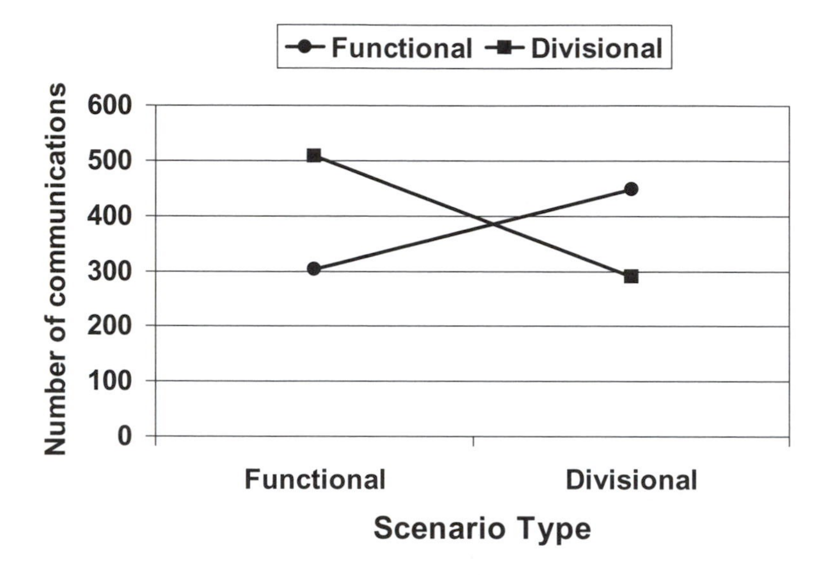

Figure 30.3 Number of communications made during the two conditions by both teams

Thus, when the scenario context did not reflect the team's underlying organisation, the patterns of communications changed in two important ways. First, some team members spoke more during the matched scenario and some spoke more during the mismatched scenario. Secondly, the overall numbers of communications made during the matched condition were significantly lower.

A final question to be asked is whether the content of the communications also changed when team organisation and mission context were incongruent. This question can be answered by analysing the schema knowledge reflected in the utterances of the teams during the scenarios. All communications were coded into six categories: the four schema knowledge components plus two additional categories of simple acknowledgement ("Roger" or "Aye") and comments ("I don't know what just happened ...") that were not addressed to any teammate. The results are shown in Figure 30.4.

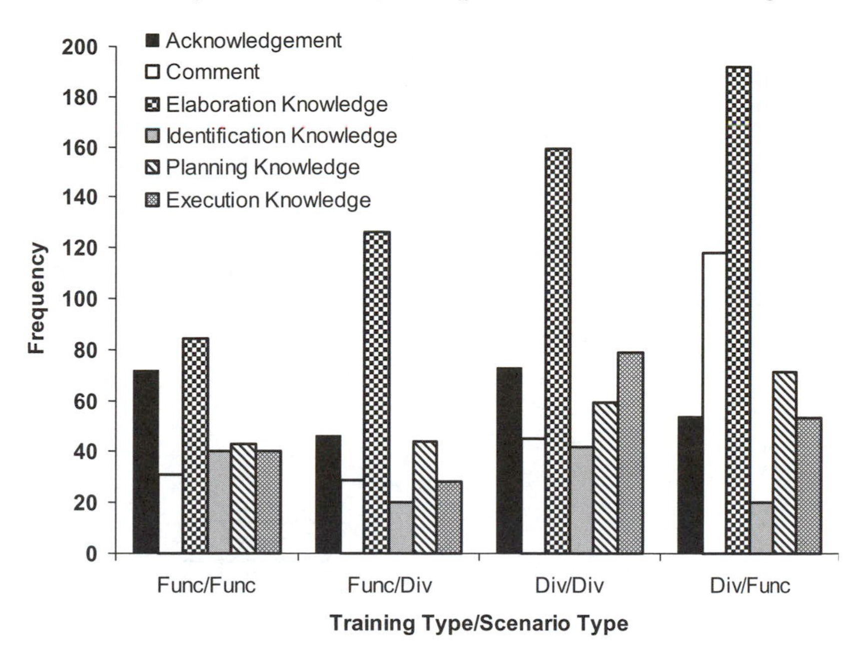

Figure 30.4 Number of times each type of communication occurred during the two scenarios, based on schema model

Consider first the two bar graphs on the left for the functional team. Several differences are evident as the team moves from the congruent to the incongruent scenario. First, the number of Acknowledgements drops as does the number of Identification and Execution Knowledge statements. Team members are not acknowledging the communications of others nor are they keeping each other up to date on new features of the situation. Secondly, there is a surprisingly large increase in the number of Elaboration statements. Most of these statements were, in fact, questions about who had specific assets and where they were located.

Now consider the two bar graphs on the right in Figure 30.4. Exactly the same patterns described for the functional team are present for the divisional team. Under the mismatched condition, Acknowledgements go down as do statements containing Identification and Execution Knowledge, while Elaborations go up. Moreover, for this team, the irrelevant comments increased significantly, which means that much of the communication occurring during the mismatched condition was not mission-directed and did little or nothing to help other team members understand the situation.

Cognitive workload

The results from the communication analysis suggest that the teams were experiencing higher cognitive workload during the mismatched condition. Their observed schema knowledge was weaker, they spoke more, and their overall performance as measured by plans and execution of tasks was lower. To confirm this hypothesis, the ICA was computed for each team in each scenario.

To facilitate interpretation, the ICA was averaged across one-minute intervals for the entire length of the 35-minute scenarios. The results are shown in Figures 30.5 and 30.6.

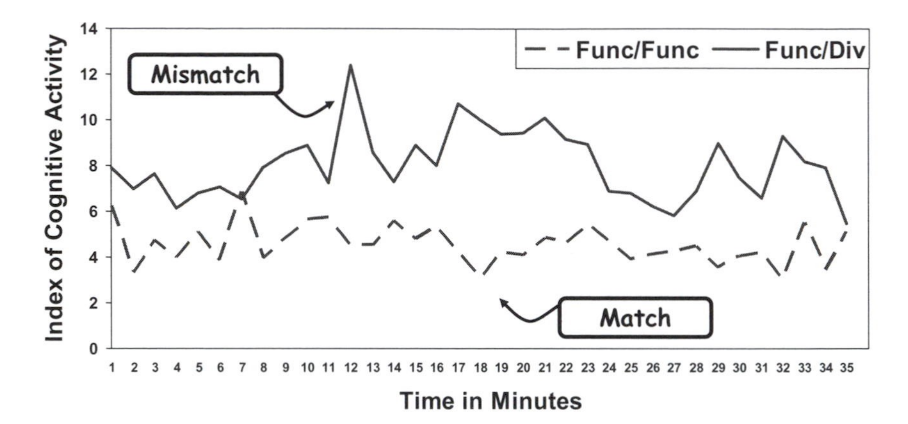

Figure 30.5 The Index of Cognitive Activity for matched and mismatched conditions for functional team member

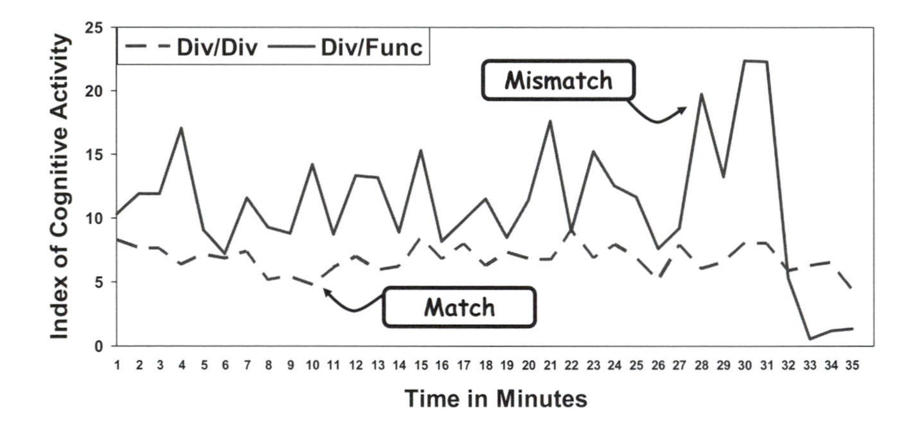

Figure 30.6 The Index of Cognitive Activity for matched and mismatched conditions for divisional team member

The results shown in the two graphs are very clear. The pattern of ICA values is relatively calm for the matched conditions for both teams. No large fluctuations occurred, and the ICA was essentially balanced across the full scenario. In contrast, on the mismatched scenarios, the ICA was higher and fluctuated quite markedly with very high peaks. The mean ICA values for the functional matched and mismatched conditions were 4.58 and 8.53 respectively. The mean ICA values for the divisional matched and mismatched conditions were 7.44 and 11.30 respectively. Paired sample t-tests for each team across the 35 one-minute intervals were significant, with $t=11.07$ ($df=34$, $p<.001$) and $t=4.63$ ($df=34$, $p<.001$) for the two sets of data.

In both cases, the ICA was significantly higher on the mismatched than on the matched condition. Moreover, on a minute-by-minute basis, the ICA for the mismatched condition exceeded the matched in virtually every instance. Thus, it was not the case that workload was high only at the beginning of the scenario when the mismatch might be the most evident to the participant. Rather, the workload measure jumped during the first few minutes of the scenario and remained elevated throughout.

Further analyses also revealed a strong correlation between communication patterns and elevated cognitive activity. The highest workload occurred during the periods of time when coordination with other team members was most urgent and when the participant being observed was tasked with making his or her assets available to the other members of the team.

For example, the highest peak for the functional team member on the mismatched condition occurred when the speaker was attempting to plan the team's mission response and was urging his team members to work quickly with him. He was obviously concerned that they would not act quickly enough to fulfil the mission.

The peaks for the divisional team member on the mismatched condition reflected her concern for her own position rather than for the team's overall mission. The first peak came when her SOF was endangered and she was unable to protect it. The last large peak occurred toward the end of the scenario when she was calling on several team members to help her. As her requests failed, her ICA value dropped, perhaps indicating that she was no longer an active participant in the mission.

Conclusions

Two approaches, schema model analysis and the ICA analysis, identified specific processes occurring during the mismatch between team structure and mission context that clarify resulting poor performance. In the mismatched condition, both teams showed weaker and possibly deteriorating schema knowledge as the nature of the communications shifted from clear and decisive Identification and Execution statements to uncertain Elaboration statements about team capabilities.

As the communications changed, so too did the cognitive workload experienced by the team members tested. The new metric used here, the Index of Cognitive Activity, successfully detected the increase in workload as a result of the incongruity between team organisations and mission context. Further, the ICA detected workload increases during critical communication and performance events when the team member was faced with possible mission failure.

Acknowledgements

This research was supported by the US Office of Naval Research under Grant No. N000140010353.

References

Beatty J. (1982). "Task-evoked pupillary responses, processing load, and the structure of processing resources". *Psychological Bulletin*, 91, 276-292.

Daubechies, I. (1988). "Orthonormal bases of compactly supported wavelets". *Communications in Pure and Applied Mathematics*, 41, 909-996.

Daubechies, I. (1992). "Ten lectures on wavelets". Philadelphia: SIAM.

Diedrich, F., Entin, E., Hutchins, S., Hocevar, S., Rubineau, B. and MacMillan, J. (2003). "When do organizations need to change (Part I)? Coping with incongruence". In *Proceedings of the 2003 Command and Control Research and Technology Symposium*, Washington, DC.

Entin, E., Diedrich, F., Kleinman, D., Kemple, W., Hocevar, S., Rubineau, B. and MacMillan, J. (2003). "When do organizations need to change (Part II)? Incongruence in action". In *Proceedings of the 2003 Command and Control Research and Technology Symposium*, Washington, DC.

Kleinman, D., Young, P. and Higgins, G. (1996). "The DDD-III: A Tool for Empirical Research in Adaptive Organizations". CCRTS.

Loewenfeld, I. E. (1993). *The pupil: Anatomy, physiology, and clinical applications* (Volume I). Ames, Iowa: Iowa State University Press and Detroit: Wayne State University Press.

Marshall, S. (1995). *Schemas in Problem Solving*. New York: Cambridge University Press.

Marshall, S. P. (2000). US Patent No. 6,090,051. Washington, DC: U.S. Patent & Trademark Office.

Marshall, S., Pleydell-Pearce, C. and Dickson, B. (2003). "Integrating psychophysiological measures of cognitive workload and eye movements to detect strategy shifts". In *Proceedings of the 36th Annual Hawaii International Conference on System Sciences*. Los Alamitos, CA: IEEE.

Ogden, R. T. (1997). *Essential Wavelets for Statistical Applications and Data Analysis*. Boston: Birkhauser.

Smith, D. E. and Marshall, S. P. (1997). "Applying hybrid models of cognition in decision aids". In Zsambok, C. and Klein, G. (eds), *Naturalistic Decision Making* (pp. 331-343). Mahwah, NJ: Erlbaum.

Chapter 31

Crew Mental Workload for the Vetronics Technology Testbed Vehicle

Christopher Smyth

Introduction

The US Army is developing combat vehicles that are smaller, lighter, more lethal, survivable, and more mobile to support the rapidly deployable forces of the Future Combat Systems (FCS). These designs, combined with an increase in vehicle and C⁴IS (Command, Control, Computers, Communications and Information Support) systems integration and performance, will assimilate and distribute more information to, from, and within the vehicle as the Army operates in the digital electronic battlefield. Consequently, the Army will use sophisticated, highly integrated crew stations for operating these future combat vehicles and to control the subordinate unmanned air and ground robotic elements.

In support of this effort, the Tank Automotive Research Development and Engineering Center (TARDEC) has developed the Crew-Integration and Automation Test-bed (CAT) Advanced Technology Demonstrator (ATD) with the Vetronics Technology Testbed (VTT) vehicle. The purpose of the CAT-ATD is to demonstrate crew interface, automation, and integration technologies that are required to operate these future combat vehicles. The technologies are being considered for use in a command vehicle with a two-person crew controlling a platoon of robotic ground vehicles, a design feature of the FCS. Of interest to designers is the effect that the automation and crew station design have on the crew interaction and the resulting mental workload.

In totality, the soldiers, crew organisation, and crew station technology may be considered as forming a military socio-technical system (Taylor and Felten, 1993). In military automated systems of a non-routine or combat nature with highly technical tasks dependent upon a continuous influx of knowledge, the technical and human-social components are tightly bound and interconnected, and the interaction determines the system performance. In turn, the social component is influenced by the attitudes, values, and behaviour styles of the soldiers, and their relations within the organisation. An effective combat system results when the technical interface between the human and machine is properly integrated with the social culture and organisation (Whitworth and deMoor, 2003). Both the technical and social domains combined influence the social-technical architecture and therefore the military system performance. The social design factors include the span of control or level of autonomy and the modalities for collaborative effort, including the crew and system responsibilities. In turn, these factors influence the system performance through the mental workload, the degree of shared mental models and situation awareness, the support of knowledge representation, and the resulting trust of the crew (Salas, Sims and Burke, 2005).

The crew work is performed at several different levels. As well as the task work performed at the automated crew station on individual assignments, the crew member supports the crew with team-work and helps plan the crew activities with meta-work (Bowers, Braun and Morgan, 1997). Crew team-work is the work done to maintain the crew functioning as a team by communicating with the other crew members, monitoring their work and providing backup, and if appropriate, coordinating the efforts and providing or supporting leadership and decision making. Meta-work is the work done to provide planning for the crew by situational assessment, defining problems, setting priorities, and scheduling crew activities. These activities impose additional demands upon the attention and cognitive resources of the crew member (Beith, 1987), because of the need to maintain knowledge of the crew objectives, functions and status and their individual roles and one's own relation to them (Orasanu and Fisher, 1991). However, the result is a multiplication of his or her effectiveness through participation in the crew as a team member.

At the request of TARDEC, the US Army Research Laboratory (ARL) is providing human factors expertise in determining the effect of these new crew station technologies on system performance through a continuing series of studies and investigations. As part of this effort, in June and September 2001, ARL participated with TARDEC in a demonstration of the VTT at the Camp Grayling Military Reservation, Michigan. During the demonstration, researchers collected subjective data with questionnaires about workload and related measures.

Experimental methodology

The vehicle, crew stations, functional display screens, participants, and research procedures that were used in this study are as follows.

VTT vehicle

In the 2001 design, the experimental apparatus is an M2 Bradley Fighting Vehicle (BFV) chassis, modified by General Dynamics Land Systems with camera arrays attached to the roof of the vehicle for indirect vision driving and target acquisition, and two experimental crew stations arranged side by side in the troop compartment (Figure 31.1).

Crew stations

The camera outputs are seen on fixed flat panel video displays that are mounted across the top of an experimental crew station (Figure 31.2). The displays for the front camera array are arranged with a central display directly in front of the operator, and left and right side displays pivoted inwards to match the angles of the array side cameras. A hand yoke controller and a foot pedal brake and accelerator are situated in the crew station for driving and target acquisition.

Figure 31.1 VTT vehicle

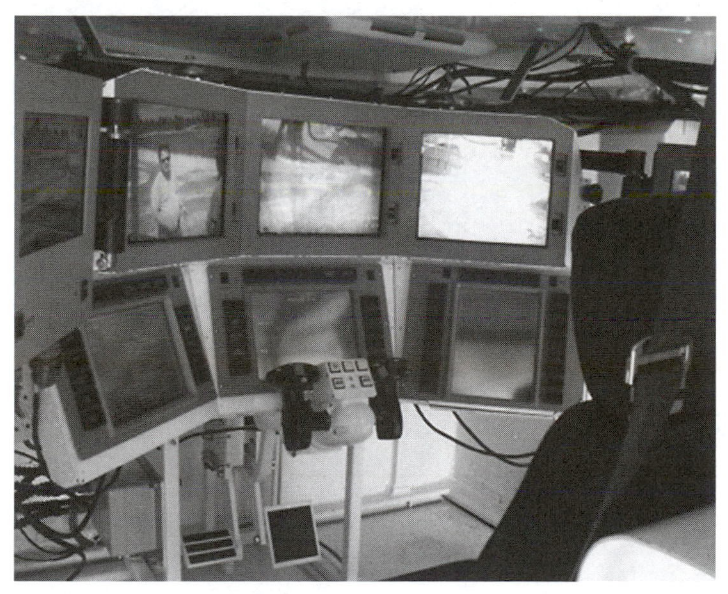

Figure 31.2 Crew station

Multifunctional display screens

In addition to the indirect vision displays, each station has a lower bank of three graphic multifunctional displays that are used for tactical maps, target acquisition, communications, and system status (Figure 31.3). The crew plans route waypoints on the tactical map using the screen touch panel and programmable display bezel switches. The driver uses the system status screen to select the crew station for driving the vehicle, setting the drive controls, and monitoring the status of the vehicle equipment. He or she drives the course with the yoke controls and foot pedals from the steer-to indicator on the indirect vision displays and the route course on the tactical map.

During the mission, the crew member searches for, acquires, and engages targets from the acquisition screen using the non-driving yoke controller to direct the acquisition scope and gun on the vehicle. As a command vehicle, the crew receives and files electronic spot, situation and field reports including obstacle, logistics, and call for fire, as well as operations and warning orders. These reports are processed from layered screen menus with the touch panel and bezel switches, and the keyboard.

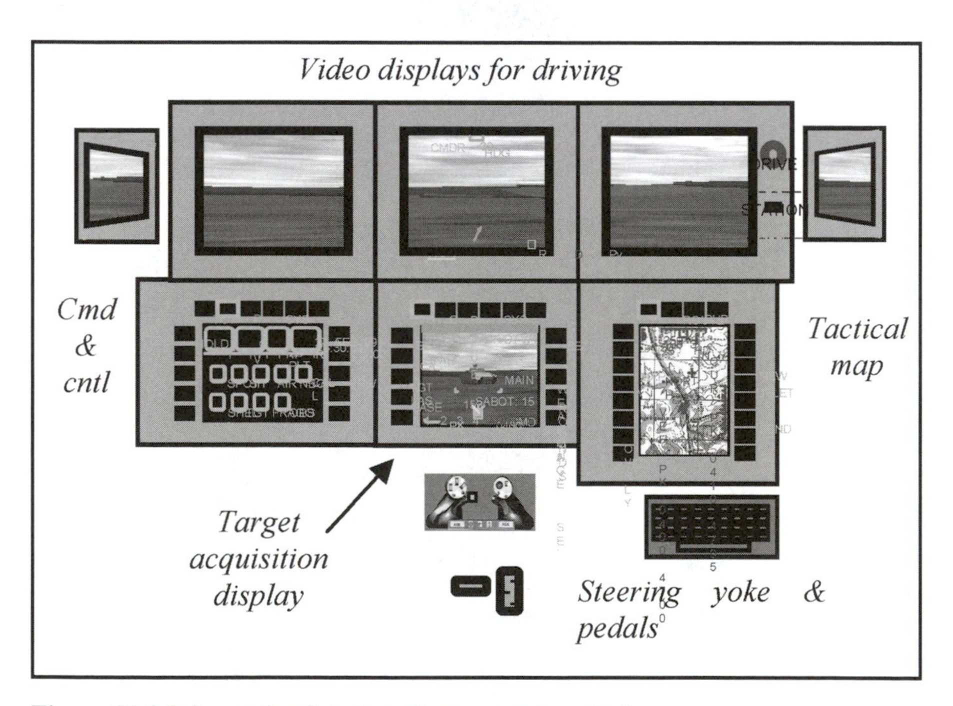

Figure 31.3 Schematic of station displays and controls

Mission scenarios

In the demonstration of the VTT as a command vehicle, the crew coordinated the advance and deployment of a virtual scout platoon including robotic elements, and engaged enemy targets when in view, for two operations orders. The scenario for the first order was to conduct a tactical road march followed by a movement to contact with the (virtual) enemy forces and seize an objective. On verbal order via radio, the force continued the movement to contact north. For the second order, the force conducted a tactical road march to establish a company defence in order to deny the enemy access to an intersection. In the planning phase, the scout platoon leader, having been provided with the operations order by the company commander, enters control measures for the phase lines, boundaries, and route into the VTT computer database and transmits an electronic Platoon Operations Overlay to the (virtual) vehicle commanders (TC) in the platoon.

Demonstration course

The participants drove on a road course connecting stationary firing points in a controlled training area at Camp Grayling Military Reservation, Michigan.

Participants

Eight military male volunteers from the armoured cavalry participated in this study. The participants, all with 20/20 – 20/30 (corrected) vision, good hearing acuity, and crew experience with a Bradley, were assigned to four two-man crews for the demonstration. The assignment was in a seemingly random manner from two populations: senior ranks as crew chief and junior ranks as crew mate. The senior rank (chief) of each crew was a Sergeant First Class (E-7); the junior rank (mate) was a Staff Sergeant (E-6), Sergeant (E-5), or Specialist (E-4). All were posted to the regular army at Fort Knox, KY, and most had extensive experience with the BFV.

Questionnaires

A battery of questionnaires was administered to determine: (1) the crew personality traits (Zuckerman et al., 1993); (2) the work social culture; (3) the task, teaming and meta-work types; (4) the NASA-TLX perceived workload (Hart and Staveland, 1988); (5) the situation awareness rating (Taylor and Selcon, 1994); and (6) task-specific workload elements consisting of task attention (McCracken and Aldrich, 1984), cognitive compatibility (Taylor, 1995), motion sickness (Kennedy et al., 1989), subjective stress (Kerle and Bialek, 1958), sleepiness (Hoddes et al., 1973), and trust rating. The work social culture and work-type questionnaires were based on the literature (Bowers, Braun and Morgan, 1997); the trust questionnaire was a seven-point, bipolar rating scale with end point verbal anchors of “little” and “complete”, for rating the amount of trust.

Statistical analysis

The statistical models were represented by mixed linear effects models (type III) in SPSS 12.0 (SPSS, Inc., Chicago); this is because participants were assigned to the crews in a seemingly random pattern. The model includes the between-subjects crew role (chief versus mate) as a fixed factor and the subject-by-crew interaction as a random effect. The covariance structure was variance components. Contrasts were conducted as planned comparison among conditions to test specific hypotheses of interest. All post-hoc pair-wise comparisons were multiple comparison Least Significant Differences tests.

Procedure

Each crew received an extensive one-week training in the VTT operations and then participated in a week-long demonstration of the vehicle. In the demonstration of the VTT, the crew performed mission scenarios under the command of a platoon leader who directed them by radio to drive to an observation point where they evaluated and engaged simulated targets and composed spot and situation status reports. Following the scenario exercises, the crew completed a set of questionnaires about the mission, functions and responsibilities, and the workload. In this process, the participants rated the effort put into the work-type activities by both himself and the other crew member for the planning and execution mission phases, the perceived mental workload for himself (self-rated) and the other crew member (other-rated) in the execution phase, and the situation awareness and task-specific workload elements of task attention, cognitive compatibility, motion sickness, stress, sleepiness, and trust rating, along with the personality questionnaire for himself. He rated his trust separately in the system, crew, and crew station.

Results

In preview, the other-rated team and meta-work types and the other-rated perceived workload are significantly different by crew role (chief versus mate); however, this is not true of the self-ratings for these or the remaining measures. Considering the small sample size, the data is presented in raw data plots where appropriate.

Personality traits

The personality traits appear to separate reasonably by military rank, with the higher rank being less impulsive, less activity prone, more aggressive, showing less sociability, and seeking acceptance less than did the lower rank. However, it appears that both members were less impulsive, less neurotic, less aggressive, and showing less sociability than the general male college population that was tested in experiments by Zuckerman, Kuhlman, Joireman, Teta and Kraft (1993).

Work assignments

Although the crews were encouraged to cooperate on tasks, depending upon their workload because of the commonality of the crew station designs, they tended to separate the tasks into fixed assignments depending upon their rank. For example, the higher ranked member of the crew had command and communication responsibilities (that is, of the crew chief) while the lower rank usually had the responsibilities (that is, of the crew mate) of driving and engaging targets when in a fixed position.

Task, teaming, and meta-work types

Statistical analysis shows insignificant differences of the self-rated work-types by crew role for both the mission planning and execution phases, but significantly more other-rated work for the chief than the mate in the planning and execution for teamwork (Planning: $F[1,6] = 17.00$, $p = .006$; Execution: $F[1,6] = 9.93$, $p = .020$), meta-work ($F[1,6] = 49.39$, $p < .001$; $F[1,6] = 37.35$, $p = .001$), and total work ($F[1,6] = 31.26$, $p = .001$; $F[1,6] = 22.67$, $p = .003$), and the planning task work ($F[1,6] = 15.21$, $p = .008$), but not the execution task work. The difference between the self-rated and the other-rated work for the same crew member is significantly by the crew role (chief versus mate). This is true for the sum of the execution team and the meta-works ($p = .046$), but not the sum for the planning works or the work-types by themselves for both mission phases. See Figure 31.4 for data plots of the total work ratings for the chief and crewmate of each crew; the data are shape-coded by rank (squares for E-7, triangle for E-6, and circles for E-5 and E-4), and numbered by crew.

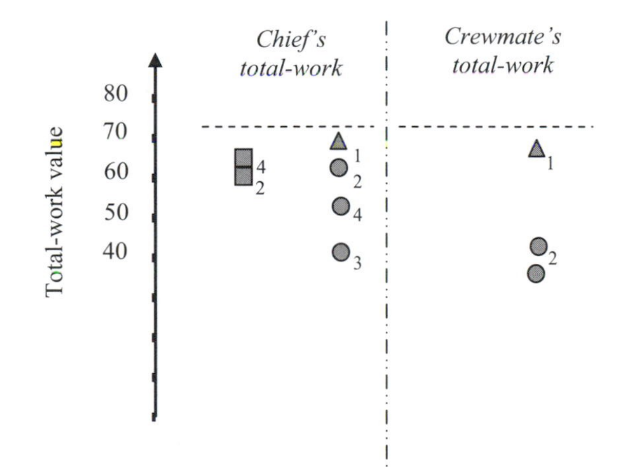

Figure 31.4 Work ratings

Perceived workload

Statistical analysis shows insignificant differences in the self-rated perceived workload (NASA-TLX; Hart and Staveland, 1988) by crew role, but significantly more other-rated work for the chief than the mate. The other-rated perceived workload is significantly greater for the chief than the mate for the global sum (F[1,6] = 9.023, p = .024), but not the dimensions or measures, although the mental demand shows a trend in the same direction as does therefore the demand dimension. The difference between the self-rated and the other-rated perceived workload for the same crew member is significantly greater for the chief than the mate, at least for the mental demand (p = .005), but not for the differences between the other measures, dimensions or global sum. The differences between the self-rated and other-rated work-type sums and those differences between the perceived workload global sums are positively correlated (Pearson Correlation: N = 8, $R^2 = 0.794$, p = .019). See Figure 31.5 for the total perceived workload data plots.

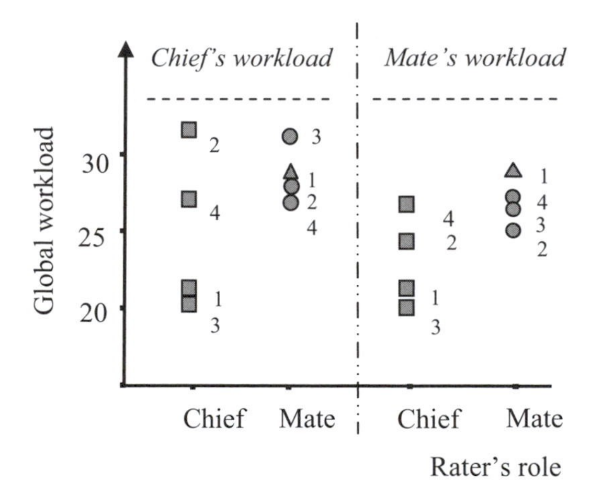

Figure 31.5 TLX total perceived workload

Task attention loading

Although insignificant, of interest are the attention loading (McCracken and Aldrich, 1984), for the driving, targeting, and menu activity tasks. The visual attention loading is about the same for the three tasks (involving "reading" to "scanning"). However, while the cognitive attention loading is about the same for the driving and menu tasks, which involve judgement, recognition, and recall, the cognitive loading tended to be higher for the targeting task, which involves estimation and evaluation. Although the motor attention loading for the driving and targeting tasks is relatively low since the tasks involve gross manipulation through the yoke handle control, the motor loading tended to be higher for the menu task since this task involves the finer motor controls needed for discrete adjustments and typing. Finally, the auditory attention loading values tended to be higher for the driving and targeting tasks since they involve interpreting verbal directives from the crew chief, while the menu task involves no speech.

Situation awareness rating

Although insignificant, the situation awareness was probably higher for the chief than the crewmate, since the Situation Awareness Rating Technique (SART) questionnaire scores (Taylor and Selcon, 1994) for the supply and understanding dimensions tended to be higher for the chief than the crewmate, while the demand scores remain about the same. The chief tended to report higher arousal, more spare mental capacity, and higher concentration and division of attention. The information quantity and quality tended to be higher for the chief, along with the familiarity of the situation. Figure 31.6 shows the data for the understanding dimension computed as the sum of the scores for the information quantity and quality, and the familiarity.

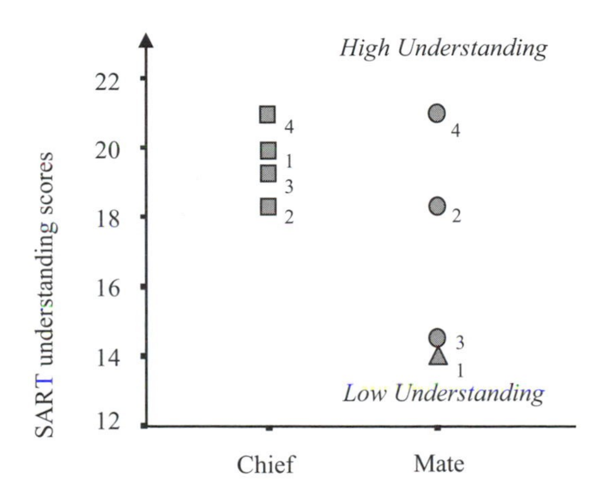

Figure 31.6 SART understanding

Cognitive compatibility

Although insignificant, the CC-SART Cognitive Compatibility questionnaire (Taylor, 1995) ratings show a tendency of higher cognitive compatibility between the tasks and displays for the chief. This is reasonable because of the increased automatic processing and activation of knowledge that resulted from his experience; however, the ease of reasoning was apparently less straightforward for both crew members because of the complex tasks that they performed. See Figures 31.7 and 31.8 for corresponding data plots.

Stress, sleepiness, and trust

Although the overall rating of subjective stress (Endsley, 1994) was comfortable, the chief tended to report slightly more stress than did the crewmate. While the chief reported himself as fully alert, the crewmate tended to report himself as slightly sleepy according to the Stanford Sleepiness scale (Hoddes et al., 1973). Finally, the chief tended to trust the system and team less than did the crewmate.

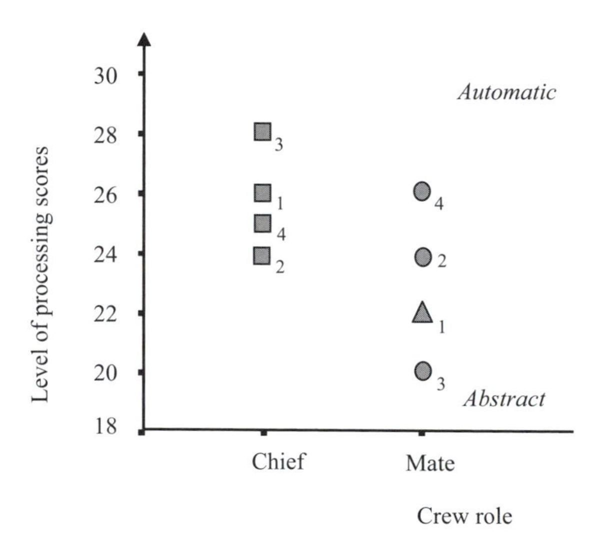

Figure 31.7 Cognitive compatibility level of processing

Motion sickness

Although some participants reported symptoms of motion sickness according to the Motion Sickness Symptoms Questionnaire (Kennedy et al., 1989), sickness was not an apparent problem during the restricted experimental conditions. Among those reporting motion sickness, the crewmate experienced stronger nausea symptoms than did the chief, while the ocular and disorientation symptoms were about the same for both.

Discussion

For this study, the knowledge that one crew member had of the work performed by the other crew member was influenced by his role in the crew as determined by his rank. The crew chief rated the amount of work for the crewmate as being less than what the mate rated for himself, and the mate rated the work for the chief as being more than what the chief rated for himself. That is, the chief tended to down-rate the work for the mate, while the mate tended to over-rate the work for the chief. This is true for the crew team-work and the meta-work, but not the task work. Crew team-work refers to those interactions between the crew members that are needed to maintain the crew functioning as a team, while meta-work is that accomplished to plan and schedule crew activities Bowers, Braun and Morgan, 1997). This result is also true of the perceived workload (NASA-TLX; Hart and Staveland, 1988) for which the mate over-rated the workload for the chief and the chief under-rated that for the mate, at least for the mental demand. The differences between the self-rated and other-rated work-type sums and those differences between the perceived workload are positively correlated. The personality traits and the work social cultural factors of organisation, responsibilities, interaction, and decision making are what would be expected for a military crew of an armoured tank or cavalry scout vehicle.

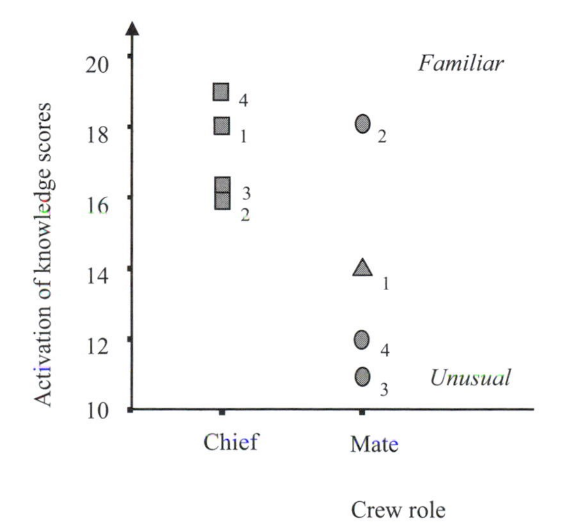

Figure 31.8 Cognitive compatibility activation of knowledge

The situation awareness and task workload elements are insignificant by crew role although differences could be expected because of the greater experience and maturity of the crew chief. This is also true of the self-rated work-types and perceived workload. While the data shows tendencies in these directions, still the insignificance statistical differences in all of these measures for the self-rating imply that the potential for maintaining mental models should be the same for both crew members.

The existence of a discrepancy between the knowledge of work distribution can impact crew interaction and therefore crew performance (Swain and Mills, 2003). A consequence is that the crew members may not provide timely support for teaming efforts or know when to expect such support. Because of the different perspectives of the effort needed to complete a task, the crewmate may not recognise when he or she needs to participate in teaming work. This may result in insufficient support on one hand and, on the other, task-interference by an offer of support when none is needed or a demand when it is not yet available.

One source of this discrepancy in workload perception may be the difference in experience of the ranks. For this reason, one crew member may not be aware of the tasks that the other crew member needs to perform or they may attribute different workload efforts to the tasks. Perhaps more crew training was needed to sensitise the crew members to the task stages at which teaming should occur and the amount of work involved in each. Another possibility is that the mental models that the crew had of each other's work are idealised representations used to maintain an awareness of crew positions in the organisational hierarchy at least during the initial crew formulation phase. For example, the mental model that was held by the chief in downgrading the work done by the crewmate may reinforce his own image of his position. Similarly, the crewmate may inflate the chief's work to justify mentally to himself his subordinate role. If so, this suggests a hierarchical component to the mental model that interferes with the workload understanding.

Still another possibility is that the automation isolates the crew from aspects of the task that they need to maintain an awareness of the teaming status. This is because the automation performs those aspects without the participation of the crew members, such a, for example, electronic communication between the crew stations. This may result in a specialised form of attention deficit that has been reported to occur with automation (Sheridan, 1992), that is, a crew-teaming attention deficit for the shared work.

For effective crew-teaming work, the members need to maintain a task prioritising strategy which is based on their situation awareness of the intra-crew relations. This situation awareness results from an additional task at the meta-cognition level for maintaining a mental model of the crew process. The crew members use their mental models to predict the mission task progress and teaming needs. To maintain this situation awareness, the crew must continually update their mental models of the situation (Minionis, Zaccaro and Perez, 1995), by observation and associations based on expectancies.

As noted by Endsley (1994), situation awareness is a precursor to optimal performance, since a loss in awareness has an impact on decision making and leads to an increased risk of error in performance. This is presumably true also of the situation awareness for crew-teaming. The maintenance of the internal model is affected by limitations on working memory and the availability of knowledge of critical features and important relationships stored in semantic and episodic memory in the form of schemas and scripts (Orasanu and Fisher, 1991).

The additional work needed to maintain crew-teaming awareness can be reduced through automation where the mission-task progress and teaming alerts are automatically displayed at the crew stations. The automation should follow good machine design principles to reduce excessive task workload and increase situation awareness (Endsley, 1994). The automation technology at the soldier-machine interface should be cognitively compatible with the tasks at the levels of processing input, reasoning, and activation of knowledge; a design supporting an automatic response by the soldier will have little need for association and therefore reduce demand on working memory (Taylor, 1995).

Conclusion

In this study, the knowledge that each crew member had of the workload was influenced by his role in the crew as determined by rank. The existence of a discrepancy between the knowledge of work distribution among the crew members could be a tactical disadvantage during high stress conditions, since the crewmate may not recognise when he or she needs to participate in teaming work, a result that could impact crew interaction and therefore crew performance. The automation technology may isolate the crew from those aspects of the task that they need for maintaining an awareness of the teaming status and this can produce a form of task attention deficit that has been reported to occur with automation. The implication is that the automation must be designed to support a knowledge base that includes the shared work.

Acknowledgements

The author would like to express appreciation for the support and interest in this experiment by the Vetronics Technology Area at the US Army Tank Automotive Research Development & Engineering Center in Warren, MI.

References

Beith, B. H. (1987). "Subjective workload under individual and team performance conditions". In *Proceedings of the Human Factors Society 31st Annual Meeting* (pp. 67-71). Santa Monica, CA: Human Factors Society.

Bowers, C. A., Braun, C. C. and Morgan, B. B. Jr. (1997). "Team workload: Its meaning and measurement". In M. T. Brannick, E. Salas and C. Prince (eds), *Team performance, assessment, and measurement* (pp. 85-105). Mahwah, NJ: Lawrence Erlbaum Associates.

Endsley, M. R. (1994). "Situation awareness in dynamic human decision making: theory". In R. D. Gilson, D. J. Garland and J. M. Koonce (eds), *Situation awareness in complex systems*. Daytona Beach: Embry-Riddle Aeronautical University Press.

Hart, S. G. and Staveland, L. (1988). "Development of the NASA Task Load Index (TLX): Results of empirical and theoretical research". In P. A. Hancock and N. Meshkati (eds), *Human mental workload* (pp. 139-183). Amsterdam: North-Holland.

Hoddes, E., Zarcone, V., Smythe, H., Phillips, R. and Dement, W.C. (1973). "Quantification of sleepiness: A new approach". *Psychophysiology*, 10, 431-436.

Kennedy, R. S., Lilienthal, M. G., Berbaum, K. S., Baltzley, D .R. and McCauley, M. E. (1989). "Simulator sickness in U.S. Navy flight simulators". *Aviation, Space, and Environmental Medicine*, 60, 10-16.

Kerle, R. H. and Bialek, H. M. (1958). *The construction, validation, and application of a Subjective Stress Scale (Staff Memorandum Fighter IV, Study 23)*. Presidio of Monterey, CA: US Army Leadership Research Unit.

McCracken, J. H. and Aldrich, T. B. (1984). *Analysis of selected LHX mission functions: Implications for operator workload and system automation goals. Technical Note ASI 479-024-84.* Fort Rucker, AL: Army Research Institute Aviation Research and Development Activity.

Minionis, D. P., Zaccaro, S. J. and Perez, R. (1995). "Shared mental models, team coordination, and team performance". In *Proceedings of the 10th Annual Meeting, Society for Industrial and Organizational Psychology*, Orlando, FL.

Orasanu, J. M. and Fisher, U. (1991). "Information transfer and shared mental models for decision making". In R. S. Jensen and D. Neumeister (eds), *Proceedings of the 6th International Symposium on Aviation Psychology* (pp. 272-232). Columbus, OH: The Ohio State University.

Salas, E., Sims, D. E. and Burke, C. S. (2005). "Is there a Big Five in Teamwork?" *Small Group Research*, 36(5), 555-599.

Sheridan, T. B. (1992). *Telerobotics, automation, and human supervisory control.* Cambridge, MA: MIT Press.

Swain, K. and Mills, V. (2003). "Implicit Communications in Novice and Expert teams". Defense Science and Technology Organisation Technical Note DSTO-TN-0474 (AR 012-550). Edinburgh, Australia: Systems Science Laboratory.

Taylor, J. C. and Felten, D. F. (1993). *Performance by design: Socio-technical systems in North America*. Englewood Cliffs, NJ: Prentice Hall.

Taylor, R. M. (1995). "CC-SART: The development of an experimental measure of cognitive compatibility in system design". *TTCP Technical Panel UTP-7, Human Factors in Aircraft Environments. Minutes of the Annual Meeting*, 13-18 June 1995, Annex I, Appendix 2 (pp. 100-112). Toronto, Canada: DCIEM.

Taylor, R. M. and Selcon, S. J. (1994). "Situation in mind: Theory, applications, and measurement of situational awareness". In R. D. Gilson, D. J. Garland and J. M. Koonce (eds), *Situational awareness in complex systems*. Daytona Beach: Embry-Riddle Aeronautical University Press.

Whitworth, B. and deMoor, A. (2003). "Legitimate by design: Toward trusted virtual community environments". *Behaviour and Information Technology*, 22(1), 31-51.

Zuckerman, M., Kuhlman, D. M., Joireman, J., Teta, P. and Kraft, M. (1993). "A comparison of three structured models for personality: The big three, the big five, and the alternative five". *Journal of Personality and Social Psychology*, 65(4), 757-768.

Chapter 32

Full Spectrum Analysis: Practical Operational Research in the Face of the Human Variable

Graham Mathieson

Introduction

This chapter is based on presentations given at the *20th International Symposium on Military Operational Research* (2003) and *NEC – The Human Dimension* (2003) combined with reflections on the presentations and discussions from *Human Factors of Decision Making in Complex Systems* (2003). It explores the need for Operational Research (OR) studies to embrace the broad range of variability arising from human participation in the operations and systems studied, and the consequent need for the human science communities to integrate and present their knowledge in forms which are amenable to exploitation by OR practitioners. Although the chapter is principally concerned with the exploitation of human sciences by OR there are clear implications for cross-disciplinary integration across the full spectrum of the human and system sciences as a prerequisite for satisfying OR needs. The chapter is, therefore, a call to action for the various scientific and technical communities relevant to OR.

The nature of OR

OR is concerned with the analysis of interventions with the operation of systems or organisations of interest to executive decision makers, who are the OR study customer. Since OR tackles real world problems of interest to human executives, and since the systems involved are usually embedded in human organisations, it can fairly be asserted that OR is principally concerned with the analysis of socio-technical systems. *(NOTE: for the rest of this chapter the terms 'system' and 'system of interest' will be assumed to refer to socio-technical systems.)*

Consequently, it is important for OR methods to be able to deal with socio-technical factors and issues. This raises several challenges for OR methods which have not, to date, received enough attention, but in the face of which steps can be taken to improve the state of practice.

Challenges for OR

The key challenges for OR can be categorised into the problems of modelling, data, prediction and intervention.

The problem of modelling

Good OR studies begin with a problem formulation stage involving the construction of a conceptual model of the system of interest. This problem model captures the joint understanding of the analysts and their client about the system (and intervention options), serves as a common description between the analyst and providers of expert knowledge, and is a key factor in selecting the analysis methods (NATO, 2002).

It is critical that the problem model be requisite, that is, that it faithfully represents real world factors, structures, processes and effects that significantly impinge on the study problem. A small study team cannot hope to have detailed knowledge of all of the disciplines required to model a socio-technical problem, and must rely upon knowledge from specialists in a range of disciplines, including: information technology, systems engineering, organisational psychology, management science, economics, cognitive psychology, anthropology, and so on. It is critical that such knowledge be trustworthy, comprehensible and usable.

This challenges providers of specialist knowledge to ensure that *their* conceptual models and theories are adequately comprehensive, comprehensible and coherent across the variety of disciplines needed to model the OR problem. For example, Farrell (see Chapter 26) emphasises the need to have a more rigorous framework to analyse complex systems. Using multiple perspectives is risky if the perspectives have inconsistent models. Although an element of post-modern thinking is useful to OR, the problem model must be kept self consistent to avoid misleading the executive decision maker.

The problem of data

To construct a requisite problem model critically depends on having data describing the key variables of the system, their likely values (or distributions) and the nature of the relationships between them, including the current operation of the system, its structures and processes. Such data are not easy to acquire, particularly if the system of interest is not in continuous operation (like a production line) but only called into action on a contingency basis (like a military capability). In the latter case even the constituents of the system may be unclear in advance of the contingency. In this context the OR practitioner needs either generic data or potential data distributions in order to transform the conceptual problem model into a generative model which can support inference about the system. Data are often less accurate, precise, reliable or available than would be ideal, but OR has evolved robust methods for dealing with such data, while still producing insights and advice which improve upon the executive's intuitive understanding. OR methods can make use of logical, descriptive or numerical data, although numerical is preferred for a variety of reasons. As with modelling, the reliability of data is paramount and this implies coherence across the domains of expertise providing it.

The problem of prediction

At the heart of OR is the presumption that the system of interest can be analysed to produce insights which the executive will use to intervene in order to produce desired effects. This, in turn, presumes some measure of predictability either in the system response to intervention or in some more abstract properties which influence future system behaviour. Even a good conceptual model with adequate data is no guarantee of predictability. Socio-technical systems tend to be complex adaptive systems (CAS), as described by Allen (1988), presenting a fundamental problem with macro-behaviour forecasting, even with complete system knowledge. CAS may exhibit simple and stable macro-behaviours despite micro-level complexity and variability, or vice versa, and they may become chaotic, responding so sensitively to minute variations in the detail that the macro behaviour appears effectively random. They may flip between different modes of behaviour in response to apparently insignificant changes or even with no apparent change at all.

This difficulty has led many analysts to declare that prediction is not possible in CAS. However, drawing useful inferences about the consequences of executive intervention does not require precise prediction of system behaviour. Executives are prepared to take risks based on general trends, or statistical forecasts; anything that gives them information to take a better gamble. OR is, so to speak, "loading the dice" in favour of success. For example, Mintzberg (1979) and others have identified relationships between organisational structure and task environment. This knowledge can be used as a generic model to give insights, in broad terms, about the likely response of an organisation to a forced change of structure or environment.

However, one must remain aware of the assumptions behind theory and be constantly sceptical about whether those assumptions hold. For example, one feature of Industrial Age organisations identified by Mintzberg is that their structure evolves in the face of environmental variety in ways which limit the demands on individual managers. In the Information Age, there is the promise of empowering managers to cope with more complex tasks, allowing greater freedom to create more complex organisations. If so, then perhaps the empirical basis of Mintzberg becomes less valid.

Historical data alone are not a sufficient basis for executive action without at least the prediction that the data used will remain valid in the timescales of the proposed intervention.

The problem of intervention

The final challenge for OR, and the providers of its underpinning domain knowledge, is the fact that socio-technical systems are all, to a greater or lesser extent, self-aware and liable to behave reflexively in the context of interventions perpetrated upon them. Such complex adaptive reflexive systems (CARS) have an extra dimension to their response in which individual and collective decision making become a critical feature. CARS can generate behaviour which either reinforces or undermines the executive's intentions, and this behaviour can be pre-emptive, driven by perceptions of the executive's intent prior to substantive executive action or its effect. Thus, bizarre situations can occur such as organisation members reacting to a false perception of a hostile executive intent and taking mitigating actions, which the executive falsely perceive as hostile, leading to the adoption of a hostile executive intent where none previously existed. Human affairs are full of such self-fulfilling prophecies and OR needs to allow for the social processes involved in them.

The human sciences (HS) have a key role to play in providing OR practitioners with the knowledge needed to take account of such reflexive behaviours and the subsidiary interventions needed to avoid or mitigate them as required. OR practitioners, for their part, need to understand the possibilities for reflexive response, on top of the other challenges, and to think of executive intervention as a multi-cycle process, with interactions between intentioned and motivated actors, rather than as an event with consequences. An essential element of the understanding required by OR practitioners is a clear concept of what a socio-technical system is and what it implies.

What is a socio-technical system?

The concept of socio-technical systems has a long history in the literature (Salvendy, 1987). However, for the purposes of this chapter, only a very basic distinction needs to be made between technical, social and socio-technical systems.

A system is as an interacting collection of parts. If all of the parts of a system are non-human technologies then one has a purely technical system, for example an autonomous robot or an unmanned production facility. For the present purpose, socially aware artificial intelligences are neglected (as are systems of non-human animals).

A purely social system is an interacting collection of humans in which non-human technologies are either not present or not significant to system operation. A community of people doing something like talking, for which technology is not really an issue, might be considered a purely social system although, in modern societies, such technology-free activity is rare.

By extension, a socio-technical system is a collection of human and non-human parts interacting in an integrated way, in which overall system behaviour arises from multiple cycles of interaction within and between the human and non-human parts. This implies that socio-technical systems are also likely to be CARS.

The assertion that all systems of interest to OR are socio-technical is especially, though not exclusively, relevant in the military domain. Military conflict is essentially a social affair, but one in which technology is deeply and inseparably embedded. Technology is so important to modern military affairs, especially with the increasing use of automation, "smart" munitions, and unmanned vehicles, that it has been tempting for military OR practitioners to consider the technical component alone. It is equally tempting, and equally misguided, to fixate on the human component of conflict to the exclusion of all else.

Whilst much about the impact of humans on systems is difficult to predict or understand, the one dependable fact is that humans bring variability to systems – they are, so to speak, the constant variable – a fact which presents challenges for systematic analysis. A consideration of human variability and its impact is, therefore, a good starting point for the dialogue between the HS and OR.

Humans – the constant variable

Any system with humans involved will change and adapt. Research clearly shows that there is significant variability between individuals and groups and also within individuals and groups over time. Understanding sources of variability will allow OR to include them in its methods and models. This chapter will, therefore, spend some time cataloguing and discussing the nature and sources of human variability and the consequences for OR's exploitation of HS.

Individual human variability

Humans differ from each other in ways which affect how they behave and perform tasks inside systems. Individual humans change over time as a consequence of learning and experience, or in response to changing context.

Human cognition and behaviour generation is still poorly understood, but many things are known with some certainty. Studies using brain scanning technology have begun to unravel some of the richness and complexity of human cognition. For example, clinical studies of patients with specific types of brain damage (Carter, 1998) have shown that the affective component of cognition is deeply implicated in higher reasoning and the formation of belief. Further, the study of left-right brain duality (Carter, 1998) provides evidence that the two halves of the human brain, far from being just parallel processors, are capable of thinking different thoughts and holding different aspirations and goals. It is logical to conclude that human reasoning is more likely the result of a complex interaction of multiple, possibly competing, thought processes, and that the coherence of behaviour that results is more akin to an emergent property of sub-conscious processes than the result of a conscious and coherent directing mind.

This role of affect on belief and reasoning has implications for the understanding of awareness and the exploitation of work such as that of Endsleigh (2003). It is important to recognise the importance of the sub-conscious component in decision-making when interpreting the work of Klein (see Chapter 2).

The pioneering OR modelling work of Moffat (2002) makes some use of human science theories such as those of Klein, but synthesises ideas from Janis and Mann (1977) to justify treating decision making as a rational process driven by coherent concepts of utility. The most recent understanding of individual human variability would indicate that such a synthesis may be inconsistent. It is important for the human science community to synthesise its own literatures to clarify current knowledge before OR practitioners can rely upon that literature to inform executives.

Variation between individuals

Given exactly the same situation, under exactly the same conditions, two typical people will react and behave differently. How differently depends on many things. For example, imagine that you are reading this chapter seated on a park bench when a man in soldier's field uniform carrying a rifle comes up and stands in front of you. How will you react? What will you make of the situation? What will you "see" standing there in front of you? One person may see a strong protector of freedom and security, and perhaps wonder if there is a threat nearby, a terrorist bomb or a riot. Another person will see the soldier himself as a threat, a menacing representative of repression and injustice. Yet another may see a young, immature fool, suckered into a dangerous profession by propaganda and the promise of a trade. Each of these views could legitimately be held by a citizen of the UK, depending on their past experiences with soldiers. A resident of certain streets in Belfast will see the soldier quite differently from the landlord of a public house near the army barracks in Bordon, Hampshire, and a quite different view again may come from a proud old war veteran.

Such different perceptions arise from a host of sources, for example, memories of past experiences, cultural norms instilled since childhood, self perceptions and the way a person "spins" their position in relation to the world. It has been shown experimentally (Malish, Mathieson and Berry, 2003) that personality can significantly affect how different military commanders choose to act given the same situation and information. Work by Sicard, Jouve and Blin (2003) suggest the intriguing possibility that risk taking behaviour is related to a need to regulate some internal risk "thermostat" which is linked to physiological responses in the brain.

At a more basic level, what each person "sees" is the result of a complex cycle of perception, attention and recognition involving the imposition of previously formed categorisations or symbolisations onto the "sensory wash" and the construction, from remembered fragments, of a story to "explain" the juxta-position of those symbolic representations. This idea is supported by the work of Klein (see Chapter 2). Indeed the need to "make sense" of the world in this way may even result in the construction of quite fictitious explanations and the neglect of countervailing perceptions in order to preserve the current "mental model". The best optical illusions work because we use a subset of the image to trigger model building and then persist in our belief in the model despite contrary evidence, even at the cost of disturbing cognitive dissonance.

This combination of history, culture, politics, psychology and physiology provides many ways for people to differ, and the interaction of the different causes can make it difficult to provide a clear pattern or distribution from which to generalise. Some clues might be had from past observations, such as in Bolia, Vidulich, Nelson and Cook (see Chapter 18), or from profiling of various sorts. If one is interested in advising soldiers on how to conduct themselves in the course of peace-keeping operations, for example, it is vital to understand what knowledge is relevant to the analysis and how the various disciplines interact.

Collective human variability

In most systems of interest to OR, humans are involved collectively, invoking a whole range of additional sources of variability. Social networking, co-operation and competition, collective self-awareness and reflex, and interactions between system structure and function produce macro-level behavioural variability which is not always clear in its origins. Systems differ in their formally appointed structures, goals, strategies and processes. They also have different histories and collective experiences, which lead to wide variety in informal structures, goals and processes, even where their formal expressions are similar.

Two teams given the same task in the same context are likely to diverge in their approach to task execution and to the many non-task-related goals and behaviours which arise in any human collective. In general, OR modellers only look at formal structures and processes and neglect to account for the informal, perpetuating the comfortable myth that those things the executive controls dominate organisational behaviour. Research, such as Siemieniuch and Sinclair (see Chapter 17), indicates that in dynamic situations role structures will adapt in ways which depend upon individual team member capabilities. Salas, Guthrie and Burke (see Chapter 21) highlights how team competence is more complex than the combination of individual competencies.

It is generally true that OR practitioners (and many HS researchers) obtain their data on organisational processes by elicitation from organisation members, which approach often tends to reproduce the formal rather than actual processes and structures. In this context, OR needs to exploit the techniques developed by disciplines such as behavioural psychology to acquire knowledge of human behaviour that does not rely entirely upon self-report.

Practical OR responses

The wide spectrum of sources of variability in systems outlined above demands a response from OR, if the discipline is to retain its credibility. One possibility is to retreat from analysis into the direct facilitation of executive decision making in complex systems. However, this would be to remove a main source of the value of the OR discipline, namely, the ability to derive insights about the consequences of intervention which add value to the executive's intuitive understanding.

OR's practical response to the challenge of human variability should include a broadening of scope, at all stages of the analysis process, to cover the full spectrum of significant factors. In particular attention needs to be paid to a balanced problem formulation, a more "natural" approach to OR modelling, the explicit treatment of uncertainty, and the synthesis of advice from analysis results.

Balanced problem formulation

The recently updated NATO Code of Best Practice for C2 Assessment (NATO, 2002) emphasises the importance of completeness in problem formulation, and the treatment of practical constraints, such as data availability, as modifiers rather than drivers of the problem model. The Code also advises an open and adaptable approach to problem bounding and assumption setting.

In considering human decision making in complex systems it is critical for OR to account for all sources of variability which might prove significant rather than, as is often done, restricting the analysis to those variables which are readily observable. Since model development is often a capital intensive project, it is also important to design adaptability into models to facilitate future development in the face of new understandings. This requires an approach to modelling which does not adhere blindly to the KISS principle ("Keep It Simple, Stupid"), but adopts the more holistic KISMET principle proposed by Maeers, Mathieson, and Rose [personal communication] in the margins of the ISMOR conference (2003). KISMET stands for "Keep It Specific, Manageable, Exploratory and Testable" and tries to evoke a more open-ended, iterative and exploratory approach. Such an approach will tend to produce more "natural" models with a more explicit treatment of system variability and its sources.

Modelling systems "naturally"

One early text (Air Ministry, 1963) defines OR as "numerical thinking about operations, with the aim of formulating conclusions which, applied to operations, may give a profitable return for a given expenditure of effort". Today, OR practitioners, particularly those in the military domain, are dominantly drawn from hard science, mathematics or engineering backgrounds. Consequently, systems are typically modelled from a rationalist perspective. Even where humans are treated explicitly a rational construct based on utility theory and choice optimisation is often used. In the military OR domain, models of decision making typically assume that command decisions are driven by the full set of information in the commander's situation display, and that multiple options for course of action are considered before the "best" course is chosen to meet the operational goal. This thinking is the usual interpretation of the ubiquitous OODA construct (Observe, Orient, Decide, Act) defined by USAF Colonel John Boyd, although Boyd's original was much more cognitively and socially contextualised than current usage implies (*Decision Making: OODA Loop*, 2004).

It is now widely accepted in cognitive psychology that a natural description of decision making based on expertise, situation recognition and a satisficing strategy is more representative (Klein et al., 1993). A natural model of the decision making process should include concepts like attention, attribution, construction, recognition, limitations in working memory (with consequences for cognitive strategies), and learning.

Current OR models (at least in the military domain) tend to assume that organisation members share formal goals, and faithfully follow formal processes. In military capability investment appraisal it is widely assumed that improving information sharing will significantly improve shared situation understanding, which will greatly enhance operational effectiveness. In the current OR models this assumed causality is usually already embedded in the model, rendering it incapable of being used to effectively question or challenge investment options involving information technologies. A more natural model of organisations would probably be centred on social networks rather than formal structures, dealing explicitly with the effects of multiple, unshared beliefs and goals, informal and *ad hoc* processes, emergent roles and rules, the interaction of multiple cultures, and organisational adaptation as outlined above.

Modelling systems more naturally will initially make models more complicated, but will enhance OR's ability to absorb human science knowledge and provide a sound basis for model evolution and adaptation in the light of new knowledge.

Some of the improvements in OR modelling implied by the "natural" approach could be implemented in the short term. Representations of perception and attention, situation recognition (already demonstrated by Moffat, 2002), satisficing strategies for decision making, the impact of internal and external moderators on cognition and, at the collective level, multiple goals and social network influences on collaboration could all be added or improved with current knowledge and in the context of current models. In the longer term, new decision making algorithms based on constructed mental models, adaptivity in organisational structures and processes and a treatment of self awareness and reflex are possible with limited additional methodological research.

Embracing uncertainty

Expanding the scope of analysis to encompass the full spectrum of variables and factors significant to the behaviour of CARS will bring in variables and factors for which solid empirical data are not available. OR practitioners already have tried and tested techniques for dealing with such unknowns. A combination of sensitivity analysis, stochastic modelling and risk based reasoning will allow credible and useful advice to be generated even in the face of high uncertainty, provided it is embraced rather than ignored or suppressed.

The treatment of uncertain knowledge is quite different in the OR and HS communities. HS research tends to be very sceptical even in the face of statistically significant experimental results, because of the imperative to be conservative in adding knowledge to the scientific canon. Often, uncertain results are the trigger for further research proposals. Conversely, OR practitioners tend to be happier to use data with low statistical significance provided only that it appears to contribute to discrimination of investment options in the context of the immediate decision problem. Time for further research is a rare luxury for OR.

Analysts and researchers need to understand each other's views on significance and usefulness of uncertain data before there can be a free flow of useful knowledge between them.

Synthesis

Synthesis is the often unregarded twin of analysis. Even those human and organisational issues which have to be excluded from the analysis through lack of capability or usable data can be re-introduced during synthesis (provided they were explicitly identified at the start). It is recognised best practice for problem formulation not only to provide problem segments amenable to analysis, but also a clear and valid mechanism for meaningful synthesis to provide coherent knowledge about the original, larger problem (NATO, 2002).

Full spectrum analysis

Dealing with socio-technical issues will challenge existing OR capabilities. The immediate response of OR practitioners should be to use multiple methods, possibly even multiple theoretical bases, to address issues raised by a properly scoped problem formulation. The longer term response must be to seek synthesis of theories so that multiple methods become compatible.

Multi-disciplinary teams are also vital to the effective treatment of human decision making in complex systems. The hard science bias in the OR community needs to be removed and the recruitment of human scientists into OR teams should be a priority in coming years. The effective integration of scientific knowledge across traditional disciplinary boundaries is a core task for the OR community, one in which the human science community should participate eagerly, since it will render much of their hard-won knowledge more exploitable.

Conclusions

This chapter has sought to explore the challenges faced by OR as it tries to support interventions in complex, socio-technical systems. It asserts that the validity (that is, fitness for purpose) of OR depends upon a balanced treatment of factors and that this means a significant broadening of the scope of models used by OR to predict the consequences of interventions in systems.

It emphasises the importance of the human sciences as a basis for understanding variability in systems, and notes that the wide range of disciplines is not integrated as a coherent body of knowledge. Specific proposals are made for the improvement of OR models and for an ongoing programme of integration to produce useable knowledge across the full spectrum of scientific disciplines. Together these developments will produce a capability for "full spectrum analysis", capable of helping executives to intervene effectively in complex systems.

Acknowledgements

Thanks are due to Dr. Malcolm Cook for his generosity in allowing this chapter into these pages.

References

20th International Symposium on Military Operational Research, a conference held in August 2003 by UK Ministry of Defence at Eynsham Hall, Oxford, UK.

Air Ministry (1963). *The Origins and Development of Operational Research in the Royal Air Force.* Air Publication 3368, HMSO, London.

Allen, P. M. (1988). Dynamics of evolving systems. *System Dynamics Review*, 4, 109-130.

Bolia, R., Vidulich, M., Nelson, W. and Cook, M. (2003). *The Use of Technology to Support Military Decision-Making and Command & Control: A Historical Perspective.* Presented at *Human Factors of Decision Making in Complex Systems* conference. Chapter 18.

Carter, R. (1998). *Mapping the mind.* London: Wiedenfeld & Nicolson.

Decision Making: OODA Loop. Retrieved 4 February 2004 from www.mindsim.com/MindSim/Corporate/OODA.html .

Endsleigh, M. (2003). *Designing For Situation Awareness In Complex Systems.* Presented at *Human Factors of Decision Making in Complex Systems* conference.

Farrell, P. S. E. and Lichacz, F. (2003). *LOE 02: Canadian Human Factors Analysis.* Poster presentation at *Human Factors of Decision Making in Complex Systems* conference. Chapter 26.

Human Factors of Decision Making in Complex Systems, a conference held in September 2003 by the University of Abertay Dundee at Dunblane Hydro, Scotland, UK.

Janis, I. L. and Mann, L. (1977). *Decision-making: A psychological analysis of conflict.* New York: Free Press.

Klein, G. (2003). *A Data/Frame Model of Sensemaking.* Presented at *Human Factors of Decision Making in Complex Systems* conference. Chapter 2.

Klein, G. A., Orasanu, J., Calderwood, R. and Zsambok, C. E. (eds) (1993). *Decision making in action: Models and methods*. NJ: Ablex Publishing Corporation.

Malish, P., Mathieson, G. and Berry, A. (2003). *Contribution of the Human Element to Command Effectiveness – The Impact of Information on Command Effectiveness*. Defence Science and Technology Laboratory, UK, reference Dstl/JA07514. Submitted to *Journal of the OR Society*, November 2003.

Mintzberg, A. (1979). *The structuring of organisations*. Englewood Cliffs, NJ: Prentice-Hall.

Moffat, J. (2002). *Command and Control in the Information Age Representing its Impact.* HMSO, London. ISBN 011 772984 1.

NATO (2002). *NATO Code of Best Practice for Command and Control Assessment.* Reprinted by US Department of Defense, Command and Control Research Programme (see http://www.dodccrp.org).

NEC – The Human Dimension, a conference held in November 2003 by the Royal Military College of Science, Shrivenham, UK, on behalf of the UK Industry/MoD Command and Digitization Group.

Salas, E. (2003). *Why Training Team Decision-Making is Not as Easy as You Think.* Presented at *Human Factors of Decision Making in Complex Systems* conference. Chapter 21.

Salvendy, G. (ed.) (1987). *Handbook of human factors*. New York: John Wiley & Sons.

Sicard, B., Jouve, E. and Blin, O. (2003). *Decision-making and Extreme Risk-taking.* Presented at *Human Factors of Decision Making in Complex Systems* conference. Chapter 5.

Siemieniuch, C. and Sinclair, M. (2003). *Changes in Organisational Roles When Disaster Strikes.* Presented at *Human Factors of Decision Making in Complex Systems* conference. Chapter 17.

Chapter 33

Effective Taxonomies in Organisational Safety

Brendan Wallace and Alastair Ross

Introduction

The development of effective taxonomies to organise large databases (whether electronic or not) has its roots in knowledge management and library science (Reardon, 1998). Knowledge management is a series of techniques for the accurate storage and retrieval of information, normally involving the use of a taxonomy, which structures the database. However, the taxonomies developed in knowledge management are normally used for generic projects (for example, libraries). Techniques for developing project-specific taxonomies (in this example, for qualitative or quantitative safety databases), and, moreover, testing the effectiveness of these taxonomies are largely lacking. We propose the phrase "Taxonomy Theory" as a general rubric to cover these issues.

One of the key tasks in safety management is to create a taxonomy, or taxonomies, with which to classify the various safety issues that have to be dealt with. In our work with the Confidential Information Reporting and Analysis System (CIRAS) for the UK Railways (Wallace, Ross and Davies, 2003), and the Strathclyde Event Coding and Analysis System (SECAS) for the UK nuclear industry (Wallace, Ross and Davies, 2002), we were involved in creating taxonomies to organise large databases of safety data. This chapter will discuss the issues that arose from the creation of these taxonomies. It will be seen that the techniques developed for their creation could be used in the creation of *any* taxonomy for *any* form of database. We will also look at techniques for assessing the effectiveness of taxonomies, looking at the specific example of a taxonomy of road accident event reports. It will be seen that many taxonomies which are generally considered adequate may not always be functioning as well as they might, and that this has implications in a wide range of fields.

Taxonomies and databases

Purposes and goals of databases

It is clear that the major goal of a database is that information should be accessible easily and quickly. Therefore it is important that it is organised effectively. To take the example of a library, it is self-evident that a well-organised library will be one in which two or more different people can access the same book by looking at the same category, at two (or more) different times. In order for this to happen, therefore, the same book must also be classified in the same place by the different classifiers (in this example, library staff). It should be noted that there is more than one way of classifying books in libraries. For example, some libraries use the Dewey system and some use the Library of Congress system. There is no "right" or "wrong" way to classify data (Ryle, 1938). However, there can be systems that function pragmatically: that is, they classify all data that it is reasonable to expect in a given situation.

It should be noted that this can be interpreted in two ways. First, it can be pointed out that, since any given field (in this example, books in a library) can be categorised via more than one taxonomy, that the key to deciding which taxonomy should be used should be *utility*, however this is defined (ease of use, comprehensiveness, and so on). The second, perhaps slightly less obvious corollary is that it is not, therefore, necessarily the case that *one* taxonomy should cover all the specific situations in which objects might be classified (so there is no necessary reason as to why one classification system should be used for all the libraries in the world above and beyond the level of pragmatics).

Therefore, even in the field of safety, there is no *logical* reason as to why there should be only one taxonomy that should be used to apply to *all* cases of (for example) "human error" (normally referred to as "cognitive failures" or something similar). Of course, as in the library example, pragmatically speaking this might be thought to be desirable, but equally it might not. For example, it might be thought that the specifics of accident "causation" are sufficiently different in a nuclear power plant from accident "causation" in the railway industry, and that the specifics of data are such in these two specific industries that perhaps it might be better to have two separate taxonomies for these two separate situations. It should be noted that in our own experience, after consultation with management and staff, we have found it easier to create taxonomies for specific industries and situations, rather than to adapt previously existing taxonomies which were developed for perhaps very different settings, and to deal with very different problems (Wallace and Ross, 2006).

Taxonomies

To return to the library example: most classification systems subdivide major elements into "smaller" elements. For example, if the library was organised by theme, then "Social Sciences" might be broken down into "Psychology", "Economics", and "Sociology". Psychology might then be broken down into Social Psychology, Cognitive Psychology, and so on. This is a hierarchical taxonomy, and it is generally agreed that hierarchical taxonomies should be Mutually Exclusive and Exhaustive (MEE) (Robson, 1993). However, this is more complex than it might seem, because in practice, most taxonomic categories are "fuzzy" (Kosko, 1994). For example, in a library context, where is a book entitled "Cognitive Social Psychology" to be categorised? It could obviously be categorised under *either* cognitive *or* social psychology. Therefore a decision must be made and codified in the form of a rule. The formal structure must be imposed on the fuzzy or qualitative field in order to ensure classificatory agreement. This is precisely the role of a good taxonomy. For example, a rule might be created such that, if a book has two "classifier" words in the title, precedence should be given to the first one. So in the above example, Cognitive Social Psychology would be classified under Cognitive Psychology, not Social Psychology. Two points must be made here. First, these rules are pragmatic, and are, to a certain extent, arbitrary. It *doesn't really matter* whether one decides that "Cognitive" should be given priority or "Social". What does matter is that this is a pragmatically useful "rule". This leads to the second point, that these rules are useless unless they are widely known. It is not enough for the classifier to know the "rules": the user must as well.

Creation of taxonomies

The creation of a taxonomy is fundamentally a social activity (Bowker and Star, 1999). Preferably, end users must collaborate in terms of creating the definitions and rules that will be necessary to understand and use the taxonomy. For example, it must be ensured that the language used is clear and understandable: vague or ambiguous phrases are a key source of taxonomic unreliability. We have discovered in our own development of taxonomies that one of the best ways of ensuring that taxonomic subcategories are mutually exclusive is to create them as being logically exclusive in an AND/OR format. To take an example from the CIRAS database, a division is made at the frontline level that communications problems are EITHER between staff, OR are between staff and managers (either from staff to managers or from managers to staff). Given the division of the workplace presumed in CIRAS, logically no other division is possible (Wallace, Ross and Davies, 2003).

This is important in order to avoid what we term the "Bucket" category or the "Other" category. We do accept that when building up a taxonomy, "Other" categories may be essential. However, it should always be the aim of a *completed* taxonomy to omit all "Other" categories. There are two reasons for this. First: "Other" is by definition impossible to define in a rule-based fashion except by default. The only possible definition is "classify the item as this if you can't classify it as anything else". However, given that categories are "fuzzy", then almost any item could, theoretically, be classed as "Other". For example, in the "psychology" situation above, should the book have been classed as "Other" because theoretically it could be classified in two different ways? Or what about a further sub-classification "educational psychology". Should a book entitled "the psychology of teachers" be put here? Without an "Other" category, the answer will probably be "yes". But with the addition of "Other" suddenly a new option presents itself, and one will discover the other categories suddenly becoming much "tighter" as everything that is not self-evidently to be put in one category will be classified as "Other". The situation gets worse where items can be classed in more than one way. This can lead to the creation of "Bucket" categories, which are, essentially, "Other" categories, but which can be chosen regardless of the situation. For example, a phrase like "safety culture" or "mental models" could function as a bucket category, given that they are not tied to specific observable processes or actions that could be reliably classified (in contrast to "communications" or "rule violation" for example).

Hierarchies

One more point should be made about hierarchies in taxonomies, which is that the arrangement of these hierarchies should be tightly nested. For example, in a biological context, if one divides "living creatures" into "animals" and "plants" then (because there is no "other" category), "animals" and "plants" "add up to" "all living creatures". "Living creatures" is, therefore, a "supercategory" and the other categorisations are "subcategories". The creation of tightly nested taxonomies is greatly facilitated with the elimination of "Other" and "Bucket" categories.

The concept of "tightly nested taxonomies" is taken, as the above example indicates, from biology (in this case, cladistics (Kitching et al., 1998)), but we have found that such taxonomies greatly facilitate the production of acceptable reliability data (as demonstrated in a "reliability" or "consensus" trial).

Data retrieval

This matters because inadequate taxonomies will "skew" the database. For example, a "Bucket" category, or even an "Other" category which gets picked all the time will turn up in analysis (for example, bar charts) as the most picked topic, and will lead to safety reports which claim that "Other" is the category into which most resources should be channelled. Of course, this negates the whole purpose of a classificatory database.

Forms of information

It should be noted that from the point of view of a classifier, all information is simply information to be classified. The specific form of the data is irrelevant. For example, a web page may be primarily textual or primarily numeric or mathematical. From the point of view of classification this "difference" is irrelevant. The same principle applies in non-Web-based databases. It is irrelevant what data is contained in a database: numeric, textual, some mixture of the two, or some other format. It does not matter for the purposes of classification if something is written 1,256 or as one thousand two hundred and fifty six. Quantitative and qualitative are taxonomic differences just like everything else, and are useful (or not). Moreover, in an *electronic* database data can be cut down into fundamental units. For example, to return to our library example: imagine not just references to the books, but the books themselves were stored electronically. It would be highly convenient not just to be able to locate the books themselves, but data within the books (for example, specific chapters, or even words). In an electronic database, the distinction between "book" and "data within the book" is arbitrary. All that matters is that the data can be classified and retrieved easily.

A number of points should be made here about the purpose of databases. The World Wide Web is, in some senses, a database. However, what "search engines" evaluate when organising the Web is *popularity*. For example, if I put the name of an actor into a search engine, I expect to be led to the most popular site which features the actor. Therefore, when I put in exactly the same phrase six months later, the "number one" site may well be completely different.

It should be stressed that this is not the kind of database to be discussed here. In a library, if one puts in "War and Peace" into the library computer on two separate occasions, one expects that the same book will be accessed both times. It is not popularity but positioning in the database that is important here, and a user will assume that this will stay the same.

It should also be stressed that again, to pursue the library analogy, not only will one wish to access the books, but to count them. For example, we will wish to know not just where *War and Peace* is on the library shelves, but how many copies there are, and how the number of copies has changed over time. Are there more or less copies than there were last year? This will influence buying strategies (do books need to be replaced or not?) Clearly for a functioning database, one will want to know how many of each item one has, and the rate of change (if any). To stay with the library example: if one normally loses a book a month in a particular section, this may be acceptable; but if the rate changes so that one month one loses one book, the next one loses two, the next one loses three and so on, this probably indicates a problem. Therefore, we not only want to be able to count elements in our database, but also rates of change in this raw data. Needless to say, this is particularly important in a safety related database.

Reliability or consensus

The efficacy of a database, therefore, can be gauged by the effectiveness by which data can be classified (given that with effective rules of classification, retrieval should then become easy). As much as possible, the same data must be classified in the same "section" of the database by different people. Bias attributed to different coders should be quantified and used to evaluate databases and to improve or refine definitions of categories. Therefore the only way to assess the effectiveness of a database is a test of reliability (or to be more precise, consensus); that is, the way in which data is entered and retrieved.

The criteria by which taxonomies are to be tested or evaluated has been called "inter-judge" (Cohen, 1960), "inter-observer" (Caro et al., 1979) or "inter-rater" (Posner et al., 1990) reliability. We more recently suggested "Inter-Rater Consensus

(IRC)" as a term which avoids confusion as to what is being assessed (Davies et al., 2003). Tests of this criterion have usually been simply called reliability studies (Grove et al., 1981). Trials can also be concerned with *Intra*-Rater Consensus (Robson, 1993); the extent to which each single classifier's coding is consistent on different occasions (this is particularly important where there is only one classifier).

We shall now briefly outline an Inter-Rater Consensus trial on a coding taxonomy. The trial was conducted prior to the codes being accepted for operational purposes. The focus here is on the implications of taxonomic work like this. More detailed reliability data for this trial can be found elsewhere (Wallace and Ross, 2005).

The trial

The taxonomy in question was a series of 53 classificatory codes used by a UK Police force to classify qualitative "road accident event reports". The codes were as follows:

Drivers and riders

1 Under The Influence Of Drink/Drugs
2 Taken Ill (for example, Heart Attack, Epileptic Fit)
3 Fatigue
4 Excessive Speed (In Excess Of Speed Limit)
5 Excessive Speed (With Regard To Prevailing Conditions)
6 Turning Round In The Road Way (U Turn)
7 Driver Inexperience
8 Cutting Into Moving Traffic From A Stationary Position
9 Swerving To Avoid A Hazard In The Roadway
10 Changing Lane Without Ascertaining That The Road Is Clear
11 Overtaking Improperly
12 Emerging Carelessly From A Side Road (Going Straight Ahead)
13 Making A Right Turn Manoeuvre When Unsafe
14 Making A Left Turn When Unsafe
15 Losing Control On A Right Hand Bend
16 Losing Control On A Left Hand Bend
17 Losing Control-Other Manoeuvre
18 Reversing Negligently
19 Failing To Maintain A Safe Distance
20 Pedal Cyclist Leaving The Footway Without Warning
21 Failing To Give Precedence To A Pedestrian
22 Disobeying Automatic Traffic Signal
23 Disobeying Traffic Sign
24 Misjudging Clearance/Distance
25 Dazzled
26 Not Displaying Lights

Pedestrians

27 Crossing Carelessly
28 Disobeying Traffic Signal
29 Crossing While Masked By A Stationary Vehicle
30 Under The Influence Of Drink/Drugs
31 Lying In The Roadway
32 Playing On The Roadway

Passengers

33 Distracting Driver
34 Stealing A Ride
35 Opening A Door While Unsafe
36 Negligence

Vehicle defect

37 Mechanical Fault
38 Tyre Defective
39 Insecure Load
40 Insecure/Defective Trailer

Other causes

41 Wet Weather Conditions
42 Icy/Snowy Weather Conditions
43 Reduced Visibility
44 Greasy Road Surface
45 Icy Road Surface
46 Defective Road Surface
47 Road Works
48 Parked Vehicle Causing Obstruction
49 Defective Traffic Signal
50 Cause Unknown

Flagged items

51 Foreign Driver
52 Stolen Vehicle

Reliability data

Thirty-one previously unclassified event reports from minor road traffic accidents were selected at random and given to six experienced event classifiers. Each coder coded the same set of 31 reports independently, being unable to discuss the reports during the coding process. One single choice of classification from the list above was applied to each of the 31 events, as in the standard operating procedure for the taxonomy in question.

Reliability was calculated by comparing each of the six coders with all others, giving 15 paired comparisons in total (coder 1 with coder 2; coder 1 with coder 3; and so on). Agreement on codes assigned to the 31 events ranged from 25.8 per cent to 61.3 per cent (average agreement 43.2 per cent). The average Kappa coefficient (Cohen, 1960) which corrects for agreement due to chance was .4 (for a discussion of Kappa, see Wallace and Ross, 2006). Agreement was clearly below acceptable levels (for example, Borg and Gall, 1989).

Discussion of trial

Low consensus between coders can be attributed to various issues, for example, differential experience of coders; training in use of the codes; presentation and instruction during the trial. However, the best predictors of failures like this tend to be a lack of logical structure in the taxonomy itself and/or poorly defined classificatory choices. A list of codes like the one above needs to be accompanied by definitions of categories which allow multiple coders to come to the same conclusions. It is obvious that without such a guide, each coder will tend to interpret each event/code in an idiosyncratic fashion, reducing reliable coding.

Taxonomic structure and mutually exclusive codes

The basic question here is whether grouped sub-codes (for example, codes 1-26 in the "Drivers and Riders" category or codes 27-32 in the "Pedestrians" category) are *mutually exclusive* (that is, MEE as discussed above). A basic examination of the codes indicates that there may be problems in this regard. There are many different dimensions to the codes. For example, the experience of the driver (code 7) and the specific driving manoeuvre undertaken (for example, code 13) are *not exclusive*. So, in the case of an inexperienced driver making an unsafe manoeuvre, a choice of codes would apply. Code 27 (Pedestrian Crossing Carelessly) and 28 (Pedestrian Disobeying Traffic Signal) would also seem to overlap, as the latter is an example of the former. Similarly, code 11 (Overtaking Improperly) and code 15 (Losing Control On A Right Hand Bend) would presumably both apply if a driver overtook on a right hand bend and lost control. It is not hard to imagine a situation where "Parked Vehicle Causing Obstruction" (Code 48) might coincide with a driver "Swerving To Avoid A Hazard In The Roadway" (Code 9) or "Changing Lane Without Ascertaining That

The Road Is Clear" (Code 10). And we hope the overlap between codes 42 and 45 is obvious.

There are many examples like this in the coding structure which might lead to unreliable categorisation unless definitions are included to make codes mutually exclusive. For example, though the taxonomy under discussion is probably beyond repair, reliable use of code 15 would require it to be defined as "Losing Control On A Right Hand Bend" which *did not involve* overtaking, or driver inexperience, or any of the other possibilities listed.

Exhaustive groups of codes

A second principle of taxonomic arrangement is that codes should be *exhaustive* for the purposes for which they are designed, leading to the general term MEE.

Mutually exclusive codes are best arranged in hierarchical groupings so that they "add up" to a super-category or "higher level" code. This code in turn is broken down into the sub-codes. Within each "box" codes should be exhaustive – that is there should be no other logical choice. For example, an implicit hierarchy in the taxonomy would cover codes 15-17 under a super-category "Losing Control". Notwithstanding problems with how vague this aspect is (it could apply to overtaking, reversing, speeding, changing lane or any number of activities which there are other coding choices for), we can see how the first two codes do not appear to be exhaustive of "Losing Control", indicated by use of a third "Other" category. Ideally, taxonomies can be designed so that groups of sub-codes cover all aspects of higher level codes without resorting to such a choice (Wallace and Ross, 2006). The problem is that the existence of "Other" codes usually leads to them being used! This defeats the very purpose of a taxonomy – to discriminate between events on the basis of distinctions which are of theoretical or practical interest.

A problem with exhaustivity here is that the supposed "higher level" code "Drivers and Riders" is too large and contains too many confounded choices. Tighter hierarchies with fewer choices should be made explicit. Another example would be the apparent super-category "Road Surface" covering codes 44-46. If this was included in a logical arrangement then we could create mutually exclusive categories and definitions that would help coding. Moreover, *even if disagreement did occur at the sub-category level* we would still have more information, because we could see that disagreement occurred at the relatively trivial level of: "what specifically was wrong with the road"? We would still know this was a "Road Surface" problem.

To reiterate: we must be careful here not to fall into the trap of stating that there is a "true taxonomy". Perhaps we might decide that "weather conditions" was the super-category and that sub-categories should be created on this basis. However, the point is that rigorous MEE hierarchies must be created for consensus to be achieved.

Implications

These are not trivial issues. It is one of the axioms of Taxonomy Theory that a consensus/reliability trial must take place on any taxonomy before it can be considered effective, and yet this is rarely done. And so data from established taxonomies is used by governments to feed policy, and yet the meaning and accuracy of this data must be considered unproven until effective reliability data is available. We have published elsewhere details on our own taxonomies which have achieved adequate data (Wallace, Ross and Davies, 2002, 2003). It should be noted that Taxonomic Theory has implications far beyond safety. We would regard a reliability trial as essential to demonstrate the efficiency of *any* taxonomy (see, for example, Ross, Wallace and Davies, 2004).

There are three further points to be stressed here. The first is that we are highlighting the taxonomic nature of classification to a much greater extent than is normally done in the literature, in which the *methodology of the initial data collection* is normally considered to be of more importance. So, for example, specifics of data collection and output are frequently considered to be the differentiating factors in terms of the data analysis methodology: that is, whether the data to be stored and classified is "quantitative" or "qualitative", of "minor" or "major" events and so on. We on the contrary are less interested in this because we are looking at the categorisation process itself. From this point of view, it is less relevant where the data "originated"; instead, it is to be considered, primarily *as data*: that is, as data to be classified, analysed, and then produced in various formats. Therefore the distinction (which is in itself a taxonomic distinction) between "reliable" and "unreliable" data is more important (Ross, Davies and Plunkett, 2005).

If this is accepted, it brings us to the next point, which is that we would suggest that once a reliable taxonomy has been established, therefore, one can act on the assumption that any data input must be tailored to feed the system (that is, adapting the data to the taxonomy rather than the other way round). This has the huge advantage of creating a criterion for the relevance or not of data produced in any investigation, interview (or whatever). So, for example, in terms of SECAS (Wallace, Ross and Davies, 2002) we created an accident investigation form that was tailored to answer the specific questions that would be asked in terms of applying appropriate "codes" (categories) to the data (which was, in this case, qualitative). We feel that this is a better way of dealing with the problems in current accident investigation techniques described by Benner (1981).

Finally there is a point that has been made before but which is worth stressing again. It is almost universally assumed in the literature that one, "off the shelf" taxonomy of a given area or situation can apply to all cases of this situation. Now, as we have argued, this might be the case, but if it is, this should be decided only at the level of pragmatics or utility. There is no logical reason why this should be the case. It is interesting in this regard, therefore, that in our own work we have found that, pragmatically, it is frequently not the case. That is, for the major projects we have been involved with, we have found that different taxonomies are required for different situations, and that only by developing "situation specific" taxonomies can the required reliability be produced. Doubtless this finding also has implications for the required locus of research. If one concentrates on the individual then perhaps one can argue that a single model of "human error" might describe all (or nearly all) such events, and that a taxonomic system can be inferred from this model. If, on the other hand, one views humans as being fundamentally situated, then this would imply that actions and behaviours will always be influenced by the situation, and that, therefore, a "universal taxonomy" is not desirable (regardless of whether it is possible) (Wallace and Ross, 2006).

Conclusion

This chapter has discussed Taxonomy Theory, which demonstrates how taxonomies are created, and how their efficiency can be demonstrated. We have argued that taxonomies are project-specific, and we should add that in our own experience it is far easier to build and use them if the staff who have to use the taxonomy are involved at an early stage.

We have argued that taxonomies should be hierarchical and "nested" as well as MEE. In terms of presentation, if codes are mutually exclusive, this makes them easier to present, as coders can be led down a "logic tree" with two or three mutually exclusive choices to make. Moreover, tight definitions should be provided of all codes.

We have also discussed the fact that in our own experience, we have found "situation specific" taxonomies tailored to specific industries (developed in collaboration with staff in these industries) to be more effective in producing the requisite reliability than pre-designed "off the shelf" taxonomies. Finally, we would argue that it is only with an effective reliability/consensus trial, similar to the one described here, that taxonomies can be demonstrated to be effective.

Acknowledgements

We would like to thank Megan Sudbery, Deborah Grossman and Rachel Herriot, for invaluable help in developing the CIRAS taxonomy (and for helping with the hard work of classifying!) We would also like to thank Sally Cathcart for help with the trial discussed in this chapter. Finally, we would like to thank Professor John Davies, for general support and help.

References

Benner, L. (1981) "Methodological biases which undermine accident investigations". In *Proceedings of the International Society of Air Safety Investigators 1981 International Symposium*. Washington, D.C., September 1981.

Borg W. and Gall, M. (1989). *Educational research.* London: Longman.

Bowker, G. and Star, S. (1999). *Sorting things out: Classification and its consequences.* London: MIT Press.

Caro, T. M., Roper, R., Young, M. and Dank, G. R. (1979). "Inter-observer reliability". *Behaviour*, 69(3-4), 303- 315.

Cohen, J. (1960). "A coefficient of agreement for nominal scales". *Educational and Psychological Measurement*, 20, 37-46.

Davies, J., Ross, A., Wallace, B. and Wright, L. (2003). *Safety management: A qualitative systems approach.* London: Taylor and Francis.

Grove, W. M., Andreasen, N. C., McDonald-Scott, P., Keller, M. B. and Shapiro, R. W. (1981). "Reliability studies of psychiatric diagnosis". *Archives of General Psychiatry*, 38, 408-413.

Kitching, I. J., Forey, P. L., Humphries, C. J. and Williams, D. M. (1998). *Cladistics.* Oxford: Oxford University Press.

Kosko, B. (1994). *Fuzzy thinking.* London: Flamingo.

Lakoff, G. (1990). *Women, fire and dangerous things.* Chicago: University of Chicago Press.

Posner, K. L., Sampson, P. D., Caplan, R. A., Ward, R. J. and Cheney, F. W. (1990). "Measuring interrater reliability among multiple raters: An example of methods for nominal data". *Statistics in Medicine*, 9, 1103-1115.

Reardon, D. (1998). "Knowledge Management: the discipline for information and library science professionals". In *Proceedings of the 64th IFLA General Conference, Amsterdam, August 16th to August 21st.* Retrieved 9 February 2004 from http://www.ifla.org.sg/IV/ifla64/017-123e.htm.

Robson, C. (1993). *Real world research.* Oxford: Blackwell.

Ross, A. J., Davies, J. B. and Clarke, P. (2004). "Attributing to positive and negative sporting outcomes: A structural analysis". *Athletic Insight*, 6, 3.

Ross, A. J., Davies, J. B. and Plunkett, M. (2005). "Reliable qualitative data for safety and risk management". *Process Safety and Environmental Protection*, 83, 3, 117-121.

Ross, A. J., Wallace, B. and Davies, J. B. (2004). "Measurement issues in taxonomic reliability". *Safety Science*, 42, 8, 771-778.

Ryle, G. (1938). "Categories". Retrieved 9 February 2004 from http://www.philosophy.ru/library/ryle/categ.html.

Wallace, B. and Ross, A. J. (2005). "Effective taxonomies for database management". *Municipal Engineer*, 158(4), 253-257.

Wallace, B. and Ross, A. J. (2006). *Beyond human error: Taxonomies and safety science.* Florida: CRC Press.

Wallace, B., Ross, A. and Davies, J. (2002). "The creation of a new minor event coding system". *Cognition, Technology and Work*, 4(1), 1-8.

Wallace, B., Ross, A. and Davies, J. (2003). "Applied hermeneutics and qualitative safety data: The CIRAS project". *Human Relations*, 56(5), 587-608.

Chapter 34

Using Signal Detection Theory to Measure Situation Awareness: The Technique, the Tool (QUASA^{TM}), the Test, the Way Forward

Graham Edgar and Helen Edgar

Introduction

Given the widespread use of the term "situation awareness" (SA) it is surprisingly difficult to define SA. Reviews of definitions of SA (for example, Dominguez, 1994 cited in Banbury and Tremblay, 2004) illustrate that there are many differing views of what the concept of SA represents. A review by Breton and Rousseau (2001, cited in Banbury and Tremblay, 2004) classified some 26 definitions of SA and found that they were divided into two main categories, treating SA as either a *state* or a *process*.

Perhaps the best known definition that treats SA as a state of knowledge, is that of Endsley (1995) which defines SA as, "the perception of elements in the environment within a volume of time and space, the comprehension of their meaning, and the projection of their status in the near future". Considering SA as a process, Sarter and Woods (1995) proposed a definition which suggests that situation awareness covers a variety of cognitive activities (perception, attention, and so on) that are crucial to maintaining awareness in a complex environment. Endsley (2000) provides a useful review of a range of cognitive processes that may be important in the acquisition of SA. One thing that becomes clear is that not all potentially available information is used in building SA. For instance, Endsley and Smith (1996) found that fighter pilots tended to direct attention to stimuli that they perceived as particularly important to the task. Endsley and Rodgers (1998), in a study of air traffic controllers, found that as workload increased attention to less important information was reduced. This suggests that a proportion of available information may be gated out by attentional processes and not used in building up SA.

Evidence suggests that SA is not built up using only information immediately available from the environment. An inherent cognitive bias may lead individuals to be influenced by what they expect to happen. Taylor, Endsley and Henderson (1996) found that individuals that were given a set of expectations were more likely to make errors if the situation did not develop as expected. Errors of this kind may underpin serious errors of judgement, for example, friendly fire (Edgar, Edgar and Curry,

2003) or, in driving, "looked but failed to see" accidents (Sabey and Staughton, 1975). Furthermore, individuals may choose not to use all information that is available to them. Kahneman and Tversky (1973) found that people often did not use all the information given to them in short vignettes when making judgements. It might be considered that using all available information would be the best strategy in building SA but this assumes, of course, both that all the available information is true and that it is possible to assimilate *all* the information. It is entirely possible that some of the information presented to, or held by, the individual could be false. Obviously, it seems unlikely that anyone would knowingly hold false information, but there are a number of ways in which false information could be available. There could (especially in conflict situations) be a deliberate effort to provide the individual with false information, the information may be true but be misinterpreted by the individual, or the information may have been true at one time but the situation has changed.

Thus, SA can be considered to be the building up of a mental model of a situation by the selection of information. The selection of information can occur early in the system (using attention) or later (by choosing which available information to use or discard). Most current techniques for measuring SA tend to measure the "endpoint" of this process, the state of knowledge of the individual compared with the actual situation they are in. Poor SA, however, could arise from either having a limited amount of information available or inappropriate selection from a larger body of information. It would be useful therefore to be able to assess not only the SA state, but also the process used to attain that state. The technique described in this and previous papers (Edgar, Smith, Stone, Beetham and Pritchard, 2000: Edgar, Edgar and Curry, 2003) is designed to assess both the state of SA and provide insights into the process by which that SA is acquired. This technique has been embodied in a tool referred to as QUASA™ (QUantitative Analysis of Situation Awareness). This approach was developed, and named, by the authors of this chapter while working for BAE Systems and is now being used by that company and its current employees (McGuinness, 2004).

Theory underlying the QUASA™ tool

The starting point for the QUASA™ tool is the notion that information pertinent to a situation has to be internalised if it is to be used effectively in decision making. It is also assumed that the storage of this information is *not* "all or nothing". This notion seems plausible as there is a large body of research that considers that there are different states of awareness associated with memory retrieval. Specifically, it has been suggested that there is distinction between "remembering" and "knowing" (Tulving, 1985). It has also been suggested that remembering and knowing may reflect different levels of confidence concerning the accuracy of the recall (Inoue and Bellezza, 1998). This confidence could be based on the putative strength of the underlying memory trace. Some information may be very strongly represented (you may be looking directly at something providing you with the information at the time) or the representation may be much weaker (it may be a representation of information gathered days or years ago). A theoretical (and arbitrary) distribution of the representation strengths for a number of true items of information is shown by the solid curve in Figure 34.1.

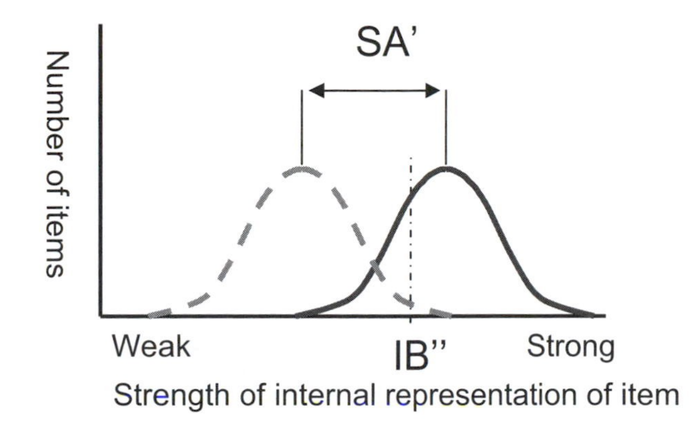

Figure 34.1 Theoretical distributions of the internal representation strengths of true (solid curve) and false (broken curve) items of information

A crucial aspect of the theory underlying the QUASA™ tool is the assumption that not only *true* information may be stored: *false* information may also be stored. It is unlikely that anybody would choose to store false information about a situation deliberately, but there are a number of ways in which false information might be present. For instance, the information given to the individual may have been false, the information presented may have been true but the individual misinterpreted it, or the information *was* true but the situation has now changed to the extent that the information is now false. This false information is *also* likely to have a range of representation strengths and a possible distribution is illustrated by the broken curve in Figure 34.1.

Obviously, the true and false information is not conveniently labelled for the individual and both may be used in building up the individual's own view of the situation. Thus, a basic premise of the QUASA™ tool is that good SA is represented by *the ability to tell true and false information apart*. So, good SA would be represented by a wide separation of the curves in Figure 34.1 (true and false information having widely different representation strengths); poor SA would be represented in the extreme by the curves overlying one another. Given, however, that the representation strengths of true and false information may well overlap (as shown in Figure 34.1) the individual also has to establish a criterion representation strength above which information may be accepted as true and below which the information may be rejected as false. This criterion level is indicated by the dashed line labelled "IB" in Figure 34.1. If the criterion is set "high" (the line moves to the right) the individual is setting a very cautious criterion – only willing to accept information that is strongly represented (and that they perhaps feel more certain of). If the individual is sufficiently cautious, it is possible that *no* false information will be accepted – but some true information may also be rejected. On the other hand, if the criterion is set low (the line moves to the left) then the individual is willing to accept more information as true – some of which may *actually* be false information. The criterion can, of course, be set independently of the separation of the underlying curves and could theoretically change from moment to moment. A strength of the QUASA™ tool is that it makes few assumptions about the shapes of the underlying distributions of the strengths of true and false information, or even that true information should have a generally stronger representation than false.

Applying the QUASA™ tool

The approach that the QUASA™ tool takes to measuring SA is to measure an individual's ability to tell true information (drawn from the situation of interest) from plausible false information. This is done by presenting the individual with a series of probe statements such as, "There is an enemy unit at position x". Some of the statements are true and some of them are false. The individual's task is to respond as to whether they *believe* the probe statement is true or false. The QUASA™ tool uses signal detection theory (for a comprehensive discussion of the area of signal detection theory, see Green and Swets, 1966) to give a measure of how well the individual can tell true from false information and *also* an indication of the individuals bias, that is, how biased they are towards believing information is true or false. There are a number of different measures that have been developed within signal detection theory. The best known is d', which is an *estimate* of the distance between the means of the two distributions (true and false information). This measure is, however, problematic if the variances of the two distributions are not equal (and there is no compelling reason to assume that they are when dealing with true and false information to build SA) and in this case a "nonparametric" measure may be more appropriate. One that is widely used in memory research is A' (Grier, 1971) and this has been demonstrated to be a better estimate of sensitivity *if* the variances are unequal (Donaldson, 1993). A' is the area under the isosensitivity (receiver or relative operating characteristic – ROC) curve (Luce, 1963), but it should be pointed out that A' is not truly nonparametric and, like d', is only an estimate of sensitivity. The data reported here were calculated using A' as a measure of sensitivity as it appears to be a generally more robust measure of sensitivity than d', although the QUASA™ tool, as used by the authors, does calculate both A' and d'. As Donaldson (1993) pointed out, both A' and d' are only estimates of sensitivity and therefore some uncertainty remains. This uncertainty can be reduced by obtaining ratings of how *confident* individuals are that the item is true or false. The collection of confidence ratings as a part of the process of applying signal detection theory is not a new one. As Macmillan and Creelman (1991) pointed out, "the user of detection theory who does not collect ratings is at risk". From a psychological point of view, collecting confidence ratings from individuals also allows an estimate of how well "calibrated" they are (for example, Lichtenstein and Fischhoff, 1977; Lichtenstein, Fischoff and Phillips, 1982); that is, how well they know what they know.

Collecting confidence ratings slows the administration of the test down, but does allow a better estimate of sensitivity. Confidence ratings *were* collected in some of the trials reported here, but these data are not discussed here. However, the output from the QUASA™ tool will be considered in more detail.

Testing the QUASA™ tool

The QUASA™ tool was tested in a simulated command and control environment. The BAE Systems terrain model facility was used (see Figure 34.2). This was a 1/300 scale model of a section of Germany 10x2.5km ground scale – just north-east of Hildesheim (near Hanover). Navigation features (churches, factories, unusual buildings, and so on) were all modelled individually. All buildings were correctly positioned (to the level of garden sheds) and contours were accurately reproduced. It was painted to simulate (when filmed in black and white and inverted) a thermal return representative of a late afternoon in September.

Figure 34.2 Testing the QUASA™ tool

A war game was conducted. Blue forces had four commands A, B, C and D which were involved in an assault scenario. A, B and C were armour-heavy and tended to lead the assault. D was an infantry-heavy reserve. Four commanders controlled the blue forces and it was to these commanders that the QUASA™ probes were presented. Red forces were controlled by umpires, according to certain pre-determined characteristics for red commanders. All communication between commanders was by written message (therefore there was a complete record of all communication). All red units were hidden unless blue had "eyes on" (that is, a blue unit on the ground was judged able to "see" the red unit). Spotting of red units was probabilistic with a random element used to ensure that there was some uncertainty in the spotting. Some sightings could not initially be identified (players were only told there was "something" there) and might later resolve into tractors, cows, and so on. SA probes were presented as situation reports to headquarters. All players appeared comfortable with answering the SA probes in this way.

The probe statements were designed to correspond to Endsley's (1995) classification of the components of SA (perception, comprehension and projection) so, for instance, perception was probed using statements of the form, "A has crossed the river at point x". Integration (used in preference to Endsley's comprehension term as it is felt to better represent the possibility of integrating information to give a *false* picture) was probed by statements of the form, "The strongest enemy force is present in area *a*" (which could only be answered by working out whether this was likely from losses to friendly forces – the enemy were not physically visible in that location). Projection was probed using statements of the form, "At time *x* you will reach point *y*". Obviously, it is not known at the time whether these projection statements are true or false – but they can be scored *post hoc*. The breakdown of the SA probes into these components has been carried out to conform with Endsley's classification – but is essentially arbitrary (the borderlines between the different categories are not always distinct). Other classifications can also be used, for example, relevant information vs. non-relevant information.

The situation being assessed can be easily appreciated by inspection of the probes used. This is a strength of the QUASA™ approach. It is therefore essential to consider carefully what aspects of the situation are relevant, and should be tested, when designing the probes. Approximately five probe statements were presented in each turn of the war game, with a minimum of 20 probes required to evaluate any aspect of SA.

The simulation was designed to provide the components of a command and control task in a controlled and replicable environment. Variants of the QUASA™ tool have, however, also been successfully applied in more "realistic" simulations of command and control both by the authors of this paper and also by other workers that have adopted the general approach (for example, McGuinness, 2004).

Output of the QUASA™ tool

Components of SA

The scenario was run a total of seven times. The data from one such run are shown in Figure 34.3. As discussed, it is possible to tailor the probes to address specific aspects of SA, namely perception, integration and projection. The data for two of the commanders (A and D) have been broken down into these components.

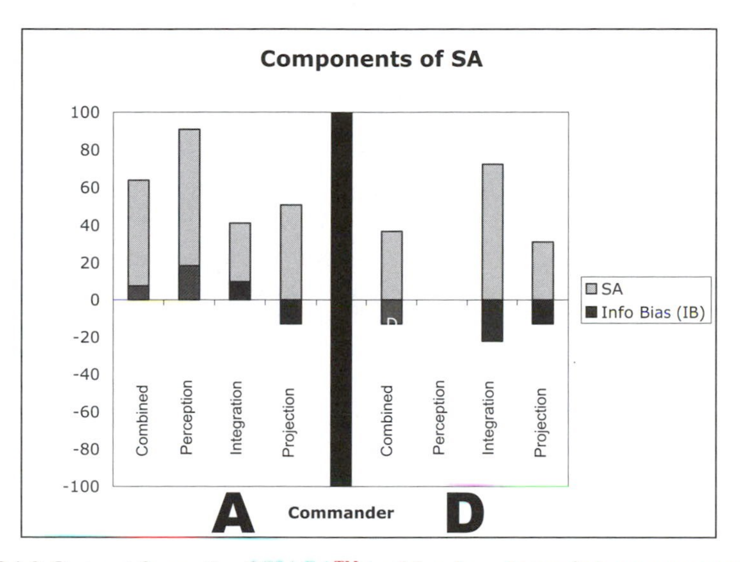

Figure 34.3 Output from the QUASA™ tool broken down into components of SA

The shaded bars give an indication of how well the individual is able to tell true from false information. A score of +100 indicates that the person was always able to tell true from false information; a score of zero means that they were unable to tell true and false information apart. Note, it *is* possible for the SA score (SA') to go negative, and this will be considered later. The cross-hatched bars give a measure of what has been termed "information bias" (IB") and this indicates whether the individual has any bias towards answering "true" or "false". If the IB" score is positive it suggests that the individual is tending to answer "false" more than is appropriate, with the implication that they are rejecting information that is true. If the IB" score is negative it suggests a bias towards answering "true" more than is appropriate, with the implication in this case that the individual is accepting information that is false.

An interesting point to note from these data is that D appears to have no "perception" component to SA. Taken in tandem with the nature of the probes used, this suggests that s/he had no awareness of where individual friendly and enemy units were. It is also interesting in that it suggests it is possible to build up awareness at the level of integration – without building up awareness at the level of perception. This suggests that the structure of SA need *not* be hierarchical, with perceptual information not necessarily being used to build up the integrated "big picture". This leads to the possibility that at least some aspects of SA may be built using "top down" information from an individual's knowledge, experience and expectations and there is support for this notion (Taylor, Endsley and Henderson, 1996; Endsley, 2000). The possibility that individuals may build up their SA based not on what they perceive, but on what they know (or believe they know), has important implications for understanding how certain situations may be misinterpreted. Individuals almost certainly do not go into a situation without some prior knowledge that they will use to build SA – and which may well affect how they perceive the situation. To use a saying that has often been attributed to Immanuel Kant, "*We see things not as they are, but as we are*."

Dynamic SA

It is also possible to use the QUASA™ tool to look at how SA might change over time. This was done by running a five-turn "window" over the data; analysing the data for turns 0-4, 1-5, 2-6, and so on. Individual data points have to be treated cautiously as, in this experiment, the number of data points in each "window" is getting rather low – but it still appears to be effective in giving an impression of the change in SA. The dynamic data for D from the same run as before are shown in Figure 34.4. All the components of SA have once again been combined, due to insufficient data to plot each component separately.

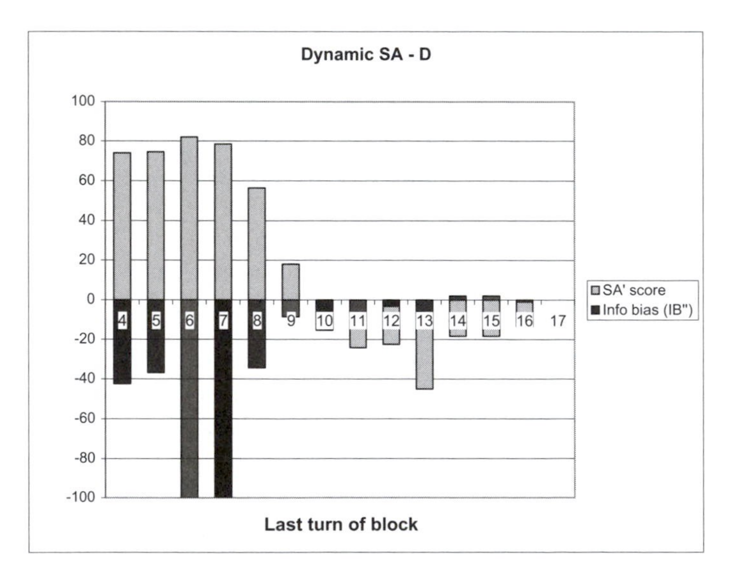

Figure 34.4 Dynamic SA for commander D

An interesting aspect of these data is that the QUASA™ SA' score goes *negative* for this commander. In terms of the representation of information, this suggests that false information is actually *more* strongly represented than true. The most reasonable explanation for this result is that this individual has an awareness of the *wrong* situation and is more likely to believe (and base decisions on) false information as opposed to true. An awareness of the wrong situation, taken together with this individual's lack of perceptual awareness (of where individual units are) suggests that this individual may make mistakes – the most catastrophic of which is likely to be "friendly fire" on his own forces. This individual was very nearly involved in a "blue-on-blue" incident in this trial, and was only prevented from firing on his own forces by the reaction of the umpires.

The way forward

Information bias

The information bias score for this individual is also of interest (darker bars in Figure 34.4). Note that the information bias goes strongly negative just before, or possibly coincident with, a drop in SA. This result was found consistently in the majority of participants in these studies. This raises the possibility that changes in bias may be associated with losses in SA. This is an important finding as measuring actual SA in "real time" is likely to be difficult, but it is theoretically possible to measure shifts in bias. Thus sudden bias shifts could be monitored and used to warn of possible difficulties with SA. The theoretical possibility of using bias shifts to predict losses of SA is currently under investigation by the authors.

Team and shared SA

An important application of this technique is assessment of team and/or shared SA. For instance, it is entirely possible to apply QUASA™ to different members of a team and see to what extent they are aware of the same information, or information from different sources. Also, one logical extension of the tool is to use it to assess an individual's knowledge of what other people in the team know. This technique was successfully applied in some of the studies described here. Participants were presented with the true/false statements and were *also* asked who else in their team they believed would get the answer correct. Signal detection theory was then applied to these answers, comparing the responses of the individuals with the actual responses of other members of the team. An example of the data is shown in Figure 34.5. The format is similar to that illustrated in Figures 34.3 and 34.4 except that the meaning of the graph is quite different. The data for Participant A are their actual SA and bias scores (as before). The data for B, C and D, however, indicate their *knowledge* of how much A knows.

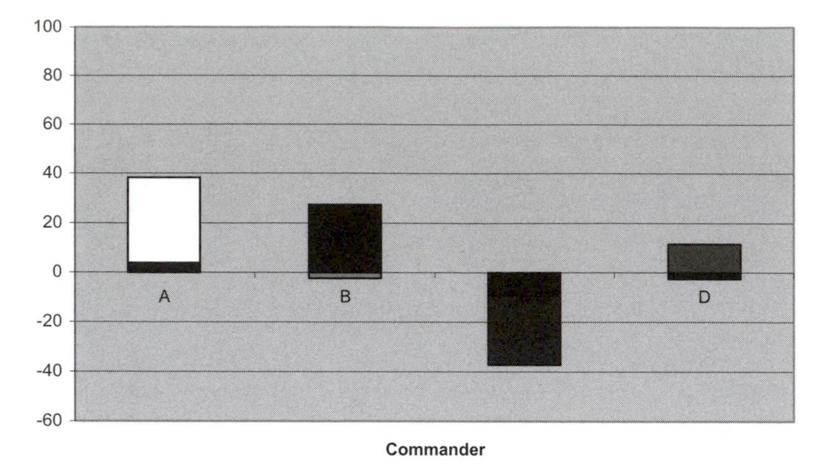

Figure 34.5 An illustration of what other members of the team believe A's level of knowledge to be, and their attitude towards that player

The dark shaded area (players B, C and D) indicates judgements of A's knowledge (a score of +100 would indicate that they were always able to judge correctly what A knows and does not know). Thus, in Figure 34.5, B has a *limited* knowledge of how much A knows, whereas C is completely *wrong* in what they believe A knows. The bias score for B, C and D (light shading) now indicates how "good" they think A is in terms of how much they know. A negative score indicates that the individual believes that A knows more than they do, a positive score that they know less.

This is an important measure in that it gives an indication of an individual's attitude to other members of the team. For instance, if they do not trust other members of the team they may show a strongly positive bias score – indicating that they believe that the person is unlikely to get the answers right! The bias scores in these trials were all relatively low. This may be an indication of the fact that the teams were formed *ad hoc* and the team members had no particular reason for believing the other team members would be either particularly knowledgeable or not. In teams where the members are familiar with each other it is possible that the bias scores may be quite different.

Meta-knowledge

The process described above gives a good indication of how aware each member of the team is of the knowledge of the other team members – and their attitude to those team members. It is also, of course, equally possible to apply the same technique to measure an individual's awareness of their *own* state of knowledge. To do this, the true/false probes are used as usual, but the individual is also asked how confident they are that *their own* answer is correct. This can either be done using a rating scale or a binary confident/not confident response. Both methods have advantages and disadvantages. In these trials, a combined technique was used, asking for confidence ratings on a four-point scale that can then be easily split to give a confident/not confident response that can be analysed using signal detection theory, based on the notion of a Type 2 ROC curve (Clarke, Birdsall, and Tanner, 1959). Although the use of a four-point scale is an efficient method of gathering data, paradoxically, it may not be quite as sensitive as asking for binary judgements of confidence (Tunney and Shanks, 2003). By using a combined technique, data were gathered in these studies that are appropriate for Type 2 analysis and the results of this analysis will be presented separately.

Conclusions

This chapter presents an original approach to measuring situation awareness that was invented and developed by the authors of this chapter. This approach uses the well-established technique of signal detection theory to assess many aspects of SA, both within individuals and teams, and across a wide range of situations. There are, however, many others applications of this approach currently being developed by the authors and also other workers in the general area of situation awareness.

Acknowledgements

The name "QUASA" is the trademark property of BAE Systems plc, UK. The underlying tool is openly available from the authors (gedgar@glos.ac.uk). The data reported in this paper were collected while Graham Edgar and Helen Edgar were employees of BAE Systems plc, UK. The authors are grateful to Andrew J. Smith, D. Lee Beetham and Ceri Pritchard for their invaluable assistance in designing and running the wargame described in this chapter, and for helpful discussion throughout. Some of the data discussed in this chapter have been presented previously (Edgar et al., 2000; Edgar, Edgar and Curry, 2003).

References

Banbury, S. and Tremblay, S. (eds). (2004). *A cognitive approach to situation awareness: Theory and application*. Aldershot, England: Ashgate Publishing Ltd.

Breton, R. and Rousseau, R. (2001). *Situation awareness: A review of the concept and its measurement* (Technical Report No. 2001-220). Defence Research and Development Canada, Valcartier.

Clarke, F. R., Birdsall, T. G. and Tanner, W. P. (1959). "Two types of ROC curves and definitions of parameters". *Journal of Acoustical Society of America*, 31, 629-630.

Comstock, J. R. and Arnegard, R. J. (1992). "The multi-attribute task battery for human operator workload and strategic behavior research". NASA Technical Memorandum No. 104174.

Dominguez, C. (1994). "Can SA be defined?" In M. Vidulich, C. Dominguez, E. Vogel and G. McMillan (eds), *Situation awareness: Papers and annotated bibliography* (pp. 5-15). Brooks Air Force Base: United States Air Force Armstrong Laboratory.

Donaldson, W. (1993). Accuracy of d' and A' as estimates of sensitivity. *Bulletin of the Psychonomic Society*, 31, 271-274.

Edgar, G. K., Smith, A. J., Stone, H. E., Beetham, D. L. and Pritchard, C. (2000). "QUASA: QUantifying and Analysing Situational Awareness". Paper presented at the *IMCD People in Digitized Command and Control Symposium*, RMCS Shrivenham, UK.

Edgar, G. K., Edgar, H. E. and Curry, M. B. (2003). "Using signal detection theory to measure situation awareness in command and control". In *Proceedings of the Human Factors and Ergonomics Society 47th Annual Meeting*. Denver, Colorado, pp. 2019-2023.

Endsley, M. R. (1995). "Toward a theory of situation awareness in dynamic systems". *Human Factors*, 37(1), 32-64.

Endsley, M. R. (2000). "Theoretical underpinnings of situation awareness: A critical review". In M. R. Endsley and D. J. Garland (eds), *Situation Awareness analysis and measurement* (pp. 3-32). Mahwah, NJ: Lawrence Erlbaum Associates.

Endsley, M. R. and Rodgers, M. D. (1998). "Distribution of attention, situation awareness, and workload in a passive air traffic control task: implications for operational errors and automation". *Air Traffic Control Quarterly*, 6(1), 21-44.

Endsley, M. R. and Smith, R. P. (1996). "Attention distribution and decision making in tactical air combat". *Human Factors*, 38(2), 232-249.

Green, D. M. and Swets, J. A. (1966). *Signal detection theory and psychophysics*. New York: Wiley.

Green, M. (1999). *Measuring situation awareness with the "ideal observer"*, [Internet. Available: http://www.ergogero.com/sitaw/sitaware.hTMl9/99].

Grier, J. B. (1971). "Nonparametric indexes for sensitivity and bias: Computing formulas". *Psychological Bulletin*, 75, 424-429.

Inoue, C. and Bellazza, F. S. (1998). "The detection model of recognition using know and remember judgements". *Memory and Cognition*, 26, 299-308.

Kahneman, D. and Tversky, A. (1973). "On the psychology of prediction". *Psychological Review*, 80, 237-251.

Kanis, H. (2000). "Questioning validity in the area of ergonomics/human factors". *Ergonomics*, 43(12), 1947-1965.

Kuchar, J. K. and Yang, L. C. (1997). "Incorporation of uncertain intent information in conflict detection and resolution". Paper presented at the *36th IEEE Conference on Decision and Control*. San Diego, CA.

Lichtenstein, S. and Fischhoff, B. (1977). "Do those who know more also know more about how much they know?" *Organizational Behavior & Human Performance*, 20, 159-183.

Lichtenstein, S. and Fischhoff, B. and Phillips, L. D. (1982). "Calibration of probablities: the state of the art to 1980". In D. Kahneman, P. Slovic & A. Tversky (eds) *Judgement under uncertainty: heuristics and biases*. Cambridge: Cambridge University Press.

Luce, R. D. (1963). "Detection and recognition". In R. D. Luce, R. R. Bush and E. Galanter (eds), *Handbook of mathematical psychology*. New York: Wiley.

MacMillan, N. A. and Creelman, C. D. (1991). *Detection theory: A user's guide*. Cambridge: Cambridge University Press.

McGuinness, B. (2004). "Quantitative analysis of situational awareness (QUASA): Applying signal detection theory to true/false probes and self-ratings". Paper presented at the *2004 Command and Control Research & Technology Symposium (CCRTS)*. San Diego, CA.

Pritchett, A. R., Hansman, R. J. and Johnson, E. N. (1996). "Use of testable responses for performance-based measurements of situation awareness". Paper presented at the *International Conference on Experimental Analysis and Measurement of Situation Awareness*. Daytona Beach, Florida.

Sabey, B. and Staughton, G. C. (1975). "Interacting roles of road environment, vehicle and road user". Paper presented at the *5th International Conference of the International Association for Accident Traffic Medicine*. London.

Sarter, N. B. and Woods, D. D. (1995). "How in the world did we ever get in that mode? Mode awareness in supervisory control". *Human Factors*, 37, 5-19.

Taylor, R. M., Endsley, M. R. and Henderson, S. (1996). "Situational awareness workshop report". In B. J. Hayward and A. R. Lowe (eds), *Applied Aviation Psychology: Achievement, Change and Challenge* (pp. 447-454). Aldershot, England: Ashgate Publishing Ltd.

Tulving, E. (1985). "Memory and consciousness". *Canadian Psychology*, 26, 1-12.

Tunny, R. J. and Shanks, D. R. (2003). "Subjective measures of awareness and implicit cognition". *Memory and Cognition*, 31(7), 1060-1071.

PART 7
A Final Comment

Chapter 35

Intelligence, Uncertainty, Interpretations and Prediction Failure

Malcolm Cook, Corinne Adams and Carol Angus

"'A hidden limitation of intelligence is its inability to transform a mystery into secret,' wrote Lord Butler in his report into the intelligence on Iraq's weapons of mass destruction. The enemy's order of battle may not be known, but it is knowable. The enemy's intentions may not be known, but they are knowable. But mysteries are essentially unknowable." Henry Porter, *The Guardian*, 14 May 2006.

"Reports that say that something hasn't happened are always interesting to me, because as we know, there are known knowns; there are things we know we know. We also know there are known unknowns; that is to say we know there are some things we do not know. But there are also unknown unknowns – the ones we don't know we don't know." Donald Rumsfeld quoted on BBC website, 2 December 2003.

Introduction

The analysis of the bombings in London on 7 July 2005 has suggested that there was a failure to make use of intelligence that, with hindsight, seems to indicate clearly the involvement of those actively involved in terrorist activities. It has been suggested that the indications were strong enough but operational difficulties prevented the relevant organisations prioritising the relevant individuals for surveillance or pre-emptive arrest under the legislation available to the police at that time. The pattern of activities undertaken by al-Quaeda and affiliated organisations following the 11 September attacks on New York and Washington had been predictable and unremarkable. Retrospective analysis of the events leading up to 11 September indicate that significant clues existed from a number of sources, including the al-Quaeda group, that identified an airborne threat to a limited number of US mainland targets (Hawthorne, 2002; Posner, 2003). This type of evidence makes the failure to pre-empt their attacks on London and Madrid all the more surprising because history is replete with examples of providing obvious clues as to future events from the past events. The British attacks on the Italian Fleet in Pearl Harbour were a major stimulus to the Japanese in their attack on the harbour, and the British use of radar in the Battle of Britain should have provided the US forces at Pearl Harbour with confidence in the new technology to locate large numbers of aircraft. It is interesting to note that after Pearl Harbour the signs were so obvious that conspiracy theorists considered it possible that the proverbial back door had been left open (Keegan, 2003). These conspiracy theories are broadly similar to those associated with the 11 September attacks, where it seems clear that intercepts would have been received by security services (Fouda and Fielding, 2003). However, even although communications intercepts have been the mainstay of intelligence activity for many years (Bamford, 2002) there is every reason to expect that the growth in telecommunications traffic represents a major search and interpretation challenge for the relevant services. Indeed, if the 11 September attacks suggest one thing, it is that the possession of the relevant intelligence information is not a guarantee of effective pre-emption or action against terrorist activities.

One of the significant concerns about the work of intelligence agencies is with regard to their ability to process effectively available information to predict accurately intent and actions of terrorist organisations (Betts, 2002; Pettiford and Harding, 2003). Information is not equivalent to knowledge and this was clearly illustrated by the events of September 11. The production of knowledge in specific areas requires knowledge and meta-knowledge to infer what is a realistic interpretation of the information available. Knowledge is crucially important in intelligence. As Shulsky and Schmitt (2002) noted, intelligence refers to the creation of knowledge, by an organisation and through an activity, with knowledge creation at the core of the intelligence process. The process of knowledge creation and use in intelligence can be divided into three parts: collection, analysis and dissemination. Failures can occur in any of the processes and the failures can be located in human factors issues, cognitive and social psychological.

Failures occur in intelligence analysis (Berkowitz and Goodman, 2000; Carter, 2001; Herman, 2001a; Herman, 2002; Odom, 2003) and there are many reasons to suspect that part of this may reflect cognitive limitations of operators, social factors shaping the handling of data and technological limitations in supporting the process. Currently the empirical evidence in the area is scant because of the limited access to the environment. The process of managing intelligence information has been revolutionised by the sheer volume of information that can be collected and submitted for analysis from secret and open source media (Shulsky and Schmitt, 2002; Treverton, 2001; Berkowitz, 2003). Electronic management of information has in turn revolutionised the dissemination of information (Sharfman, 1996) making the propagation of inappropriate interpretations more problematic. Herman (2002) has identified a number of issues that specifically relate to psychological and social aspects of information sharing and usage.

In intelligence, a delicate balance must be struck between revealing information in aiding the process of collection and guarding intelligence to protect the sources of information. If one accepts that the ebb and flow of information may vary in speed and quality the level of shared situation awareness amongst the potential users will vary. Allowing for retention of information at one time and rapid sharing of information at other critical times, a new format of information storage must be created. The danger in using technology alone to solve the problem is the ability to create large warehouses of information that are inaccessible, unintelligible and unusable. Two issues should be considered with regard to an intelligence warehouse.

First, the ease of using the methods for encoding and retrieving information to develop intelligence briefs will be examined. The process of working with information should effectively create the product because any requirement to transcribe, translate or summarise information could distort information as effectively as the process of serial reproduction discussed by Sir Frederick Bartlett in 1932. Bartlett (1932) investigated the distortions that took place in the repetition of reproduction to understand better the way that memories can be distorted over time. That work suggests that various processes of normalisation and alignment to pre-existing memory schema occur as encoding and reproduction take place. The more frequent translation and repetition that occurs, the more distortion that may occur and the greater the resulting misunderstandings. It has been suggested that the development of intelligence briefings is a major performance indicator in the intelligence community and a significant factor in career progression, that is, the more reports the greater the progression. It might be assumed that this would produce higher quality output, but it is more likely that this will polarise inputs into conservative estimates producing no surprises or exaggerated estimates that will never be qualified by experience, as pre-emptive action is taken. The evidence from history suggests that both types of failure have occurred in the recent past.

Secondly, the appropriateness of the knowledge structure, implicit in an interface to an intelligence information warehouse, needs to be considered with regard to the conceptual requirements of intelligence. Previous work with high-level decision makers in command and control teams (Macklin et al., 2002) suggests that it may be possible to construct more effective interfaces by using a conceptual structure derived from critical incident debriefing of practitioners. Critical incident debriefing has been extensively used in human factors research to acquire knowledge structure implicit in information for use in system design of complex socio-technical systems in command and control (Klein, 2000b).

One candidate knowledge structure for effective storage and retrieval is a narrative or storyboard format that inter-relates level 1 situation awareness (perception of events), with level 2 situation awareness (comprehension or interpretations of events), and level 3 situation awareness (prediction of future events). The codification of information in terms of these levels of situation awareness and in terms of a narrative format (with temporal and spatial codes) allows agent-based representation of searches and inquiries to be executed on behalf of human operators on a continual basis, by other human and computer software agents. Thus, a new format for information storage and retrieval could simultaneously improve encoding of information, subsequent retrieval, re-use of information by other agencies and integration of all-source intelligence material into a single integrated framework. The three levels of explicit situation awareness would give an internal metric for the quality of total situation awareness as projection situation awareness would imply that the intelligence was ahead of the events but perception situation awareness would indicate that the intelligence agencies were simply reacting to events as they occurred. This new approach could lead to improvements in intelligence functions considered by a number of authors (Berkowitz and Goodman, 2000; Treverton, 2001) as a result of the open-source availability of information and the technology afforded by the information revolution.

Knowledge craft

The events of 11 September made it clear that intelligence lapses needed further investigation to understand the mechanisms and processes that had perhaps failed to capture and use the relevant information that was available before the events occurred (Herman, 2001b). In many respects, intelligence functions could potentially benefit from the same kinds of strategic human factors analysis that has enhanced other types of complex social and cognitive processes in other knowledge environments from dealing rooms to news environments.

The attack in Madrid on 11 March 2004 consisted of near simultaneous attacks on trains where bombers triggered bombs carried in backpacks, and while the attacks were small the impact that they had was very significant in potentially influencing the subsequent election process in Spain. Transport and communication systems represent a key feature of modern secular societies that promote individual freedom, and the ability to navigate freely throughout the society is perceived as a basic human right. Transport systems are highly vulnerable to terrorist attacks but the cost of providing pre-emptive protection within transport systems is prohibitive. Instead, surveillance may simply provide the opportunity to identify the perpetrators post-event as it did with the follow-on abortive London bombings on 21 July 2005. As Normal Mailer wryly observed with regard to Presidential security, it is not always an effective deterrent: "All the security around the American President is just to make sure that the man that shoots him gets caught." *Sunday Telegraph*, 1990.

Uncertainty

The central issue of this chapter is to consider how intelligence is collected, used and what expectations can be attached to the patterns of information with regard to prediction. The quotations that prefix this chapter indicate that there is a great concern with the unknown, unknowable and the uncertain, and yet the current trends are towards greater and greater information dominance through the use of technology. It is argued that this concern and the actions taken to resolve uncertainty may never ever be satisfactorily resolved and the more significant issue of working with uncertainty needs to be considered. In particular, the core issue is how the human decision maker responds psychologically to uncertainty and the manner in which they justify their actions with regard to incomplete information. There are psychological models that have examined exactly this problem in social psychology and which suggest that the information processing strategy can change in response to the meta-cognitive cues that indicate familiarity with a type of event sequence or the novelty of an event sequence not previously experienced (Forgas, 1995). This type of issue needs to be examined in a group decision making context where high stakes decisions are taken under conditions of uncertainty.

Recent years have revealed the propensity for intelligence to over-estimate apparently the veracity and accuracy of information, with the prime example being the second Gulf War, which was tied officially to the existence of Weapons of Mass Destruction (WMD), which at the time of writing have still remained highly elusive. Intelligence failures such as this are not new phenomena. Indeed, post 9-11 many were quick to compare the attacks of al-Quaeda on New York and Washington with those of the Japanese on Pearl Harbour (Griffin, 2004). There is no doubt that Pearl Harbour shared many of the same problems that later became obvious in the post 9-11 inquiries. The historical accounts (Slackman, 2001) and the military analyses of failures (Cohen and Gooch, 2003) suggest that enough of the information was available, as it was in the case of the al-Quaeda attacks on New York, but the critical decision makers failed to act. There is general agreement that significant opportunities to pre-empt the attacks of 11 September 2001 were missed by many, among them political commentators (Hersh, 2004), by governmental committees (National Commission on Terrorist Attacks upon the United States, 2004) and former senior figures in the US administration who had attempted to focus attention on the clear and present danger (Clarke, 2004). This indicates unsurprisingly that group processes identified by earlier researchers (see Brown, 2000) can still occur within groups of very senior decision makers such that group think, the risky shift, the conservative shift, and other pathological processes can undermine the effective use of the available information.

Group processes are equally evident in the inter-group rivalries between organisations charged with the defence of the countries concerned. The potential for so-called turf wars within US intelligence agencies, which may have prevented effective information sharing, can be traced back to the years immediately after the war when consideration was given to the structure of the agencies responsible for preventing another unforeseen attack like that at Pearl Harbour (Trento, 2001). There have been suggestions that the FBI should bear the brunt of the responsibility (Lance, 2003), because, like MI5 in the UK, the FBI (Federal Bureau of Investigation) was largely responsible for homeland security in the United States of America. Some might suggest that it is difficult to decide what the most significant intelligence failure was with regard to the 9-11 attacks, but it seems likely the that failure of various agencies to collaborate, cooperate and share information across institutional boundaries (see Hersh's foreward in Ritter, 2005) meant that the bigger picture, or the Total Information Awareness was not achieved. The bigger picture was apparently lost as a piecemeal jigsaw across various agencies and in this manner it shares very strong similarities with the Pearl Harbour attacks where the periscopes at the harbour entrance, the radar contacts from the incoming Japanese aircraft and, critically, the intercepts from the Japanese government were not used to full effect to provide a warning of the coming attack.

There are clearly psychological issues to be considered throughout the intelligence process related to the assigned intelligence tasks, the information collected, the analysis process and the use made of intelligence by the consumer. The processes are cognitive and social psychological processes but what makes them challenging is the way they potentially interact. More complex cognitively challenging problems may result in decision makers seeking simpler approaches that are more easily explained to others in order to convince them of the desired course of action. One immediate concern is that the intelligence consumer can distort the intelligence process and there have been suggestions that this could have been the case with regard to the decisions to invade Iraq in the second Gulf War. As Jonathan Freedland (2004, p. 359) suggested:

> We saw, for example, that the intelligence agencies – so energetically promoted as an independent and therefore trustworthy source of guidance on the Iraqi menace – had been effectively co-opted by the current government.

Freedland's concerns are shared by Runciman (2004) and Oborne (2005). Some have chosen to see the analysis of intelligence reports to be deliberately coloured in a manner that distorts the truth or the meaning of the reports with the expressed intention to deceive (Oborne, 2005). In actuality it may be extremely difficult to distinguish dissimulation from genuine error because of the practical use of plausible deniability so frequently used in matters of intelligence, spycraft, assassinations and covert operations. There have been a number of distinguished commentators who have, however, indicated their concern at the wordsmiths' manipulation of the documents and the lack of attempts to correct significant *misunderstandings* that led the country to the war in Iraq (Runciman, 2004). Equally challenging to the view that intelligence assessments were at the root of the misunderstandings prior to the second Gulf War are the reports on Lord Butler's Report (Coates, 2004). Others have expressed concern about the evaluation of the operation of the intelligence agencies because of the central role that the internal security services, such as MI5, must play with regard to counter-terrorism activities (Hollingsworth and Fielding, 2003). Some models of the intelligence cycle make explicit the user or consumer's reactions and their impact on future activities, as indicated in Figure 35.1.

Another concern is that the intelligence consumers can effectively re-direct attention away from genuine areas of concern towards false threats. To some extent Clarke (2004) suggests that this was the case with the balance of concern in the US administration for Iraq and al-Quaeda, such that the former was emphasised and the latter viewed dismissively. The politicisation of the intelligence process and the distorting influence of the intelligence consumer's policy on collection, interpretation and dissemination have been outlined by other commentators (Gill, 2005).

To be fair to the decision makers, history has many examples of asymmetric conflicts where the more powerful body has seriously underestimated the threat of the *weaker* power. The arrogance with which the al-Quaeda forces were viewed may be a contributory factor in misdirecting the efforts of an intelligence analysis and this type of underestimation has figured in military decision making. Arrogant or dismissive assessments as contributors to military operational failures are still a frequent occurrence even though the technology of intelligence has changed (Regan, 2000; Keegan, 2003). Clearly one would expect that the success of the attacks on the US Embassies in Africa could have been viewed as a prelude to the attacks on mainland America. It is perhaps unusual that the US has a number of recent historical cases, such as Vietnam, where the lessons learned from the past have not been completely absorbed or have been forgotten. Finally, there is the possibility that the intelligence information provided is accurate and presented clearly in a manner that indicates that the threat is not significant, but the intelligence consumer seeks to present information selectively in a manner that creates the impression of a clear and present danger.

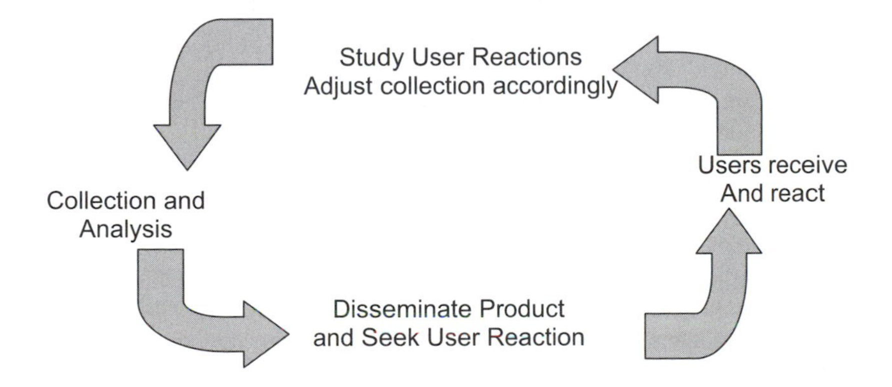

Figure 35.1 A simplified outline of the intelligence process

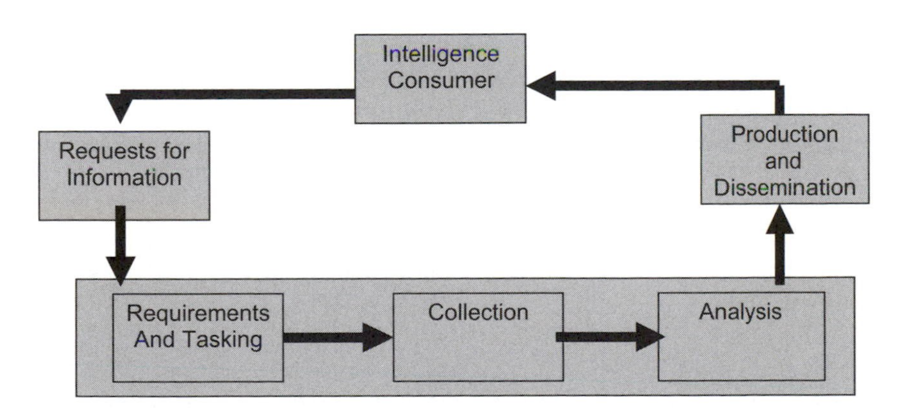

Figure 35.2 Intelligence cycle after Berkowitz and Goodman (2000)

A significant issue that can exist in the intelligence domain is the absence of human intelligence, otherwise known as HUMINT, which comes directly from individuals within the opposing forces network, who can provide a direct insight into the intent and action of the opposing force. Friedman (2004) has identified that the most significant problem influencing the war in Iraq and the following counterinsurgency was the lack of intelligence from local sources, and it is clear that this was an issue in relation to al-Quaeda. In Baer's (2002) book there are suggestions that the intelligence collection process of field officers was largely disconnected from consumer's influence indicated in Figure 35.2. The change in the size of the organisations in the post-Cold War era, where a significant draw down occurred, meant that more limited resources were more directly influenced by policy decisions and may have been less responsive to the observations of field agents. Thus, the intelligence process outlined in Figure 35.2 may have become a positive feedback loop where the desires of the consumer amplified the material collected on one threat and this led to a detriment in attention paid to developing threats. It is possible that disqualified individuals, like Lt. Col. Oliver North, who identified the al-Quaeda threat, were not given the credit for their knowledge and understanding as a result of other indiscretions regarding the sales of arms to Iran. Baer (2002) indicated that this might have been the case with his own expressed concerns about developing threats.

Studying intelligence failures

Kuhns (2003) has identified intelligence failures as one of the highly developed areas of academic study of intelligence. Other commentators have supported the existence of intelligence failures with potentially consistent factors as their contributors and human factors are a significant part of the problem at an organisational and individual level (Herman, 2002). The existence of intelligence errors at first seems the most obvious and easily verifiable of events. However, it is possible and plausible that attacks may be allowed to progress in order for agents working on behalf of the government to gain further access to terrorist organisations. In military operations during the Second World War care was taken in the use of the decoded Enigma intercepts to prevent the axis forces from becoming aware that their security was compromised. The reason for intelligence failures is often more difficult to discern because it is likely to be classified at a very high level.

It can be argued that intelligence failures can be analysed in a manner similar to accidents, with a sequence of contributory causes leading up to significant events in a manner similar to that identified by James Reason for organisational contributions to accidents (Reason, 1990; 1997). Reason has proposed that any error or failure in the operation of a system is normally not a result of a single cause but rather it is a consequence of a concatenation of errors that results in degraded performance that at some point contributes to failed responses.

Intelligence processes are normally segmented into collection, analysis and dissemination as indicated in Figure 35.2 (c.f. Berkowitz and Goodman, 2000) and in the past collection and analysis were often identified as problematic areas contributing to intelligence failure (Herman, 2002; Kuhns, 2003). The emphasis for many agencies is naturally on superior collection (Combs, 2000) because there is a belief that this would diminish uncertainty associated with decision making but it is argued that analysis is often weak, so that even with the right information the wrong decisions are made. In the final analysis, it is very unlikely that critical elements of the intelligence picture would be captured, and as a consequence intelligence will always rely upon incomplete, uncertain and confused images of the operational environment. There is also the possibility that opposing forces will use deception and feints in an attempt to mislead the intelligence agencies. This is certainly a process outlined in UK (Dorril, 2000) and Russian (Andrew and Mitrokhin, 2005) military and intelligence operations of the past. The investigative guesswork of actual operations is well captured in Baer's (2002) book that describes his pursuit of terrorists in the Middle East. While Baer was in the Directorate of Operations and not the Directorate of Intelligence, his insights as a field officer suggest that the image of the intelligence problems are rarely complete. In addition, Baer indicates a very important role for HUMINT as a special source and one of the most effective in corroboration of the current working hypothesis. Intelligence analysis does not make use of effective information technology and the interface to the knowledge interface is weak in supporting or obstructive to searching. This is surprising as the information technology revolution has been identified as a potential revolution in military affairs (O'Hanlon, 2000; Hall, 2003) and it would be not unreasonable to expect that the same might be the case for intelligence operations. Indeed, some authors have specifically identified the information age as a unique opportunity for re-thinking the manner in which intelligence operations are conducted (Berkowitz and Goodman, 2000). The visibility of the intelligence failures has in recent years become something that has been a matter for Congressional Intelligence Committees in the US because of the failures in intelligence predictions prior to the events of 11 September 2001 (Johnson, 1996; Posner, 2003).

The problems with intelligence (Benjamin and Simon, 2002; Powers, 2002) were already a matter for subject debate before the release of US Governmental evidence and Congressional judgements. The failure of intelligence to grasp what was a fairly clear footprint, if somewhat diverse (see Gunuratna, 2002), for al-Quaeda was identified in more popular reviews of intelligence function (Farren, 2003). The tactical surprise of the al-Quaeda attacks can be set alongside other attacks like that on Israeli athletes at the 1972 Olympic Games and the Aum Shinri Kyo gas attacks on the Tokyo underground (Murakami, 1997; Henderson, 2001), even though the scale of the assault by al-Quaeda was far greater. With more information available in the public domain, it has become clear that a significant body of information existed and further data collection would only have corroborated the potential method of attack, place of attack and time of attack (see Fouda and Fielding, 2003), indicating a post-collection failure in analysis or dissemination. The transparency of the connections between specific individuals to al-Quaeda has been made clear by the hijackers that were caught and imprisoned (Moussaoui, 2002). The links between individuals are indicated in detail in Fouda and Fielding's (2003) account. These failures in insight strongly support the view that there was a failure to exploit intelligence in an information age knowledge management system that suggests that the proposals for more effective processes designed to exploit information technology (Berkowitz and Goodman, 2000) have largely been ignored. The body of evidence on the attackers was sufficient to introduce measures that would have mitigated and pre-empted the attacks, even though the opposing organisation itself was not attacked.

Intelligence failures are not new and the frequent comparison of the events of 9-11 to Pearl Harbour has some basis in fact, but in an attempt to diminish 9-11 it has been suggested that it was simply a tactical surprise. It was recognised that cooperation between and within organisations was weak in fusing this intelligence that was reminiscent of the failures prior to Pearl Harbour (McNeilly, 2001). The failure to fuse information across organisations has led some to point at the need for fusion centres to significantly improve the exploitation of information available. However, even if the information was made available in a single organisation, it is likely that the thematic linkages between the individual items of information could not have been successfully exploited as a consequence of procedural, technological and organisational limitations (Benjamin and Simon, 2002). In an era of global terrorism it is clearly necessary to overcome these difficulties and particularly to address the arbitrary divide between national and international security issues. The financial and economic impact of 9-11 has been global and strategic, with the airline industry the most visible casualty, so that the surprise attacks on 9-11 should not be dismissed. Intelligence failures at Pearl Harbour resulted from critical areas of information capture that were neither exploited nor circulated to effectively exploit the critical information. The psychological issues involved in effective exploitation of intelligence are dependent on human factors. The importance of analysis supported by effective information technology and subject matter expertise is highlighted as likely to be very important in the continuing war on terrorism.

Knowledge and assimilation

The intelligence services require a sophisticated group of knowledge workers able to collate, analyse and interpret complex patterns of information to make predictions about the future course of events. The intelligence services need to transfer their knowledge to other groups and this multi-agency collaboration is used to create policy and justify actions. Thus, there is a need to store information in a manner that a very specialised community can use it but in a way in which it can easily be transformed into a format that is easily assimilated by other agencies, where cooperation is required. As Herman (2002) notes, the vast majority of intelligence failures are associated with various types of human factors issues in which the role played by the individuals within the intelligence community with regard to failure is critical and this is evident in the analysis phase in particular. Psychological models have been used previously in evaluating the risk of bias in intelligence preparation (Cremeans, 1971; Heuer, 1978) but organisational, technological and economic factors have radically re-shaped intelligence services and processes in the period of time following these investigations. Herman (2002) uses dated models of human psychological process to explain the mistakes observed in intelligence and it is not clear if the same types of error will propagate into future intelligence operations dominated by information technology and organisational change. It is proposed that a more detailed analysis by appropriately qualified human factors and domain experts could provide valuable insights to enhance the transitional process because of the wide range of social and cognitive issues associated with the use of information technology as a mediating system. A future aim might be the development and validation of a socio-cognitive model of intelligence functions using a combination of observational and empirical research based on quantitative and qualitative measures.

Consider the basic knowledge available on the use of information and the development of comprehension. It is known that pre-exposure to similar information, in order to develop a schema, facilitates assimilation of subsequent information to achieve comprehension more rapidly. The literature on text comprehensions and retrieval of information from long-term memory suggests that effective cues at the encoding stages of learning facilitate the later recall and use of information, as well as improving the immediate understanding. In the information-rich intelligence environment small improvements in the initial response to information and better recall of information may be vital in developing total situation awareness. If the volume of information is too great then the ability to search easily and effectively for information that will help with the process of sensemaking is critical. The process of sensemaking in complex environments, described by Wieck (2001), is the transformation of information into knowledge and understanding.

It is generally recognised that many information search technologies currently operate poorly because the user is not able to apply their conceptual understanding of the domain of interest via the interface, without significant effort. Thus, the current knowledge warehouses may not structure or collect knowledge in a manner that meets the needs of intelligence functions (Odom, 2003) and in combination with potential information overload, this will result in inefficient use of critical information.

Theoretical basis for the research strategy

Human factors approaches to the development of computer supportive technology, in decision-aiding and information analysis, have developed rapidly over the last 15 years or so. There is now a need for more sophisticated measures for evaluation of the technology and theoretical models to help conceptualise design problems. One aim of the current research is to identify human factors models suitable for application in the field of intelligence gathering and knowledge creation. One of the key models applied to individual cognition in complex information systems is the model of situation awareness (Endsley, 2000) and the application of this to system design has already been discussed (Endsley, 2003). This model can be applied to descriptions of the technology, systems and processes for intelligence to determine if the emphasis in current intelligence is weighted towards supporting level 1 situation awareness, the perception of events. Current analyses of intelligence functions suggest that intelligence information collection is adequate but the analysis of information is not. This observation is in direct contrast to situation awareness errors in real time systems management, where the failures are usually related to missing significant events. If one accepts that the cognitive weighting of current systems inadequately supports the development of level 2 or 3 situation awareness, it is easy to interpret the shortcomings with regard to recent terrorist incidents.

One approach taken from the applied psychology literature relates to the manner in which decision making processes occur, where it is suggested that decision making is more correctly described as a pattern recognition process where environmental cues are associated with schematic knowledge of previous events. This process of recognition-primed decision making (Klein, 1993b) (also termed naturalistic decision making by Klein (1993a)) has been used to aid the designers of new information management systems in real time control systems. It is likely that the same models of decision making, given their reliance on knowledge (explicit and implicit) and on expertise are applicable to the intelligence community operators. While many knowledge workers do not consider themselves decision makers, their role as filters of information and intelligent observers of events has strong similarities to the properties of decision makers in command and control. The information management process is essentially a socio-technical filtering operation whereby the information deluge is narrowed and shaped into a manageable stream of relevant data. This process of narrowing is subject to type 1 and type 2 errors of marking as relevant information or discarding irrelevant information. In addition, intelligence operations must manage decoys, deceptions and bluffs.

In addition to the models outlined above human factors research has identified useful methodologies for the development of new technology, called cognitive task analysis or cognitive work analysis (see Schraagen, Chipman and Shalin, 2000; Vicente, 1999; Hollnagel, 2003). While not true equivalents, both methodologies have been successful in gaining insight into complex socio-cognitive technologies where individual cognitive and group psychology factors influence performance. Cognitive task analysis is well described by Schraagen, Chipman and Shalin who suggested that it is an extension of traditional task analytic techniques to include information about knowledge, thought processes, and goal structures that underlie observable task performance. Thus, it is clearly applicable in an area such as intelligence operations, which involves the use of knowledge and critical thinking to create the intelligence product. Cognitive work analysis attempts to understand the nature of the operational domain by attempting to identify the semantics of the relevant domain (Vicente, 1999). In simple terms work only makes sense within a context and abstract representations of work can create misleading indications for system developers and process management. It has been argued that work analysis is an important method for developing computer-based systems that effectively support human work within a complex socio-technical system. Again the emphasis with these modern approaches is not description but explanatory appreciation of what work is done, the demands on the human operator and how they are best supported. Recent reviews of intelligence have already identified the significance of the analysis process and of the information revolution in intelligence there is clearly a need to appreciate the nature of the work with an appropriate methodology, such as Cognitive Work Analysis (CWA) or Cognitive Task Analysis (CTA). Similar concerns are found in Wieck's (2001) work on making sense in complex socio-technical organisations because sensemaking emphasises both the social and cognitive elements of the cooperative enterprise. The significance of social context, personal identity, salient cues, ongoing projects, plausibility and enactment can be easily identified in intelligence communities. Indeed, there is no reason to expect intelligence operations to be sterile because the human and organisational factors will cause the process to deviate from optimal function. Historically it has been found that governments can influence the craft, individuals can undermine the process with malicious intent or as a way of influencing their career progression and theories of enemy intent can be upheld in the face of incontrovertible and antagonistic evidence. Any analysis of intelligence can only explain a proportion of the data if it does not address the multifaceted web of influence on the process.

To understand the human factors issues in intelligence it is necessary to outline the steps whereby information makes sense and information is dismissed from the system. Most models of human cognition propose three major types of memory: a very short-term sensory memory that gives us access to all the environmental information; a much more limited short-term or working memory in which information is processed; and a long-term memory that retains all the products of experience. The capacity, speed and organisation of each type of memory are different and this shapes the way in which information is processed. Working memory is relatively small and the main danger is information overload where the amount of information exceeds the capacity of the memory. Working memory is critical because effective processing of information results in transfer of processed information to long-term memory and the development of experience (Carlson, 1997). Long-term memory is much slower to access and a major problem is retrieval, where information is available but inaccessible. Long-term memory does not have capacity problems but humans can mislay information so that they fail to retrieve information. Access to long-term memory can change in expert individuals but only when the information accessed is repeatedly and exhaustively used and the expertise is highly limited and situation specific (Ericsson and Delaney, 1999; Proctor and Dutta, 1995). It is clear that even after short periods of training intelligence analysts will change their methods for processing information and the type of structure they impose on the knowledge. However, their sophistication may be the meta-knowledge about which sources, which type of information and what types of corroborative evidence are significant in specific analyses. Current unpublished experimental research (Cook, Cameron, and Adams, unpublished) on processing and sensemaking with very large unfamiliar datasets for decision making indicates two critical issues. First, the initial exposure to an unfamiliar dataset or event sequence limits the capability to manipulate the material before the process of assimilation has taken place. Secondly, there appear to be a number of meta-cognitive strategies for dealing with a new corpus of information prior to full assimilation that involve the use of heuristics and biases identified in earlier research.

Having considered briefly the ways in which the different elements of memory inter-relate, one might consider why a human analyst is considered more appropriate than machine intelligence. First is the sparse nature of the information in intelligence analysis that requires conjectural developments using experience and going beyond the scope of current inferential logic driven by machine intelligence. Correctly segregating patterns embedded in unfamiliar noise masks is a human characteristic that has been exploited to bar portals to information bots that traverse the web. Secondly, the presence of misleading information in the database designed to draw attention away from or mask the intent of the terrorist group under scrutiny may generate false impressions. Thirdly, the consideration of intangible and qualitative qualifications of the sources, methods and coverage of the information collected. The accomplished intelligence analyst needs to use implicit knowledge of the information, often described as gut instinct, to qualify the judgements made. This is strength and weakness of intelligence preparation by human analysts because feelings of uncertainty associated with complexity of the information can be confused with the interpretation of analysis, to produce an uncertain or qualified interpretation.

Psychologists examining information processing strategies have suggested that affect is an integral part of how we manage the world and it impacts judgements and reasoning (Bower and Forgas, 2000; Forgas, 1995, 2000). Accepting that this is the case, technology should be designed to help the user explore their uncertainties and to protect against errors of judgement driven by decision-related anxiety. However, the need for certainty, to sanction actions, and the uncertain nature of the judgements in intelligence represents a conflict that is intrinsic to the process and would not be eliminated completely by the use of technology. Thus, the solution requires training, technology and processes to prevent erroneous judgement.

The limited analysis of the intelligence community that currently exists makes the recommended analysis a subject for exploratory investigation. There are a wide range of analytic techniques for the analysis of human-machine system design and an even wider range of techniques for task description (Beevis et al., 1999). What makes the area of intelligence somewhat unique is the focus largely on the support of interpretative analysis on information to generate knowledge or comprehension without some form of direct or immediate feedback from the real world. In effect, the plausibility or accuracy of the model proposed is unknown at least until further events occur and further evidence is accrued; as such it resembles science in only finding supporting evidence that is relatively accurate and not absolute evidence that is unquestionable. Intelligence analysis is an open system and as such it is important to develop metrics which assess both the process and the product of intelligence activity, as the value of the latter may never be totally without doubt.

The focus of any research programme should be geared towards the practical implementation of an improved intelligence process by socio-cognitive improvements in information sharing techniques. The programme of work would enable an appreciation of culture and its impact in intelligence circles, as it has been suggested that this may be destructive and undermine the exploitation of new technology (Berkowitz and Goodman, 2000). Some attempt should be made to understand the organisational culture as a factor influencing work-related activities and for this reason the type of interpretative analysis used by Wieck (2001) and the work analysis approach (Vicente, 1999) should be used. Some consideration of the more detailed issues in collaborative and coordinated working mediated by computer (see Olson, Malone and Smith, 2001) have been examined in the computer science literature, but many of the studies conducted have failed to look at mature organisations with subject matter experts, typical of intelligence services.

It seems that the time has come for the revolution in information technology to be developed to meet the requirement of the intelligence services more adequately than currently is the case. A simple technological fix will not improve the analysis process because there is currently a knowledge gap with regards to the actual appreciation of the process, and at the same time the task is changing to focus on non-state threats such as terrorist groups. A superficial and subject matter-led analysis has not taken the process far and the absence of a human factors approach to analysing and aiding the intelligence process will mean that future attempts at improvement are more likely to fail. In recognising that intelligence is knowledge craft but accepting that knowledge is not impartial, and the processes creating it are influenced by a myriad of causes, one accepts the central part of the human operator. Machines do not think and currently do not discern intent; it is the human operator that must do this. As intelligence operations against terrorism are the discernment of intent then human issues are the key to any future improvements.

Conclusion

History suggests that foreign policy interventions can fail spectacularly to achieve the outcomes that decision makers desire, and the more recent US military and Central Intelligence Agency (CIA) interventions provide numerous examples of such events (Blum, 2004). However, when the UK was a colonial power and it sought to extend its influence across the globe, it frequently found it could seek to exercise influence, only to find that the consequences were the converse of what was intended and in many cases it failed to estimate the enemy's capability, with fatal or catastrophic consequences. In recent commentaries concerns have been raised that intervention in Iraq has increased the possibility of attacks on mainland UK. In this respect it is interesting to note that intelligence assessments supporting the view that attacks on the UK mainland could have been motivated by the invasion of Iraq and actions in the Middle East have been dismissed by the same government that stood firmly behind the intelligence assessments that were used to justify the war. Even before formal investigations were released into the public domain the transparency of this position was clear to the public and commentators on the events of 7 July 2005 (Black, 2005).

Intelligence blunders are not a new concern (Hughes-Wilson, 2004) and it is likely that the governmental bodies concerned will be guarded about the issues that give rise to the failures, even in an open and democratic society, as the weaknesses and vulnerabilities in the system may be exploited by their opposition. The popular bookshelves provide ample evidence of the collection process from past operations drifting into the public domain with many recounting tales from even the more recent missions. The spying (Smith, 2003) and espionage (Bennett, 2002) issues from the past are not as sensitive as they once might have been because they describe a craft that was more relevant to a past age when Cold War tensions were high. There are encyclopedic analyses of the CIA (Smith, 2003). There are encyclopedic analyses of terrorists (Sharpe, 1997; Combs and Slann, 2003) and a plethora of other books, as well as internet sources, some from the organisations themselves. The open source material provides ample evidence of the potential activities that may be undertaken by terrorists and should be a basis for identifying the terrorist footprint. The terrorist challenges of the present have limited potential for HUMINT activity and less tangible sources where high technology can overwhelm the asymmetric foe. The terrorist footprint is smaller and more uncertain than the opposition of the past and it might be argued that this changes the way that intelligence organisations need to work with the information they have. The one significant strength that intelligence organisations have is the record of the past because it is likely that the attacks of the future have already been tried or considered in one form or another in the past.

Some of the commentators on the War on Terrorism are deeply concerned that the banner of protection might be successfully used to manipulate further the public's awareness of actual events and to restrict the hard won liberties of the past (Bovard, 2003). It is interesting to note that the economic cost of achieving this total surveillance seems hopelessly unrealistic but it does not mean that decision makers will not reach for this as an obvious solution to the problems faced by intelligence agencies. However, in military parlance it may be difficult to use surveillance to create this information superiority and achieve little else but bolting the stable door after catastrophic events. Some commentators have observed that in the US, the government may have rewarded failure by increasing the budget for security services and failing to consider effectively the manner in which the same services did not meet the capability expected from them in protecting against terrorist threats.

One conclusion from decision making on terrorism is that decisions are not the rational bounded processes that some might suggest because they are compromised by factors outside the events that people seek to control and manipulate. This is equally true for decisions in commercial operations such as nuclear power plants, chemical process industries, oil and gas production, transport operations, air traffic control and in almost every other sphere of complex decision making where economic factors impinge on the decision making process and risks are high. It is clear that even in safety critical systems, decision making can be subtly influenced by factors outside the rational boundaries appropriate to a purely technical decision making process, the cultural environment can provide a fertile environment for pathological decision making to appear (Vaughan, 1997). The unknown unknowns can sometimes be the unexpressed and implicit forces that influence decision making as much as the gaps in knowledge and limitations of the information available. The next generation of information systems will need to manage the uncertainty of the unknown more effectively by making the presumptive choices explicit and the risks of errors transparent.

References

Andrew, C. and Mitrokhin, V. (2005). *The Mitrokhin archive II: The KGB and the world*. London: Penguin Books.

Bamford, J. (2002). *Body of secrets: How America's NSA and Britain's GCHQ eavesdrop on the world*. London: Arrow Books.

Baer, R. (2002) *See no evil: The true story of a ground soldier in the CIA's war on terrorism*. Bristol, UK: Three Rivers.

Bartlett, F. C. (1932). *Remembering: A study in experimental and social psychology*. Cambridge: Cambridge University Press.

Beevis, D., Rost, R., Döring, B., Nordø, E., Oberman, F., Papin, J. P., Schuffel, H. and Streets, D. (1999). *Analysis techniques for human-machine system design*. Wright Patterson, Ohio, USA: Crew Systems Ergonomics/Human Systems Technology Information Analysis Center.

Benjamin, D. and Simon, S. (2002). "A failure of intelligence?" In R. B. Silvers and B. Epstein (eds) *Striking terror: America's new war*. New York: New York Review Books.

Bennett, R. M. (2002). *Espionage: Spies and secrets*. London: Virgin Books.

Berkowitz, B. (2003). *The new face of war: How war will be fought in the 21st century*. London: Free Press.

Berkowitz, B. D. and Goodman, A. E. (2000). *Best truth: Intelligence in the information age*. London: Yale University.

Betts, R. K. (2002). "Fixing intelligence". *Foreign Affairs*, 81(1). Available at: http://www.foreignaffairs.org/2002/1.html.

Black, C. (2005). *7-7 The London bombs: What went wrong?* London: Gibson Square.

Bovard, J. (2003) Available at: http://www.fff.org/issues/listJXB.asp.

Blum, W. (2004). Killing hope: US military and CIA interventions since Word War II. London: Zed Books.Bower, G. H. and Forgas, J. P. (2000). "Affect, memory and social cognition". In E. Eich, J. F. Kihlstrom, G. H. Bower, J. P. Forgas and P. M. Niedenthal (eds), *Cognition and Emotion*. Oxford: Oxford University Press.

Boyatzis, R. E. (2000). *Transforming qualitative information*. London: Sage Publishing.

Brown, R. (2000). *Group processes*. 2nd Ed. Oxford: Blackwell.

Carlson, R. A. (1997). *Experienced cognition*. London: Lawrence Erlbaum Associates.

Carter, A. B. (2001). "Keeping the Edge: Managing defense for the future". In A. B. Carter and J. P. White (eds) *Keeping the Edge: Managing defense for the future*. Oxford: MIT Press.

Clarke, R. (2004). *Against all enemies: Inside America's War on Terror*. New York: Free Press.

Coates, T. (2004). *Lord Butler's report: Espionage and the Iraq war*. London: Coates.

Cohen, E. A. and Gooch, J. (2003) *Military misfortunes: The anatomy of failure in war*. New York: Anchor Books.

Combs, C. C. (2000). *Terrorism in the twenty-first century*. Upper Saddle River, NJ: Prentice Hall.

Combs, C. C. and Slann, M. (2003). *Encyclopedia of terrorism*. New York: Checkmark Books.

Cook, M. J., Cameron, D. and Adams, C. S. G. (unpublished). *Narrative Accounts of Command and Control: Experimentation Version 5*. Supported on EOARD Grant #032052.

Cremeans, C. D. (1971). "Basic psychology for intelligence analysts". Studies in Intelligence 15(1). Reproduced in H. Bradford Westerfield (ed.), *Inside the CIA's private world* (1995).

Davenport, T. H. and Prusak, L. (2000). *Working knowledge*. Boston, MA: Harvard Business School Press.

Davies, N. (2003). *Terrorism: Inside a world phenomenon*. London: Virgin Books.

Dorril, S. (2000). *MI6: Inside the covert world of her majesty's secret intelligence service*. New York: Touchstone.

Endsley, M. R. (2000). "Theoretical underpinnings of situational awareness: a critical review". In M. R. Endsley and D. J. Garland (eds), *Situation awareness analysis and measurement*. Mahwah, NJ: Lawrence Erlbaum Associates.

Endsley, M. (2003). *Designing for situation awareness*. London: Taylor and Francis.

Ericsson, K. A. and Delaney, P. F. (1999). "Long-term working memory as an alternative to capacity models of working memory in everyday skilled performance". In A. Miyake and P. Shah (eds), *Models of working memory*. Cambridge: Cambridge University Press.

Farren, M. (2003). *CIA: The secrets of the "company"*. London: Chrysalis Press.

Forgas, J. P. (1995). "Mood and judgement: The affect infusion model (AIM)". *Psychological Bulletin*, 117(1), 39-66.

Forgas, J. P. (2000). "Affect and information processing strategies". In J. P. Forgas (ed.), *Feeling and thinking: The role of affect in social cognition*. Cambridge: Cambridge University Press.

Fouda, Y. and Fielding, N. (2003). *Masterminds of terror*. London: Mainstream Publishing.

Freedland, J. (2004). "Tugging back the veil". In S. Rogers (ed.), *The Hutton Inquiry and its impact* (pp. 353-361). London: Politico's Guardian Books.

Friedman, G. (2004). *America's secret war: Inside the hidden worldwide struggle between the United States and its enemies*. London: Little Brown.

Gill, P. (2005). "The politicization of intelligence; Lessons from the invasion of Iraq". In H. Born, L. K. Johnson and I. Leigh (eds), *Who's watching the spies: establishing intelligence accountability*. Washington DC: Potomac Books.

Griffin, D. R. (2004). *The new Pearl Harbour: Disturbing questions about the Bush Administration and 9/11*. Gloucester: Arris Books.

Gunuratna, R. (2002). *Inside al-Qaeda: Global network of terror*. London: Hurst and Company.

Hall, W. M. (2003). *Stray voltage: war in the information age*. Annapolis, MD: Naval Institute Press.

Hawthorne, S. (2002). "Knowledge related to purpose: Data-mining to detect terrorism". (unpublished article).

Heath, C. and Luff, P. (2000). *Technology in action*. Cambridge: Cambridge University Press.

Henderson, H. (2001). *Global terrorism: The complete reference guide*. New York: Checkmark Books.

Herman, M. (2001a). "Keeping the edge in Intelligence". In A. B. Carter and J. P. White (eds), *Keeping the Edge: Managing defense for the future*. Oxford: MIT Press.

Herman, M. (2001b). *Intelligence services in the information age*. London: Frank Cass Publishers.

Herman, M. (2002). *Intelligence power in peace and war*. Cambridge: Cambridge University Press.

Hersh, S. M. (2004). *Chain of command*. London: Penguin Books.

Hollingsworth, M. and Fielding, N. (2003). *Defending the Realm: Inside MI5 and the war on terrorism*. London: Andre Deutsch.

Hollnagel, E. (ed.) (2003). *Handbook of cognitive task design*. London: Lawrence Erlbaum Associates.

Heuer, R. J. (1978). "Cognitive biases: Problems in hindsight analysis". Studies in Intelligence 22(2) 21-28. Reproduced in H. Bradford Westerfield (ed.) *Inside the CIA's private world* (1995).

Hughes-Wilson, J. (2004) *Military intelligence blunders and cover-ups*. New York: Carroll and Graf.

Johnson, L. K. (1996). *U.S. intelligence in a hostile world: Secret agencies*. London: Yale University Press.

Keegan, J. (2003). *Intelligence at war: Knowledge of the enemy from Napoleon to Al-Qaeda*. London: Hutchinson Press.

Klein, G. (1993a). *Naturalistic decision making*. Wright Patterson Air Force Base, Ohio: Crew Systems Ergonomics Research Information Analysis Centre (CSERIAC).

Klein, G. (1993b). "A recognition-primed decision (RPD) model of rapid decision making". In G. A. Klein, J. Orasanu, R. Calderwood and C. E. Zsambok (eds), *Decision making in action: Models and methods* (pp. 138-147). Norwood, NJ: Ablex.

Klein, G. (1997a). "An overview of naturalistic decision making applications". In C. E. Zsambok and G. Klein (eds), *Naturalistic Decision Making* (pp. 49-59). Mahwah, NJ: Lawrence Erlbaum Associates.

Klein, G. (1997b). "The current status of naturalistic decision making framework". In R. Flinn, E. Salas, M. Strub and L. Martin (eds) *Decision making under stress: Emerging themes and applications*. Aldershot: Ashgate Publications.

Klein, G. (2000a). "Cognitive Task Analysis of Teams". In J. M. Schraagen, S. F. Chipman and V. L. Shalin (eds), *Cognitive Task Analysis* (pp. 417-430). London: Lawrence Erlbaum Associates.

Klein, G. (2000b). "Analysis of situation awareness from critical incident reports". In M. R. Endsley and D. J. Garland (eds), *Situation awareness analysis and measurement*. Mahwah, NJ: Lawrence Erlbaum Associates.

Kuhns, W. J. (2003). "Intelligence failures: Forecasting and the lessons of epistemology". In R. K. Betts and T. G. Mahnken (eds), *Paradoxes of Strategic Intelligence*. London: Frank Cass Publishers Ltd.

Lance, P. (2003). *1000 years for revenge: International terrorism and the FBI*. London: Harper Collins.

Luff, P., Heath, C. and Greatbatch, D. (1994). "Work, interaction and technology: The naturalistic analysis of human conduct and requirements analysis". In M. Jirokta and J. Goguen (eds), *Requirements Engineering: Social and Technical Issues*. London: Academic Press.

Macklin, C. M., Cook, M. J., Angus, C. S., Adams, C. S. G., Cook, S. and Cooper, R. (2002). "Qualitative analysis of visualisation requirements for improved campaign assessment and decision making in command and control". In C. Johnson (ed.), *21st European annual conference on Human decision making and control: GIST Technical Report G2002-1* (pp. 169-177). University of Glasgow.

McNeilly, M. (2001). *Sun Tzu and the Art of Modern Warfare*. Oxford: Oxford University Press.

Moussaoui, A. S. (2002). *Zacarias Moussaoui: The making of a terrorist*. London: Serpent's Tail Publishing.

Murakami, H. (1997). *Underground: The Tokyo gas attack and the Japanese psyche*. London: Harvill Press.

National Commission on Terrorist Attacks upon the United States (2004). *The 9/11 commission report*. London: W.W. Norton Company.

Oborne, P. (2005). *The rise of political lying*. London: Free Press.

Odom, W. E. (2003). *Fixing intelligence: For a more secure America*. London: Yale University Press.

O'Hanlon, M. (2000). *Technological change and the future of warfare*. Washington DC: Brookings Institution Press.

Olson, G. M., Malone, T. W., and Smith, J. B. (eds), (2001). *Coordination theory and collaboration technology*. Mahwah, NJ: Lawrence Erlbaum Associates.

Pettiford, L. and Harding, D. (2003). *Terrorism: The new world war*. Slough, UK: Arcturus.

Posner, G. (2003). *Why America slept: The failure to prevent 9/11*. New York: Random House.

Powers, T. (2002). "The trouble with the CIA". In R. B. Silvers and B. Epstein (eds), *Striking Terror*. New York: New York Review Books.

Proctor, R. W. and Dutta, A. (1995). *Skill acquisition and human performance*. London: Sage Publications.

Reason, J. (1990). *Human error*. Cambridge: Cambridge University Press.

Reason, J. (1997). *Managing risks of organisational accidents*. Aldershot: Ashgate Publications.

Regan, G. (2000). *Great military blunders*. Basingstoke: Channel 4 Books (Imprint of Macmillan).

Ritter, S. (2005). *Iraq confidential: The untold story of America's intelligence conspiracy.* London: I.B. Taurus.

Robson, C. (2002). *Real world research*, 2nd Ed. Oxford: Blackwell.

Runciman, W. G. (2004). "What we know now". In W. G. Runciman (ed.), *Hutton and Butler: Lifting the lid on the workings of power.* Oxford: Oxford University Press.

Schraagen, J. M., Chipman, S. F. and Shalin, V. L. (2000). "Introduction to cognitive task analysis". In J. M. Schraagen, S. F. Chipman and V. L. Shalin, (eds), *Cognitive task analysis*. London: Lawrence Erlbaum Associates.

Sharfman, P. (1996). "Intelligence analysis in an age of electronic dissemination". In D. A. Charters, S. Farson and G. P. Hastedt (eds), *Intelligence analysis and assessment*. London: Frank Cass Publishers.

Sharpe, M. E. (1997). *International encyclopedia of terrorism*. London: Fitzroy and Dearborn Publishers.

Shulsky, A. N. and Schmitt, G. J. (2002). *Silent warfare: Understanding the world of intelligence*. London: Brassey's Inc.

Slackman, M. (2001). *Target Pearl Harbour*. London: Pimlico.

Smith, M. (2004). *The spying game*. London: Pimlico's Publishing.

Smith, W. T. (2003). *Encyclopedia of the Central Intelligence Agency*. New York: Checkmark Books.

Trento, J. J. (2001). *The secret history of the CIA*. New York: Caroll and Graf.

Treverton, G. F. (2001). *Reshaping national intelligence in the information age*. Cambridge: Cambridge University Press.

Vaughan, D. (1997). *The Challenger launch decision: Risky technology, culture and deviance at NASA*. London: University of Chicago Press.

Vicente, K. J. (1999). *Cognitive work analysis*. London: Lawrence Erlbaum Associates.

Wieck, K. E. (2001). *Making sense of the organization*. Oxford: Blackwell.

Index

Note: page numbers in *italics* refer to figures and tables.

RECEIVED
FEB 28 2008